Y0-BXR-735

Principles of Underwater Sound for Engineers

Principles of Underwater Sound for Engineers

Robert J. Urick

Research Physicist
U. S. Naval Ordnance Laboratory
White Oak, Silver Spring, Maryland

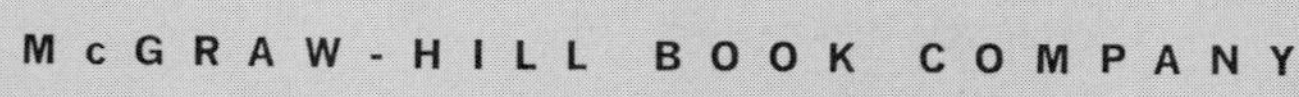

New York San Francisco Toronto London Sydney

PRINCIPLES OF UNDERWATER SOUND FOR ENGINEERS

 Library of Congress Catalog Card Number 67-15040

66085

1234567890MP72106987

Preface

Underwater sound, as a specialized branch of science and technology, has seen service in two world wars. Although it has roots deep in the past, underwater sound, as a quantitative subject, may be said to be only a quarter of a century old. Its modern era began with the precise quantitative studies undertaken with great vigor during the days of World War II. In subsequent years, its literature has grown to sizable proportions, and its practical uses have expanded in keeping with man's continuing exploration and exploitation of the seas.

This book attempts to summarize the principles of underwater sound from the viewpoint of the engineer and the practical scientist. It lies squarely in the middle of the spectrum—between theory at one end and sonar technology at the other. Its intent is to provide a summary of the principles, effects, and phenomena of underwater sound and to give numerical quantitative data, wherever possible, for the solution of practical problems.

The framework of the book is the sonar equations—the handy set of relationships that tie together all the essential elements of underwater sound. The approach is, after an introductory chapter, to state the equations in a convenient form and then to discuss in subsequent chapters each one of the quantities occurring in the equations. The final chapter is largely devoted to problem solving, in which the use of the equations is illustrated by hypothetical problems taken from some of many practical applications of the subject.

In the desire to keep the book within sizable bounds, some aspects of underwater sound have had to be slighted. One is the subject of transducers—the conversion of electricity into sound and vice versa. Although transducer arrays are discussed, only one sound source—the underwater explosion—is dealt with at any length. It is felt that the design of electroacoustic transducers for generating and receiving sound is truly an art in itself, with a technology and theoretical background that deserves a book of its own. In addition much of the basic theory of underwater sound is confined to references to the literature, and engineering matters of sonar hardware are omitted entirely. Although the book will therefore appeal neither to the theoretician nor to the hardware builder, it hopes to cover the vast middle ground between them and be of interest both to the design engineer and practical physicist. It is based on a course given for several years at the Catholic University, Washington, D.C., and at Westinghouse Electric Co., and the Martin Co. in Baltimore.

The book owes a great deal to my colleagues at the Naval Ordnance Laboratory for many discussions and helpful criticism. In particular, Mr. T. F. Johnston, chief of the acoustics division, has been a constant encouragement and stimulant in the long and arduous task of writing the book. My students have been helpful too in providing me with a receptive and critical audience for whatever is original in the presentation.

Robert J. Urick

Contents

Preface v

ONE **The Nature of Sonar** 1

1.1 Historical Survey 2
1.2 Uses of Underwater Sound 7
1.3 Basic Concepts and Units 12
References 15

TWO **The Sonar Equations** 16

2.1 Basic Considerations 17
2.2 The Active and Passive Equations 18
2.3 Names for Various Combinations of Parameters 22
2.4 Echo, Noise, and Reverberation Level as Functions of Range 22
2.5 Transient Form of the Sonar Equations 23
2.6 Statement and Examples 26
References 27

THREE **Properties of Transducer Arrays: Directivity Index** . . . 28

3.1 Transducer Responses 30
3.2 Calibration Methods 31
3.3 Reciprocity Calibration 37
3.4 Beam Patterns 39
3.5 Product Theorem 44
3.6 Shading and Superdirectivity 45
3.7 Receiving Directivity Index 47
3.8 Array Gain 49
3.9 Multiplicative Arrays 54
References 58

FOUR Generation of Underwater Sound: Projector Source Level . 60

4.1 Relation between Source Level and Radiated Acoustic Power 61
4.2 Limitations on Sonar Power 64
4.3 Explosions as Sources of Underwater Sound 69
References 80

FIVE Propagation of Sound in the Sea: Transmission Loss, I . 82

5.1 Introduction 82
5.2 Spreading Laws 83
5.3 Absorption of Sound in the Sea 86
5.4 Velocity of Sound in the Sea 93
5.5 Velocity Structure of the Sea 100
5.6 Propagation Theory and Ray Tracing 103
5.7 The Sea Surface 109
5.8 The Sea Bottom 114
References 119

SIX Propagation of Sound in the Sea: Transmission Loss, II . 122

6.1 The Mixed-layer Sound Channel 122
6.2 The Deep Sound Channel 132
6.3 The Shallow-water Channel 141
6.4 Fluctuation of Transmitted Sound 149
6.5 Deep-sea Paths and Losses: A Summary 154
References 156

SEVEN The Noise Background of the Sea: Ambient-noise Level . 160

7.1 Sources of Ambient Noise in Deep Water 161
7.2 Deep-water Spectra 166
7.3 Shallow-water Ambient Noise 168
7.4 Variability of Ambient Noise 170
7.5 Intermittent Sources of Ambient Noise 171
7.6 Effect of Depth 175
7.7 Amplitude Distribution 177
7.8 Noise in Ice-covered Waters 178
7.9 Directional Characteristics of Deep-water Ambient Noise 180
7.10 Spatial Coherence of Ambient Noise 182
7.11 Summary 183
References 184

EIGHT Scattering in the Sea: Reverberation Level 187

8.1 Types of Reverberation 187
8.2 The Scattering-strength Parameter 188
8.3 Equivalent Plane-wave Reverberation Level 190
8.4 Volume-reverberation Theory 190
8.5 Surface-reverberation Theory 194

8.6 *Target Strength and Scattering Strength* 196
8.7 *Surface Scattering by a Layer of Volume Scatterers* 197
8.8 *Reverberation Level for Short Transients* 198
8.9 *Air Bubbles in Water* 199
8.10 *Volume Reverberation* 204
8.11 *Sea-surface Reverberation* 209
8.12 *Theories and Causes of Sea-surface Scattering* 213
8.13 *Sea-bottom Reverberation* 217
8.14 *Underice Reverberation* 225
8.15 *Characteristics of Reverberation* 226
8.16 *Reverberation Prediction* 229
References 230

NINE **Reflection and Scattering by Sonar Targets: Target Strength** 234

9.1 *The Echo as the Sum of Backscattered Contributions* 236
9.2 *Geometry of Specular Reflection* 237
9.3 *Target Strength of a Small Sphere* 240
9.4 *Target Strength of Simple Forms* 243
9.5 *Target-strength Measurement Methods* 247
9.6 *Target Strength of Submarines* 249
9.7 *Target Strength of Surface Ships* 254
9.8 *Target Strength of Mines* 254
9.9 *Target Strength of Torpedoes* 255
9.10 *Target Strength of Fish* 255
9.11 *Echo Formation Processes* 256
9.12 *Echo Characteristics* 259
9.13 *Summary of Numerical Values* 261
References 261

TEN **Radiated Noise of Ships, Submarines, and Torpedoes: Radiated-noise Levels** 263

10.1 *Source Level and Noise Spectra* 263
10.2 *Methods of Measurement* 264
10.3 *Sources of Radiated Noise* 266
10.4 *Summary of the Sources of Radiated Noise* 275
10.5 *Total Radiated Acoustic Power* 277
10.6 *Radiated-noise Levels* 278
References 284

ELEVEN **Self-noise of Ships, Submarines, and Torpedoes: Self-noise Levels** 286

11.1 *Self-noise Measurements and Reduction* 288
11.2 *Sources and Paths of Self-noise* 289
11.3 *Flow Noise* 293
11.4 *Domes* 298
11.5 *Self-noise Levels* 300
References 303

TWELVE **Detection of Signals in Noise and Reverberation: Detection Threshold** 304

12.1 Definition of Detection Threshold 305
12.2 The Threshold Concept 306
12.3 Input Signal-to-noise Ratio for Detection 310
12.4 Estimating Detection Threshold 313
12.5 Effect of Duration and Bandwidth 314
12.6 Example of a Computation 315
12.7 Tabular Summary 317
12.8 Auditory Detection 317
References 324

THIRTEEN **Design and Prediction in Sonar Systems** 326

13.1 Sonar Design 326
13.2 Sonar Prediction 327
13.3 The Optimum Sonar Frequency 328
13.4 Applications of the Sonar Equations 332
References 337

Index 339

ONE

The Nature of Sonar

Of all the forms of radiation known to man, sound travels through the sea the best. In the turbid, saline water of the sea, both light and radio waves are attenuated to a far greater degree than is that form of mechanical energy known as sound.

Because of its relative ease of propagation, underwater sound has been applied by man to a variety of purposes in his use and exploration of the seas. These uses of underwater sound constitute the engineering science of *sonar*, and the systems employing underwater sound in one way or another are *sonar systems*.

Sonar systems, equipments, and devices are said to be *active* when sound is purposely generated by one of the system components called the *projector*. The sound waves generated by the projector travel through the sea to the *target*, and are returned as sonar *echoes* to a *hydrophone*, which converts sound into electricity. The electrical output of the hydrophone is amplified and processed in various ways and is finally applied to a control or display device to accomplish the purpose for which the sonar set was intended. Active sonar systems are said to *echo-range* on their targets.

Passive or *listening* sonar systems use sound (usually unwittingly) radiated by the target. Here only *one-way* transmission through the sea is involved, and the system centers around the hydrophone used to listen to the target sounds. Communication, telemetry, and control applications employ a hybrid form of sonar system using a projector and hydrophone at *both* ends of the acoustic communication path.

1.1 Historical Survey

Although the "modern age" of sonar may be said to date back a quarter of a century to the start of World War II, sonar has its origins deep in the past. One of the earliest references to the fact that sound exists beneath the surface of the sea, as well as in the air above, occurs in one of the notebooks of that versatile, archetypal engineer, Leonardo da Vinci. In 1490, two years before Columbus discovered America, he wrote (1):* "If you cause your ship to stop, and place the head of a long tube in the water and place the outer extremity to your ear, you will hear ships at a great distance from you." Although this earliest example of a passive sonar system has the enviable merit of extreme simplicity, it does not provide any indication of direction and is insensitive as a result of the great mismatch between the acoustic properties of air and water. Yet the idea of listening to underwater sounds by means of an air-filled tube between the sea and the listener's ear had widespread use as late as World War I, when, by the addition of a second tube between the other ear and a point in the sea separated from the first point, a direction could be obtained and the bearing of the target could be determined.

Perhaps the first quantitative measurement in underwater sound occurred in 1827, when a Swiss physicist, Daniel Colladon, and a French mathematician, Charles Sturm, collaborated to measure the velocity of sound in Lake Geneva in Switzerland. By timing the interval between a flash of light and the striking of a bell underwater, they determined the velocity of sound to a surprising degree of accuracy.

Later on in the nineteenth century, a number of famous physicists of the time indirectly associated themselves with underwater sound through their interest in the phenomenon of "transduction"—the conversion of electricity into sound and vice versa (2). Jacques and Pierre Curie are usually credited with the discovery in 1880 of piezoelectricity—the ability of certain crystals, when stressed, to develop an electric charge across certain pairs of crystal faces. Other physicists had dabbled in the subject before this. Charles Coulomb is said to have speculated on the possibility of producing electricity by pressure, and Wilhelm Röntgen wrote a paper on the electric charge appearing on the various faces of crystals under stress. The counterpart of piezoelectricity as a transduction process is magnetostriction, wherein a magnetic field produces a change in the shape of certain substances. The earliest manifestation of magnetostriction was the musical sounds that were heard when, about 1840, the current in a coil of wire was changed or interrupted

* The parenthetical numbers throughout the text denote numbered references to the literature in a list of references at the end of each chapter.

near the poles of a horseshoe magnet. James Joule, in the 1840s, carried out quantitative measurements on the change of length associated with magnetostriction, and is commonly credited with being the discoverer of the effect.

These studies, and those of others in the 1840s and 1850s, were the foundation for the invention of the telephone, for which a long-disputed patent was issued in 1876 to A. G. Bell. Another nineteenth-century invention that was the mainstay of sonar systems before the advent of electronic amplifiers was the carbon-button microphone, a device which became the earliest, and still probably the most sensitive, hydrophone for underwater sound.

About the turn of the century there came into being the first practical application of underwater sound. This was the submarine bell, used by ships for offshore navigation. By timing the interval between the sound of the bell and the sound of a simultaneously sent blast of a foghorn, a ship could determine its distance from the lightship where both were installed. This system was the impetus for the founding of the Submarine Signal Company (now part of Raytheon Mfg. Co.), the first commercial manufacturer of sonar equipment in the United States. The method was never in widespread use and was soon replaced by navigation methods involving radio—especially radio direction finding.

Another pre-World War I achievement was the embryonic emergence of the first schemes for the detection of underwater objects by echo ranging. In 1912, five days after the "Titanic" collided with an iceberg, L. F. Richardson filed a patent application with the British Patent Office for echo ranging with airborne sound (2). A month later he applied for a patent for its underwater analog. These ideas involved the then-new features of a directional projector of kilohertz-frequency sound waves and a frequency-selective receiver detuned from the transmitting frequency to compensate for the Doppler shift caused by the motion of the echo-ranging vessel. Unfortunately, Richardson did nothing at the time to implement these proposals. Meanwhile, in the United States, R. A. Fessenden had designed and built a new kind of moving-coil transducer for both submarine signaling and echo ranging and was able, by 1914, to detect an iceberg at a distance of 2 miles. Fessenden "oscillators" operating at frequencies near 500 and near 1,000 Hz are said (3) to have been installed on all United States submarines of the World War I period to enable them to signal one another while submerged. They remained in use until recently as powerful sinusoidal sound sources for research purposes.

The outbreak of World War I in 1914 was the impetus for the development of a number of military applications of sonar. In France a young Russian electrical engineer, Constantin Chilowsky, collaborated with a distinguished physicist, Paul Langevin, in experiments with a condenser (electrostatic) projector and a carbon-button microphone placed at the focus of a concave mirror. In spite of leakage and breakdown troubles caused by the high voltages needed for the projector, by 1916 they were able to obtain echoes from the bottom and from a sheet of armor plate at a distance of 200 m. Later, in 1917, Langevin turned to the piezoelectric effect and used a quartz-steel sandwich to replace the condenser projector. He also employed one of the newly developed vacuum-tube amplifiers—probably the first application of electronics to underwater sound equipment. For the first time, in 1918, echoes were received from a submarine, occasionally at distances as great as 1,500 m. Parallel British investigations with quartz projectors were carried out by a group under R. W. Boyle. The word "asdic" was coined at the time to refer to their then highly secret experiments.* World War I came to a close, however, before underwater echo ranging could make any contribution to meet the German U-boat threat.

In the meantime, extensive use had been made of Leonardo's air tube for passive listening, improved by the use of two tubes to take advantage of the binaural directional sense of a human observer. The MV device consisted (5) of a pair of line arrays of 12 air tubes each, mounted along the bottom of a ship on the port and starboard sides and steered with a special compensator. Surprising precision was achieved in determining the bearing of a noisy target; an untrained observer could find the bearing of a distant target to an accuracy of ½°. Another development of the time (5) was a neutrally buoyant, flexible line array of 12 hydrophones called the "eel," which could easily be fitted to any ship and could be towed astern away from the noisy vessel on which it was mounted. All in all, some three thousand escort craft were fitted with listening devices of various kinds in World War I. By operating in groups of two or three and using cross bearings, they could obtain a "fix" on a suspected submarine contact.

The years of peace following World War I saw a steady, though extremely slow, advance in applying underwater sound to practical needs. Depth sounding by ships under way was soon developed, and by 1925, fathometers, a word coined by the Submarine Signal Company for their own equipment, were available commercially in both the United States and Great Britain. The search for a practical means of echo ranging on submarine targets was carried on in the United States by a handful of

* According to A. B. Wood (4), the word "asdic" was originally an acronym for "*A*nti-*S*ubmarine *D*ivision—*ics*" from the name of the group which did the work. The suffix had the same significance as it does in the words "physics," "acoustics," etc. For many years thereafter, the word "asdic" was used by the British to refer to echo ranging and echo-ranging sonar systems generally.

men under H. C. Hayes at the Naval Research Laboratory. The problem of finding a suitable sound projector in echo ranging was solved by resorting to magnetostrictive projectors for generating the required amount of acoustic power. Also, synthetic crystals of Rochelle salt began to replace scarce natural quartz as the basic piezoelectric material for piezoelectric transducers. During the interwar period sonar received a great practical impetus by advances in electronics, which made possible vast new domains of amplifying, processing, and displaying sonar information to an observer.

Ultrasonic frequencies, that is, frequencies beyond the region of sensitivity of the unaided human ear, came to be used for both listening and echo ranging and enabled an increased directionality to be obtained with projectors and hydrophones of modest size. A number of small, but vital, components of sonar systems were added during this period, notably the development by the British of the range recorder for echo-ranging sonars to provide a "memory" of past events and the streamlined dome to protect the transducer on a moving ship from the noisy, turbulent environment of water flow past a moving vessel. By 1935, several fairly adequate sonar systems had been developed, and by 1938, with the imminence of World War II, quantity production of sonar sets started in the United States. By the time the war began, a large number of American ships were equipped for both underwater listening and echo ranging. The standard echo-ranging sonar set for surface ships was the QC equipment. The operator searched in bearing with it by turning a handwheel and listening for an echo with headphones or loudspeaker. If an echo was obtained, its range was noted by the flash of a rotating light or from the range recorder. Submarines were fitted with JP listening sets, consisting of a rotatable horizontal line hydrophone, an amplifier, a selectable bandpass filter, and a pair of headphones. The cost of this equipment with spares was $5,000! With such primitive sonar sets, the Battle of the Atlantic against the German U-boat was engaged and, eventually, won.

But from a scientific standpoint, perhaps the most notable accomplishment of the years between World War I and World War II was the obtaining of an understanding of the vagaries of sound propagation in the sea. Early shipboard echo-ranging sets installed in the late twenties and early thirties were mysteriously unreliable in performance. Good echoes were often obtained in the morning, while poor echoes, or none at all, were obtained in the afternoon. When it became clear that the sonar operators themselves were not to blame and that the echoes were actually weaker in the afternoon, the cause began to be sought in the transmission characteristics of the seawater medium. Only with the use of special temperature-measuring equipment did it become evident that slight thermal gradients, hitherto unsuspected, were capable of refracting sound deep into the depths of the sea and could cause the target to lie in what is now known as a "shadow zone." The effect was called by E. B. Stephenson the "afternoon effect." As a means to indicate temperature gradients in the upper few hundred feet of the sea, A. F. Spilhaus built the first bathythermograph in 1937; by the start of World War II, every naval vessel engaged in antisubmarine work was equipped with the device. During this period also, a clear understanding was gained of absorption of sound in the sea, and remarkably accurate values of absorption coefficient were determined at the ultrasonic frequencies 20 to 30 kHz then of interest.

On both sides of the Atlantic, as in World War I, the World War II period was marked by feverish activity in underwater sound. In the United States, a large group of scientists organized by the National Defense Research Committee (NDRC) began investigations of all phases of the subject.* Most of our present concepts as well as practical applications had their origins in this period. The acoustic homing torpedo, the modern acoustic mine, and scanning sonar sets were wartime developments. Methods for quick calibration of projectors and hydrophones began to be used, and an understanding of the many factors affecting sonar performance that are now summarized in the sonar equations was gained. Factors such as target strength, the noise output of various classes of ships at different speeds and frequencies, reverberation in the sea, and the recognition of underwater sound by the human ear were all first understood in a quantitative way during the years of World War II. Indeed, in retrospect, there is little of our fund of underwater acoustic knowledge that cannot be traced, in its rudiments, to the discoveries of the wartime period.

The Germans must be given credit for a number of unique accomplishments. One was the development of Alberich, a nonreflecting coating for submarines. It consisted of a perforated sheet of rubber cemented to the hull of a submarine and covered by a solid, thin sheet of rubber to keep water out of the air-filled perforations. This coating was effective only over a limited range of depth and frequency, and could not be kept bonded to the hull for long periods under operating conditions. Another innovation was the use of flush-mounted arrays for listening aboard surface ships. An array of this kind—given the designation GHG for *grüppen-hört-gerät* or "array listening equipment"—was installed on the cruiser "Prinz Eugen" and used with some success.

According to Batchelder (6), the word "sonar" was coined late in the

* At the end of the war the findings of that part of NDRC engaged in underwater sound were summarized in an admirable series of some 22 reports called the *NDRC Division 6 Summary Technical Reports.*

war as a counterpart of the then-glamorous word "radar" and came into use later only after having been dignified as an acronym for *so*und *na*vigation and *r*anging!

In the post-World War II years, continued improvement in sonar systems has occurred. In echo ranging this has been brought about by the use of higher acoustic powers and lower frequencies and by the incorporation of signal processing to supplement the previously almost unaided human observer. Similarly, passive sonar systems have tended to migrate to lower frequencies, where both acoustic output of the target vessel is greater and absorption of sound in the sea is less. Lower frequencies and higher powers have necessitated larger arrays containing, by earlier standards, enormous numbers of hydrophone elements. In recent years, arrays have been placed on the deep ocean floor to take advantage of the quiet environment and the good propagation conditions existing at such locations. In the meantime, continued measurements and research at sea have added to our quantitative and numerical knowledge of the underwater acoustic parameters that is needed for engineering purposes.

A striking development of the postwar period, and one that is still taking place, is the expansion of the applications of underwater sound to peaceful purposes. Beyond its traditional use in depth sounding, underwater sound is being used in acoustic navigation beacons, subbottom geologic mapping in shallow and deep water, fish finding, and in telemetry and control applications. With the expanding vistas of man's explorations of the seas, it is likely that the number of such peaceful uses of underwater sound will continue to increase in future years.

1.2 Uses of Underwater Sound

The useful spectrum of underwater sound covers some five or six decades of frequency. Beginning at the ultra-low-frequency end of the application spectrum, one finds the *pressure mine*, a device actuated by the reduction in pressure caused by the motion of a ship in its vicinity. This reduction of pressure is a hydrodynamic (Bernoulli) effect rather than a sound effect; it does not require an elastic medium for its existence and is not propagated to a distance. Yet in the mine it is apt to be sensed by a pressure transducer, to which it appears no different from an acoustic wave. The pressure disturbances that are produced by a ship under way contain frequencies generally below 1 Hz. *Acoustic mines* sense the true acoustic radiation of ships and explode when the acoustic level in their passband reaches a certain value. In *mine sweeping*, such mines are swept, or purposely exploded, by the sound of a powerful source towed behind the mine-sweeping vessel. Mines and mine sweeping

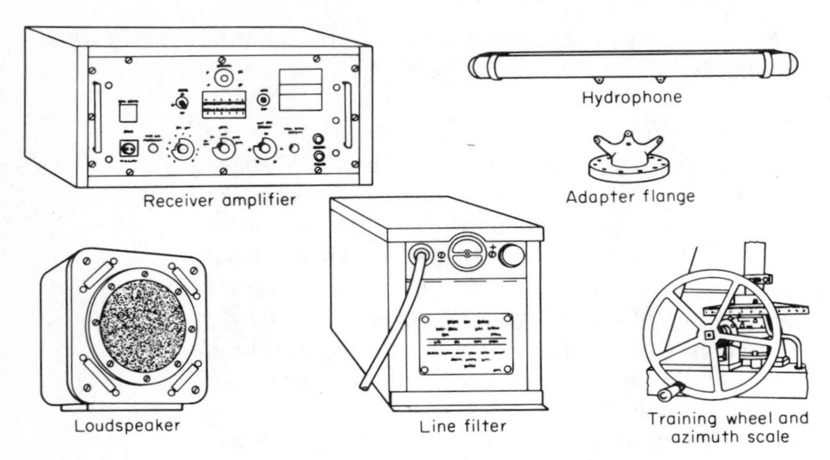

***FIG. 1.1.** The JP listening equipment for submarines. This device was in general use on United States submarines during World War II.* (*Ref.* 7.)

commonly utilize frequencies of the spectrum where the acoustic output of vessels that are the targets for acoustic mines is at a maximum. The acoustic radiation of ships and submarines is also employed for *passive detection* by a hydrophone array on another vessel or on the bottom, a long distance away. Because the receiving array must be directional, so as to be able to determine the direction to the target, somewhat higher frequencies are employed in passive detection. The common passive listening set on United States submarines in the days of World War II—the JP equipment—is illustrated in Fig. 1.1. It utilized a mechanically trained horizontal line hydrophone 3 ft long; the operator listened with earphones or a loudspeaker to the output of selected frequency bands between 70 and 12,000 Hz. At still higher frequencies occur sonar systems for the *active detection* of submarine targets, more particularly those designed for installations on destroyer-like naval vessels. An example of an echo-ranging sonar for shipboard installation may be seen in Fig. 1.2, which shows the components of the QHB scanning sonar. In scanning sonars, a short pulse of sound is sent out uniformly by the cylindrically shaped projector; on reception, the same unit is used as a hydrophone, but successive portions of the transducer are selected by an electronic scanning switch to form a rapidly rotating directional beam synchronized to the rotating sweep of a ppi (plan position indicator) oscilloscope display. The QHB sonar operates at 25.5 kHz. *Homing torpedoes* utilize moderately high frequencies because of the necessarily small size of their transducers and the consequently higher frequencies needed to form an adequately directional beam and to reduce noise. Homing torpedoes detect and steer toward their targets and may be either *active homers* or *passive homers*, depending on whether they echo-range or home on the radiated noise of the target. *Mine-hunting sonar systems* and *small-object locators* operate at high kilohertz frequencies by echo ranging with a highly directional transducer and a short pulse duration so as to be able to detect the mine or the small object in the background of clutter (reverberation) in which it appears.

A number of special-purpose sonar equipments may be mentioned. The *underwater telephone* is a device for communicating between a surface ship and a submarine or between two submarines. The UQC-1 equip-

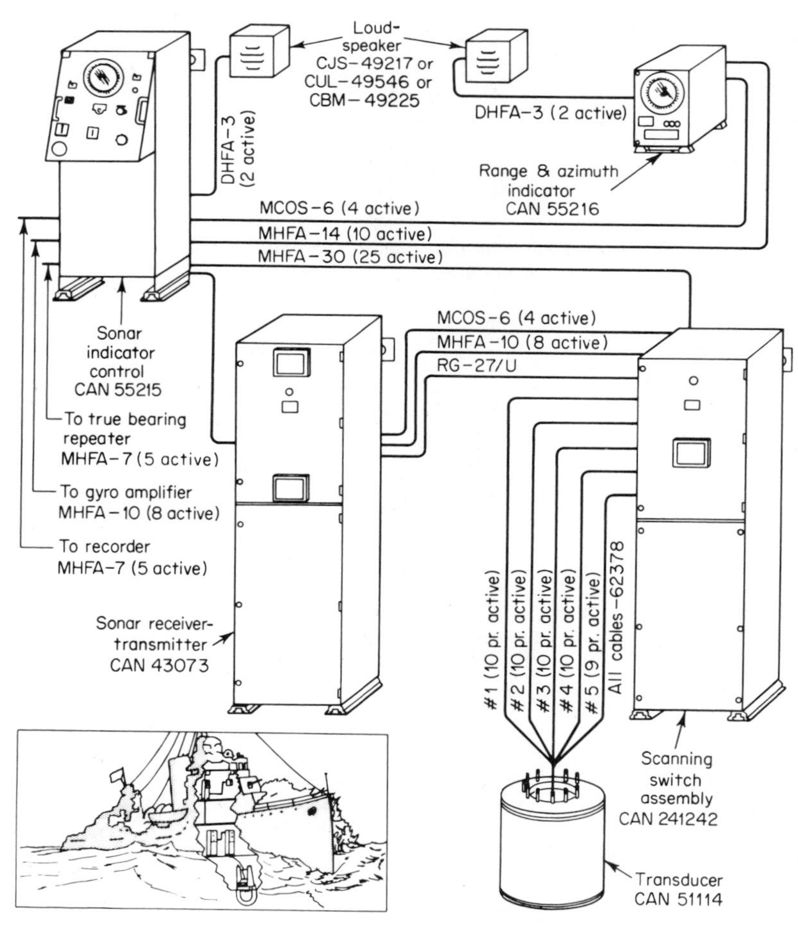

FIG. 1.2. ***A scanning sonar, Type QHB, for surface-vessel installation.*** ***(Ref. 7.)***

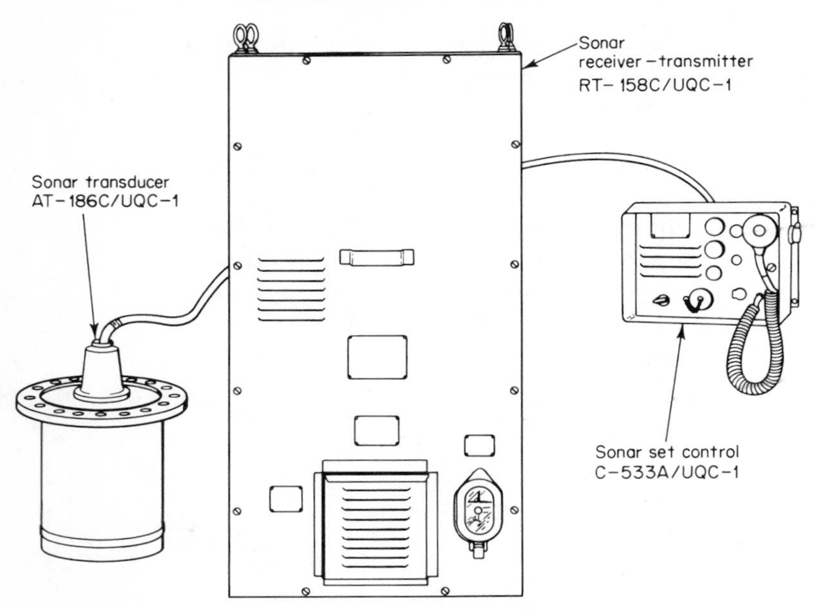

FIG. 1.3. The underwater telephone UQC-1 for communication with submarines. (Ref. 7.)

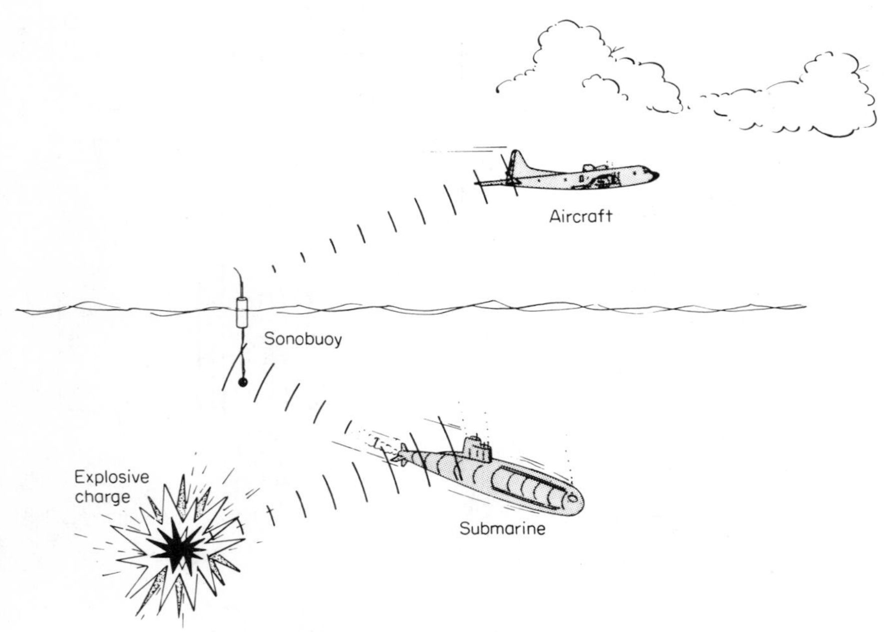

FIG. 1.4. Submarine hunting from an aircraft. The sonobuoy is the link between sound underwater and the aircraft above. The explosive charge is a sound source for echo ranging.

ment shown in Fig. 1.3 uses the single-sideband voice modulation of an 8-kHz carrier and is the underwater analog of a radio transmitter-receiver. *Sonobuoys* are small sonar sets dropped by an aircraft for underwater listening or echo ranging. This compact, expendable device contains a miniature radio transmitter for relaying signals picked up by its hydrophone. A pictorial view of an aircraft, its sonobuoy, and a submarine target is seen in Fig. 1.4. By means of a specially packaged explosive charge also dropped by the aircraft, *explosive echo ranging*, wherein the sonobuoy is used to receive and transmit echoes, can be done. One of the oldest applications of sonar is *depth sounding* (fathometry) in which, in effect, echo ranging is done against the bottom. Figure 1.5 shows the components of the UQN-1 sounding set used for obtaining the depth of the bottom in the deepest waters of the ocean. It operates at 12 kHz and prints a permanent record of the depth along the track of the vessel on which it is installed. More compact depth sounders for small craft are available commercially for use in shallow waters. Many depth sounders are capable of being used as *fish finders* when a school of fish lies beneath the vessel. Special active sonar sets for searching for fish *ahead* of the vessel have been developed commercially for fishing-boat installations.

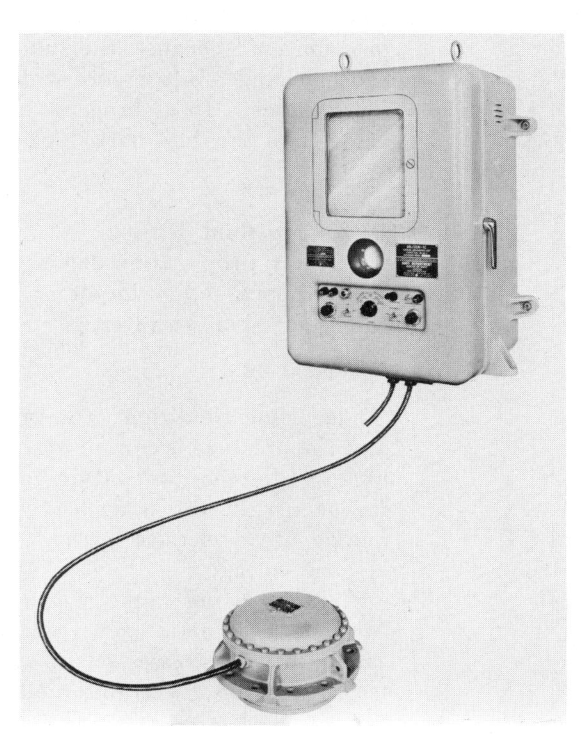

FIG. 1.5. A depth sounder, Type UQN-1, for depth finding in deep water. (Photograph by permission of Edo Corporation, College Point, New York)

Finally, the *acoustic marine speedometer* should be mentioned as an interesting device for navigational use (8). This device, intended during World War II to be an improvement on the pitometer log for finding the speed of a vessel, exploits the Doppler shift of volume reverberation produced by the vessel's motion. The principle is similar to that used in CW (Continuous Wave) Doppler radar; indeed, the underwater analog of Doppler radar, but using the acoustic backscattering of the sea bottom, has been developed as a navigational aid for ships. All these last-mentioned devices generally operate at frequencies in the vicinity of 100 kHz. Except when extremely high resolution is required, frequencies greater than a few hundred kilohertz have not been employed in seagoing applications because of the high absorption of sound at these frequencies.

1.3 Basic Concepts and Units

Sound consists of a regular motion of the molecules of an elastic substance. Because the material is elastic, a motion of the particles of the material, such as the motion initiated by a sound projector, communicates to adjacent particles. A sound wave is thereby propagated outward from the source at a velocity equal to the velocity of sound. In a fluid, the particle motion is to and fro, parallel to the direction of propagation; because the fluid is compressible, this to-and-fro motion causes changes in pressure which can be detected by a pressure-sensitive hydrophone. In a plane wave of sound the pressure p is related to the velocity of the fluid particles u by

$$p = \rho c u$$

where ρ = fluid density
c = propagation velocity of wave

The proportionality factor ρc is called the *specific acoustic resistance* of the fluid. For seawater, $\rho c = 1.5 \times 10^5$ g/(cm²)(sec); for air,

$$\rho c = 42 \text{ g/(cm}^2\text{)(sec)}$$

Under some conditions, the proportionality factor between the velocity and the pressure is complex and is then called the *specific acoustic impedance* of the medium containing the sound wave. This is the Ohm's law for acoustics; the particle velocity u may be viewed as the acoustic analog of an electric current, and the pressure p as the analog of an electric voltage.

A propagating sound wave carries mechanical energy with it in the form of the kinetic energy of the particles in motion plus the potential energy of the stresses set up in the elastic medium. Because the wave is propagating, a certain amount of energy per second will flow across a unit area oriented normal to the direction of propagation. This amount of energy per second (power) crossing a unit area is called the *intensity* of the wave. If the unit area is taken in an arbitrary direction, the intensity becomes a vector quantity analogous to the *Poynting vector* of electromagnetic propagation. In a plane wave, the instantaneous intensity is related to the instantaneous acoustic pressure by

$$I = \frac{p^2}{\rho c}$$

When the squared pressure is averaged over an interval of time, the average intensity for that interval is

$$I = \frac{\overline{p^2}}{\rho c}$$

where the bar indicates a time average. In practice, there is always some integration time inherent in the system being used, and the average intensity, rather than the instantaneous intensity, will be of practical interest. If p is in dynes per square centimeter, and ρ and c are in units of grams per cubic centimeter and centimeters per second, then the intensity I will have the units of ergs per square centimeter per second. Since 1 watt equals 10^7 ergs/sec, the intensity in units of watts per square centimeter is

$$I = \frac{\overline{p^2}}{\rho c} \times 10^{-7}$$

With $\rho c = 1.5 \times 10^5$, a plane wave in water of rms pressure 1 dyne/cm^2 has an intensity of 0.64×10^{-12} watt/cm^2.

For transient signals, and in problems where considerable signal distortion takes place in propagation or in encounters with the target, it is more meaningful to refer to the *energy flux density* of the acoustic wave. The energy flux density is the time integral of the instantaneous intensity, or

$$E = \int_0^\infty I\,dt = \frac{1}{\rho c}\int_0^\infty p^2\,dt$$

where the integral, in a practical case, is taken over the duration of the wave. The units of energy flux density are ergs per square centimeter.

In underwater sound, *the basic unit of intensity is the intensity of a plane wave having an rms pressure equal to* 1 *dyne/cm*^{2}. This unit of intensity equals 0.64×10^{-12} watt/cm^2. Similarly, the unit of energy flux density is the energy-density of a plane wave of rms pressure equal to 1 dyne/cm^2 integrated over a time interval of 1 sec. This unit amounts to 0.64×10^{-12} watt-sec/cm^2. These are the units implied when a particular intensity or energy-density is said to be "relative to 1 dyne/cm^2" (abbreviated "re 1 dyne/cm^2"). Since 1 dyne/cm^2 equals 1 microbar (μb), the abbreviation "re 1 μb" is sometimes used for this unit. For the radiated noise of ships, a reference pressure equal to 0.00204 dyne/cm^2 is commonly employed—a unit taken over from air acoustics, where it is the threshold sensitivity of the human ear at 1,000 Hz. In order to be consistent, however, we will use the 1 dyne/cm^2 reference as a single pressure for all the sonar quantities in the chapters to follow.

For other quantities in underwater sound, other units are in engineering usage. Horizontal distances (ranges) are commonly expressed in yards, kiloyards (1 kyd = 1,000 yd), or nautical miles (1 nm = 2,027 yd); depths are expressed in feet or fathoms (1 fathom = 6 ft); the speed of sound is stated in feet per second, whereas the speed of ships, submarines, and torpedoes is stated in knots (1 knot = 1 nm/hr); temperatures are specified in degrees Fahrenheit. These diverse units appear in much of the underwater sound literature and reflect the strong orientation of the subject toward naval applications. In current scientific publications, the metric system is making rapid headway.

Acousticians have long used the *decibel* for expressing intensity and energy-density because it is a convenient way to handle large changes in variables and, in addition, permits quantities to be multiplied together by the more simple process of adding their decibel equivalents. The decibel is a ratio of intensities (or energy-densities) expressed in logarithmic units. If I_1 and I_2 are two intensities, the number N denoting the ratio I_1/I_2 is

$$N = 10 \log_{10} \frac{I_1}{I_2} \qquad \text{db}$$

I_1 and I_2 are said to differ by N db. The intensity *level* of a sound wave is the number of decibels by which its intensity differs from the reference intensity as just described. That is, if I_2 in the above expression is the reference intensity, then a sound wave of intensity I_1 has a *level* equal to N db re 1 dyne/cm^2, where "re 1 dyne/cm^2" is used to indicate the reference intensity employed. For example, a sound wave having an intensity 100 times that of a plane wave of rms pressure 1 dyne/cm^2 has a level equal to 10 log (100/1) = 20 db re 1 dyne/cm^2. Energy flux densities may be compared with a reference energy flux density in a similar way through a specification of energy-flux-density level.

The term *spectrum level* refers to the level of a sound wave in a frequency band 1 Hz wide. It has meaning only for sounds having a "continuous" spectrum, that is, those for which there is some sound, however small, in any frequency band. *Band level* refers to the level in a frequency band greater than 1 Hz wide. For a flat (white) spectrum, or as an approximation for any continuous spectrum if the band is not too wide, the band level BL is related to the spectrum level SPL at the middle of the band by

$$\mathrm{BL} = \mathrm{SPL} + 10 \log W$$

where W is the bandwidth in hertz. This expression implies that the intensities in all the adjacent 1-Hz bands can be added together to give the intensity in the W-hertz band.

REFERENCES

1. Bell, T. G.: Sonar and Submarine Detection, *U.S. Navy Underwater Sound Lab. Rept.* 545, 1962.
2. Hunt, F. V.: "Electroacoustics," Chap. 1, John Wiley & Sons, Inc., New York, 1954.
3. Listening Systems, *Natl. Defense Res. Comm. Div. 6 Sum. Tech. Rept.* 14, chap. 2, 1946.
4. Wood, A. B.: From the Board of Invention and Research to the Royal Naval Scientific Service, *J. Roy. Naval Sci. Serv.*, **20**(4):185 (1965).
5. Hayes, H. C.: Detection of Submarines, *Proc. Am. Phil. Soc.*, **59**:1 (1920).
6. Batchelder, L. B.: When Sonar Was Called Submarine Signalling, *J. Acoust. Soc. Am.*, **31**:832(A) (1959).
7. Directory of Sonar Equipment, *U.S. Dept. Navy Bur. Ships* NAVSHIPS 94200.3, 1962.
8. Acoustic Marine Speedometer-completion Report, *Harvard Univ. Acoust. Res. Lab. Rept.* 6.1-ser. 287-2074, 1945.

TWO

The Sonar Equations

The many phenomena and effects peculiar to underwater sound produce a variety of quantitative effects on the design and operation of sonar equipment. These diverse effects can be conveniently and logically grouped together quantitatively in a small number of quantities called the *sonar parameters*, which, in turn, are related by the *sonar equations*. These equations are the working relationships that tie together the effects of the medium, the target, and the equipment, and they are among the design and prediction tools available to the engineer for underwater sound applications.

The sonar equations were first formulated during World War II (1) as the logical basis for calculations of the maximum range of sonar equipments. In recent years, they have seen increasing use in the optimum design of sonars for new applications. Essentially the same relationships are used in radar (2), though with linear instead of logarithmic units and with slightly different definitions of the parameters.

The essentially simple sonar equations serve two important practical functions. One is the *prediction of performance* of sonar equipments of known or existing design. In this application the design characteristics of the sonar set are known or assumed, and what is desired is an estimate of performance in some meaningful terms such as detection probability or search rate. This is done in the sonar equations by a prediction of range through the parameter transmission loss. The equations are solved for transmission loss, which is then converted to range through some assumption as to the propagation characteristics of the medium.

The other general application of the equations is in *sonar design*, where a preestablished range is required for the operation of the equipment being designed. In this case the equation is solved for the particular troublesome parameter whose practical realization is apt to cause difficulty. An example would be the directivity required, along with other likely values of sonar parameters, to yield a desired range of detection in a detection sonar or the range of actuation by a passing ship of an acoustic mine mechanism. After the directivity needed to obtain the desired range has been found, the design continues through the "trade offs" between directivity index and other parameters. The design is finally completed through several computations using the equations and the design engineer's intuition and experience.

2.1 Basic Considerations

The equations are founded on a basic equality between the desired and undesired portions of the received signal at the instant when some function of the sonar set is just performed. This function may be detection of an underwater target, or it may be the homing of an acoustic torpedo at the instant when it just begins to acquire its target. These functions all involve the reception of acoustic energy occurring in a natural acoustic background. Of the total acoustic field at the receiver, a portion may be said to be *desired* and is called the *signal*. The remainder of the acoustic field is *undesired* and may be called the *background*. In sonar the background is either *noise*, the essentially steady-state portion not due to one's own echo ranging, or *reverberation*, the slowly decaying portion of the background representing the return of one's own acoustic output by scatterers in the sea. The design engineer's objective is to find means for *increasing* the overall response of the sonar system to the signal and for *decreasing* the response of the system to the background—in other words, to *increase* the signal-to-background ratio.

Let us imagine a sonar system serving a practical purpose such as *detection*, *classification* (determining the nature of a target), *torpedo homing*, *communication*, or *fish finding*. For each of these purposes there will be a certain signal-to-background ratio that will depend on the functions being performed and on the performance level that is desired in terms of percentages of successes and "false alarms," such as an apparent detection of a target when no target is present. If the signal is imagined to be slowly increasing in a constant background, the desired purpose will be accomplished when the *signal level equals the level of the background which just masks it*. That is to say, when the sonar's purpose is *just* accomplished,

Signal level = background masking level

The term "masking" implies that not all of the background interferes with the signal, but only a portion of it does so—usually that portion lying in the frequency band of the signal. The word "masking" is borrowed from the theory of audition, where it refers to that part of a broadband noise background that masks out a pure tone or a narrow-band signal presented to a human listener.

2.2 The Active and Passive Equations

The next step is to expand the basic equality in terms of the *sonar parameters* determined by the *equipment*, the *medium*, and the *target*. We will denote these parameters by two-letter symbols in order to avoid Greek and subscripted symbols as much as possible in the writing of the equations. These parameters are levels in units of decibels relative to the standard reference intensity of a 1 dyne/cm^2 plane wave. They are as follows:

Parameters Determined by the *Equipment*
- Projector Source Level: SL
- Self-noise Level: NL
- Receiving Directivity Index: DI
- Detection Threshold: DT

Parameters Determined by the *Medium*
- Transmission Loss: TL
- Reverberation Level: RL
- Ambient-noise Level: NL

Parameters Determined by the *Target*
- Target Strength: TS
- Target Source Level: SL

Two pairs of the parameters are given the same symbol because they are essentially identical. It should be mentioned in passing that this set of parameters is not unique. Others, which could be employed equally well, might be more fundamental or might differ by a constant. For example, sound velocity could be adopted as a parameter, and TS could be replaced by the parameter "backscattering cross section" expressed in decibels, as is done in radar. The chosen parameters are therefore arbitrary; those used here are those conventionally used in underwater sound.* It should also be noted that they may all be expanded in terms of fundamental quantities like frequency, ship speed, and bearing—a subject that will be of dominant importance in the

* There is, however, no conventional symbolism for many of the parameters.

descriptions of the parameters that will follow. The units of the parameters are decibels, and they are added together in forming the sonar equations.

The meaning of these quantities can best be illustrated through some simple considerations for an active (echo-ranging) sonar (Fig. 2.1). A sound source acting also as a receiver (a *transducer*) produces by some means a *source level* of SL decibels at a unit distance (1 yd) on its axis. When the radiated sound reaches the target (if the axis of the sound source points toward the target), its level will be reduced by the *transmission loss*, and becomes SL − TL. On reflection or scattering by the target of target strength TS, the reflected or backscattered level will be SL − TL + TS at a distance of 1 yd from the acoustic center of the target in the direction back toward the source. In traveling back toward the source, this level is again attenuated by the *transmission loss* and becomes SL − 2TL + TS. This is the echo level at the transducer. Turning now to the background and assuming it to be isotropic noise rather than reverberation, we find that the *background level* is simply NL. This level is reduced by the *directivity index* of the transducer acting as a receiver or *hydrophone* so that at the terminals of the transducer the relative noise power is NL − DI. Since the axis of the transducer is pointing in the direction from which the echo is coming, the relative echo power is unaffected by the transducer directivity. At the

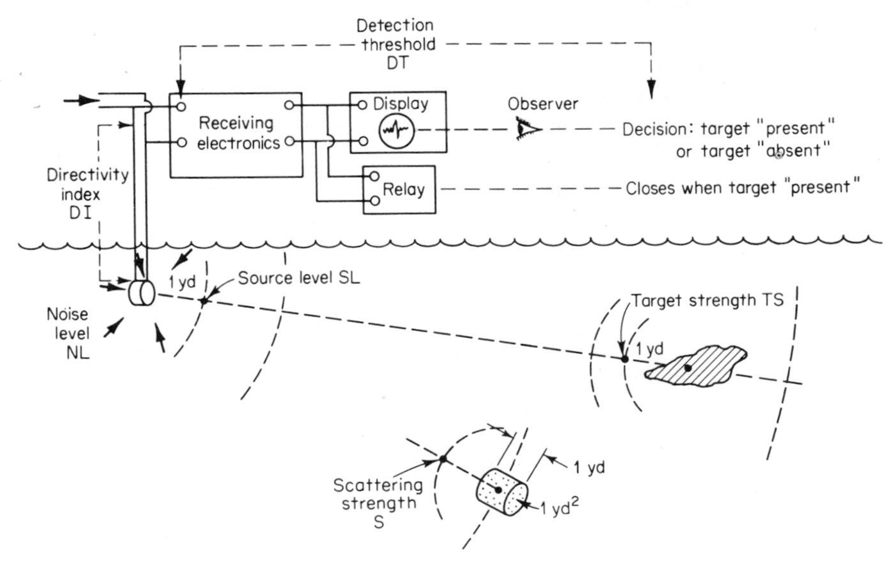

FIG. 2.1. Diagrammatic view of echo ranging, illustrating the sonar parameters.

transducer terminals, therefore, the echo-to-noise ratio is

$$SL - 2TL + TS - (NL - DI)$$

Let us now assume that the function that this sonar is called upon to perform is *detection*, that is, that its principal purpose is to give an indication of some sort on its *display* whenever an echoing target is present. When the input signal-to-noise ratio is above a certain detection threshold fulfilling certain probability criteria, a decision will be made by a human observer that a target is *present;** when the input signal-to-noise ratio is less than the detection threshold, the decision will be made that the target is *absent.* When the target is *just* being detected, the signal-to-noise ratio *equals* the detection threshold, and we have

$$SL - 2TL + TS - (NL - DI) = DT$$

We have here written the active-sonar equation as an equality in terms of the *detection threshold,* called in audition and in much of the older underwater sound literature *recognition differential.* In terms of the basic equality described above, we could equally well consider that only that part of the noise power lying above the detection threshold masks the echo, and we would have

$$SL - 2TL + TS = NL - DI + DT$$

a more convenient arrangement of the parameters, since the echo level occurs on the left-hand side and the noise-masking background level occurs on the right.

This is the active-sonar equation for the *monostatic case* in which the source and receiving hydrophone are coincident and in which the acoustic return of the target is back toward the source. In some sonars, a separated source and receiver are employed and the arrangement is said to be *bistatic;* in this case, the two transmission losses to and from the target are not, in general, the same. Also in some modern sonars, it is not possible to distinguish between DI and DT, and it becomes appropriate to refer to the sum DI − DT as the increase in signal-to-background ratio produced by the entire receiving system of transducer, electronics, display, and observer (if one is used).

A modification is required when the background is reverberation instead of noise. In this case, the parameter DI, defined in terms of an isotropic background, is inappropriate, inasmuch as reverberation is by no means isotropic. For a reverberation background we will replace

* If the human observer is replaced by a relay at the output of the detector, then the detection threshold is the input signal-to-background ratio at the transducer terminals which *just* closes the relay to indicate "target present."

the terms NL − DI by an *equivalent plane-wave reverberation level* RL observed at the hydrophone terminals. The active-sonar equation then becomes

$$\text{SL} - 2\text{TL} + \text{TS} = \text{RL} + \text{DT}$$

In the passive case, the target itself produces the signal by which it is detected, and the parameter source level now refers to the level of the radiated noise of the target at the unit distance 1 yd. Also, the parameter target strength becomes irrelevant, and one-way rather than two-way transmission is involved. With these changes, the *passive-sonar equation* becomes

$$\text{SL} - \text{TL} = \text{NL} - \text{DI} + \text{DT}$$

Table 2.1 is a list of parameters, reference locations, and short definitions in the form of ratios. More complete definitions of the parameters will

TABLE 2.1 The Sonar Parameters, Their Definitions, and Reference Locations

Parameter	symbol	Reference location	Definition
Source level	SL	1 yd from source on its acoustic axis	$10 \log \dfrac{\text{intensity of source}}{\text{reference intensity}^*}$
Transmission loss	TL	1 yd from source and at target or receiver	$10 \log \dfrac{\text{signal intensity at 1 yd}}{\text{signal intensity at target or receiver}}$
Target strength	TS	1 yd from acoustic center of target	$10 \log \dfrac{\text{echo intensity at 1 yd from target}}{\text{incident intensity}}$
Noise level	NL	At hydrophone location	$10 \log \dfrac{\text{noise intensity}}{\text{reference intensity}^*}$
Receiving directivity index	DI	At hydrophone terminals	$10 \log \dfrac{\text{noise power generated by an equivalent nondirectional hydrophone}}{\text{noise power generated by actual hydrophone}}$
Reverberation level	RL	At hydrophone terminals	$10 \log \dfrac{\text{reverberation power at hydrophone terminals}}{\text{power generated by signal of reference intensity}^*}$
Detection threshold	DT	At hydrophone terminals	$10 \log \dfrac{\text{signal power to just perform a certain function}}{\text{noise power at hydrophone terminals}}$

* The reference intensity is that of a plane wave of rms pressure 1 dyne/cm^2.

TABLE 2.2 Terminology of Various Combinations of the Sonar Parameters

Name	Parameters	Remarks
Echo level	SL − 2TL + TS	The intensity of the echo as measured in the water at the hydrophone
Noise masking level	NL − DI + DT	Another name for these two combinations is *minimum detectable echo level*
Reverberation masking level	RL + DT	
Echo excess	SL − 2TL + TS − (NL − DI + DT)	Detection just occurs, under the probability conditions implied in the term DT, when the echo excess is zero
Performance figure	SL − (NL − DI)	Difference between the source level and the noise level measured at the hydrophone terminals
Figure of merit	SL − (NL − DI + DT)	Equals the maximum allowable one-way transmission loss in passive sonars, or the maximum allowable two-way loss for TS = 0 db in active sonars

be given near the beginnings of the chapters dealing with each parameter.

2.3 Names for Various Combinations of Parameters

In practical work it is convenient to have separate names for different combinations of the terms in the equations. Methods exist for measuring some of these on shipboard sonars as a check on system operation. Table 2.2 is a listing of these names and the combination of terms each represents.

2.4 Echo, Noise, and Reverberation Level as Functions of Range

The sonar equations just written are no more than a statement of an equality between the desired portion of the acoustic field called the *signal*—either an echo or a noise from a target—and an undesired portion, called the *background* of noise or reverberation. This equality, in general, will hold at only one range; at other ranges, one or the other will be the greater and the equality will no longer exist.

This is illustrated in Fig. 2.2, where curves of echo level, noise masking level, and reverberation masking level are shown as a function of range. Both the echo and reverberation fall off with range, whereas the noise remains constant. The echo-level curve will generally fall off more rapidly with range than the reverberation-masking-level curve and will intersect it at the *reverberation-limited range* r_r given by the sonar equation for reverberation. The curve of echo level will also intersect the noise masking level at the range of the sonar equation for noise r_n. If the reverberation is high, the former will be less than the latter, and the range will be said to be *reverberation-limited.* If for any reason the noise masking level rises to the level shown by the dashed line in the figure, the echoes will then die away into a background of noise rather than reverberation. The new noise-limited range r_n' will then be less than the reverberation-limited range r_n, and the range will become *noise-limited.* Both ranges are given by the appropriate form of the sonar equation.

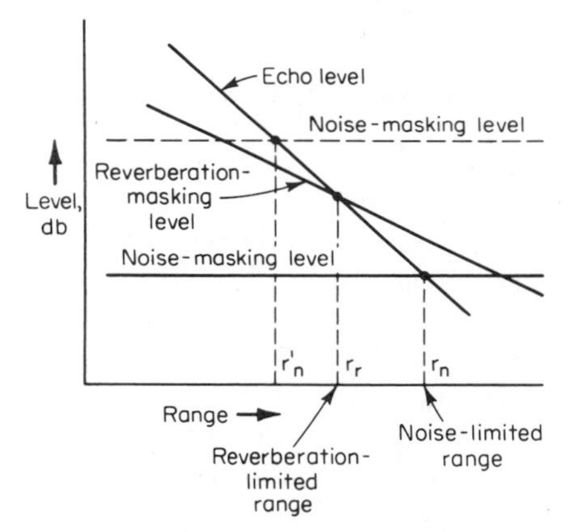

FIG. 2.2. Echo, noise, and reverberation as functions of range.

A knowledge as to whether a sonar will be noise- or reverberation-limited is necessary for both the sonar predictor and the sonar designer. In general the curves for echo and reverberation will not be straight lines because of complications in propagation and in the distribution of reverberation-producing scatterers. For a new sonar system, such curves should always be drawn from the best information available for the conditions most likely to be encountered in order to demonstrate visually to the design engineer the behavior of his signals and background with range.

2.5 Transient Form of the Sonar Equations

The equations thus far have been written in terms of *intensity,* or the average acoustic power per unit area of the sound emitted by the source or received from the target. The word "average" implies a time interval over which the average is to be taken. This time interval causes uncertain results whenever short transient sources exist or, generally, whenever severe distortion is introduced by propagation in the medium or by scattering from the target.

A more general approach is to write the equations in terms of *energy flux density,* defined as the acoustic energy per unit area of wavefront (see Sec. 1.3). If a plane acoustic wave has a time-varying pressure

$p(t)$, then the energy flux density of the wave is

$$E = \frac{1}{\rho c} \int_0^\infty p^2(t)\,dt$$

If the units of pressure are dynes per square centimeter and the acoustic impedance of the medium is in cgs units (for water, $\rho c \approx 1.5 \times 10^5$), then E will be expressed in ergs per square centimeter. The intensity is the mean-square pressure of the wave divided by ρc and averaged over an interval of time T, or

$$I = \frac{1}{T} \int_0^T \frac{p^2(t)}{\rho c}\,dt$$

so that over the time interval T

$$I = \frac{E}{T}$$

The quantity T is accordingly the time interval over which the energy flux density of an acoustic wave is to be averaged to form the intensity. For long-pulse active sonars, this time interval is the duration of the emitted pulse and is very nearly equal to the duration of the echo. For short transient sonars, however, the interval T is often ambiguous, and the duration of the echo is vastly different from the duration of the transient emitted from the source. Under these conditions, however, it can be shown (3) that the intensity form of the sonar equations can be used, provided that the source level is defined as

$$\text{SL} = 10 \log E - 10 \log \tau_e$$

where E is the energy flux density of the source at 1 yd and is measured in units of the energy flux density of a 1 dyne/cm^2 plane wave taken over an interval of 1 sec, and τ_e is the duration of the echo in seconds for an active sonar. For explosives, E is established by measurements for a given charge weight, depth, and type of explosive (Sec. 4.3). For pulsed sonars, emitting a flat-topped pulse of constant source level SL′ over a time interval τ_0, then, since the energy density of a pulse is the product of the average intensity times its duration,

$$10 \log E = \text{SL}' + 10 \log \tau_0$$

Combining the last two equations, the effective source level SL for use in the sonar equations is therefore

$$\text{SL} = \text{SL}' + 10 \log \frac{\tau_0}{\tau_e}$$

Here τ_0 is the duration of the emitted pulse of source level SL′, and τ_e is the echo duration. For long-pulse sonars, $\tau_0 = \tau_e$ and SL = SL′.

For short-pulse sonars, $\tau_e > \tau_0$, and the effective source level SL is reduced by the amount 10 log (τ_0/τ_e). In effect, the pulse emitted by the source is stretched out in time and thereby reduced in level by the multipath effects of propagation and by the processes of target reflection. The appropriate values of other sonar parameters in the equations, such as TS and TL, are those applying for long-pulse or CW conditions, in which the effects of multipaths in the medium and on the target are added up and accounted for.

For active short-pulse sonars, the echo duration τ_e is, accordingly, a parameter in its own right. Figure 2.3 illustrates a pulse as a short exponential transient at the source, as a distorted pulse at the target, and as an echo received back in the vicinity of the source. An exponential pulse, similar in form to the shock wave from an explosion of about 1 lb of TNT and having an initial duration of 0.1 msec, becomes distorted into an echo 1,000 times as long. Two actual examples of explosive echoes are shown in Fig. 2.4.

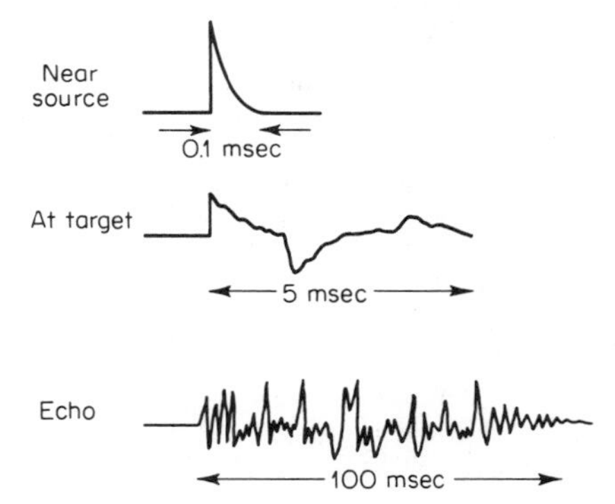

FIG. 2.3. Diagram of the pressure of an explosive pulse near the source, on arrival at an extended target, and as an echo back in the vicinity of the source.

The echo duration can be conceived as consisting of three components: τ_0, the duration of the emitted pulse measured near the source; τ_m, the additional duration imposed by the two-way propagation in the sea; and τ_t, the additional duration imposed by the extension in range of the target. In this view, the echo duration is the sum of the three components, or

$$\tau_e = \tau_0 + \tau_t + \tau_m$$

Typical examples of the magnitude of these three components of the echo duration under different conditions are given in Table 2.3. Thus,

TABLE 2.3 Components of Echo Duration

Component	Typical values, msec
Duration of the emitted pulse at short ranges	Explosives: 0.1 Sonar: 100
Duration produced by multiple paths	Deep water: 1 Shallow water: 100
Duration produced by a submarine target	Beam aspect: 10 Bow-stern aspect: 100

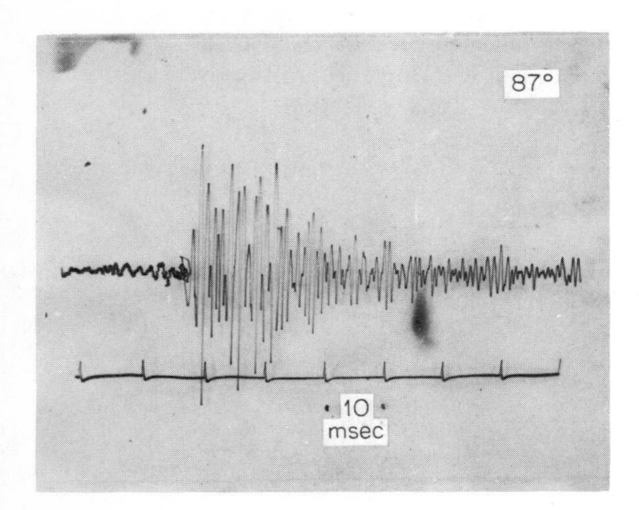

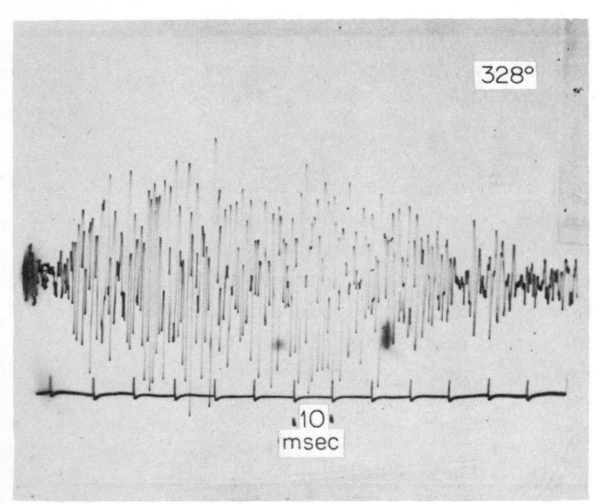

FIG. 2.4. Oscilloscope photographs of explosive echoes from a submarine at aspect angles 87° (near beam aspect) and 328° (near bow aspect). The time ticks are 10 msec apart.

with these values, the time duration of an explosive echo from a bow-stern aspect submarine in shallow water would be $0.1 + 100 + 100 = 200.1$ msec.

2.6 Statement and Examples

A condensed statement of the equations and two simple examples of their use are as follows:

Active sonars (Monostatic)

Noise background

$$SL - 2TL + TS = NL - DI + DT$$

Reverberation background

$$SL - 2TL + TS = RL + DT$$

Passive sonars

$$SL - TL = NL - DI + DT$$

1. *Range Prediction.* Given an active sonar having SL = 120 db and DI = 18 db and using processing giving a DT of +3 db, find the detection range r against a target of TS = 15 db when NL = −10 db.

$$TL = \tfrac{1}{2}(SL + TS - NL + DI - DT) = \tfrac{1}{2}(120 + 15 + 10 + 18 - 3) = 80 \text{ db}$$

The conversion to range depends on assumed propagation conditions. If spherical spreading without absorption is assumed, TL = 20 log r = 80 and r = 10,000 yd.

2. *Design.* Find the receiving directivity index of a passive sonar required to detect a noise target of SL = 50 db under conditions such that TL = 70 db and NL = 0 db and with processing such that DT = −10 db.

$$DI = NL + DT - SL + TL = 0 - 10 - 50 + 70 = +10 \text{ db}$$

Other examples will be given in the last chapter.

REFERENCES

1. Principles of Underwater Sound, *Natl. Defense Res. Comm. Div. 6 Sum. Tech. Rept.* 7, pp. 175–199 (1946).
2. Kerr, D. E. (ed.): "Propagation of Short Radio Waves," M.I.T. Radiation Laboratory Series, pp. 18–21, McGraw-Hill Book Company, New York, 1951.
3. Urick, R. J.: Generalized Form of the Sonar Equations, *J. Acoust. Soc. Am.*, **34**:547 (1962).

THREE

Properties of Transducer Arrays: Directivity Index

A human being requires a great deal of assistance in the perception and localization of underwater sound. The unaided human ear is for many reasons completely useless for the purpose. Underwater sound equipments include a variety of devices of varying degrees of complexity to make evident to an observer the existence of an underwater sound wave. These devices normally include a *hydrophone array* that transforms or *transduces* acoustic energy to electric energy, followed by some form of *signal processing* to feed some kind of aural or visual display suitable for the human observer.

Devices which convert sound and electric energy into one another—or, more generally, any two forms of energy—are called *transducers,* a hybrid Latin word meaning to "lead across." A transducer which converts sound into electricity is called a receiver or *hydrophone;* a transducer which converts electric energy into sound is called a *projector.* Some sonars use the same transducer for generating and receiving sound; others use a separate projector and hydrophone.

The ability to transduce, or interconvert, these two forms of energy rests upon the peculiar properties of certain materials called *piezoelectricity* (and its variant, electrostriction) and *magnetostriction.* Some crystalline substances, like quartz, ammonium dihydrogen phosphate (ADP), and Rochelle salt, acquire a charge between certain crystal surfaces when placed under pressure; conversely, they acquire a stress when a voltage is placed across them. These crystalline substances*

* V. A. Bazhenov has written a book on the piezoelectric properties of wood, a noncrystalline material (37).

are said to be piezoelectric. Electrostrictive materials exhibit the same effect, but are polycrystalline ceramics that have to be properly polarized by subjecting them to a high electrostatic field; examples are barium titanate and lead zirconate titanate. A magnetostrictive material is one which changes dimensions when placed in a magnetic field, and, conversely, which changes the magnetic field within and around it when it is stressed. Magnetostrictive materials are also polarized in order to avoid frequency doubling and to achieve a higher efficiency.

FIG. 3.1. A plane and a cylindrical transducer array. Array elements are 45° Z-cut crystals cemented to a steel backing plate. Resonant frequency near 24 kHz. (*Ref. 2, p. 237.*)

In air, other kinds of transducers are commonly used. Among these are moving-coil, moving-armature, and electrostatic types. In water, piezoelectric and magnetostrictive materials are particularly suitable because of their better impedance match to water. Ceramic materials have become increasingly popular in underwater sound because they can be readily molded into desirable shapes.

The ability to use these properties of matter for the reception or generation of underwater sound is based on the art of transducer design. It is a specialized science and technology of its own, which is too vast to be described here. The interested reader may find the basic principles of electroacoustic transduction in a book by Hunt (1). The design and construction of actual transducers are well described in two volumes of the NDRC Summary Technical Reports (2).

Single piezoelectric or magnetostrictive elements are normally used in hydrophones for research or measurement work. Nearly all other uses of hydrophones require hydrophone *arrays* in which a number of spaced elements are employed. The advantages of an array over a single hydrophone element are severalfold. First, the array is more sensitive, since a number of elements will generate more voltage (if connected in series) or more current (if connected in parallel) than a single element exposed to the same sound field. Second, the array possesses directional properties that enable it to discriminate between sounds arriving from different directions. Third, the array has an improved signal-to-noise ratio over a single hydrophone element since it discriminates against isotropic or quasi-isotropic noise in favor of a signal arriving in the direction that the array is pointing. These advantages dictate the use of hydrophone arrays in most practical applications of underwater sound. The first and second of the above three benefits exist for projectors as well as for hydrophones. Examples of a plane and a cylindrical array are shown in Fig. 3.1.

3.1 Transducer Responses

Since a hydrophone is a device which (normally) linearly transforms sound into electricity, there exists a proportionality factor, called the *response* of the hydrophone, that relates the generated voltage to the acoustic pressure of the sound field.

The *receiving response* of a hydrophone is the voltage across its terminals produced by a plane wave of unit acoustic pressure (before the introduction of the hydrophone into the sound field). It is customary to express the receiving response as the *open-circuit response* obtained when the hydrophone works into an infinite impedance. The receiving response is usually stated as the number of decibels relative to 1 volt produced by an acoustic pressure of 1 dyne/cm^2 and written db re 1 volt/(dyne)(cm^2). Thus a response stated as −80 db re 1 volt means that the hydrophone generates an open-circuit rms voltage of 10^{-4} volt when placed in a plane-wave sound field having an rms pressure of 1 dyne/cm^2.

The *transmitting-current response* of a projector is the pressure produced at a point 1 m from the projector in the direction of the axis of its beam pattern by a unit current into the projector. The transmitting response is usually stated as the number of decibels relative to 1 dyne/cm^2 as measured at the reference distance, produced by a current of 1 amp into the electrical terminals of the projector, and expressed as a certain number of decibels relative to 1 dyne/(cm^2)(amp). Transmitting responses are customarily referred to a point distant 1 *meter* from the source; for a reference distance of 1 *yard*, a correction of 20 log 39.4/36 or +0.87 db is required to convert the transmitting response to the source level required for the sonar equations expressed in yards. As an example, a projector having a response of 100 db re 1 dyne/(cm^2)(amp) (referred to 1 m) will produce an rms pressure of 10^5 dynes/cm^2 at 1 m when driven with a current of 1 rms amp. The corresponding source level would be 100.87 db re 1 dyne/(cm^2)(amp) at 1 yd.

3.2 Calibration Methods

The determination of the response as a function of frequency and direction constitutes the calibration of a transducer. A large number of methods for determining transducer responses are available. Table 3.1 is a summary of calibration methods as compiled from the literature by T. F. Johnston of the Naval Ordnance Laboratory. These methods have different desirable and undesirable features depending upon transducer size and frequency range.

Comparison Method The most simple and straightforward of these is the comparison method, wherein the output of the unknown transducer is compared with that of a transducer that has been previously calibrated. This method is easily and rapidly carried out and is the most commonly used of all methods, but requires a comparison transducer of known and reliable calibration.

Reciprocity Method At the other extreme is the reciprocity method of transducer calibration, which requires no reference standard, but requires a series of measurements on several transducers. Because of its basic importance, it will be described in some detail.

The practical importance of electroacoustic reciprocity was first pointed out by MacLean (3) and was applied to underwater sound calibration by Ebaugh and Meuser (4). The method was extended by

TABLE 3.1 Comparison of Frequency Response Calibration Methods

No.	Method	Advantages	Disadvantages	References (List at foot of table)
1	Reciprocity	Standards not required; absolute method	Lengthy; complex; requires reciprocal transducer	
	(a) Spherical			1
	(b) Cylindrical			2
	(c) Plane			3
	(d) Diffuse			4
	(e) Self-reciprocity			5
	(f) General theory			6–10
2	Substitution or Comparison	Simple; rapid	Not absolute method; requires calibrated standard transducer	11
3	Static	Simple	Indirect computation required	12, 13
4	Hydrostatic	Large body of water not required	Low frequency only	14, 25
5	Two-projector null method	Absolute method	Low frequency only; requires sealed tank	1, 15
6	Near calibration of large transducers	Eliminates need for large body of water	Special instrumentation; complex data analysis	16–18
7	Pulse calibration	Reduces effect of reflections and standing waves	Elaborate equipment required	19–21
8	Noise calibration	Continuous wideband coverage in frequency	White (or "pink") noise generator required	22
9	Impedance calibration	Tank or body of water not required	Complex data reduction	23
10	Explosions	Wideband	Results are complex; noisy; unsafe	24
11	Pressure tube	For high-pressure calibrations	Special tube needed	26

1. Procedures for Calibration of Electroacoustic Transducers Particularly Those for Use in Water, Z24.24-1957, USA Standards, New York.
2. Bobber, R. J., and G. A. Sabin: Cylindrical Wave Reciprocity Parameter, *J. Acoust. Soc. Am.*, **33**:446 (April, 1961).
3. Simmons, B. D., and R. J. Urick: The Plane Wave Reciprocity Parameter and Its Application to the Calibration of Electroacoustic Transducers at Close Distances, *J. Acoust. Soc. Am.*, **21**:633 (1949).
4. Diestel, H. G.: Reciprocity Calibration of Microphones in a Diffuse Sound Field, *J. Acoust. Soc. Am.*, **33**:514 (April, 1961).
5. Carstensen, E. G.: Self-reciprocity Calibration of Electroacoustic Transducers, *J. Acoust. Soc. Am.*, **19**:702 (1947).

6. Ballantine, S.: Reciprocity in Electromagnetic, Mechanical, Acoustical and Interconnected Systems, *Proc. Inst. Radio Eng.*, **17**:929 (1929).
7. MacLean, W. R.: Absolute Measurement of Sound without a Primary Standard, *J. Acoust. Soc. Am.*, **12**:140 (1940).
8. Foldy, L. L., and H. Primakoff: General Theory of Passive Linear Electroacoustic Transducers and the Electroacoustic Reciprocity Theorem: I, *J. Acoust. Soc. Am.*, **17**:109 (1945).
9. Primakoff, H., and L. L. Foldy: General Theory of Passive Linear Electroacoustic Transducers and the Electroacoustic Reciprocity Theorem: II, *J. Acoust. Soc. Am.*, **19**:50 (1947).
10. McMillan, E. M.: Violation of the Reciprocity Theorem in Linear Passive Electromechanical Systems, *J. Acoust. Soc. Am.*, **18**:345 (1946).
11. Albers, V. M.: "Underwater Acoustics Handbook," 1st ed., chap. 18, p. 243, The Pennsylvania State University Press, University Park, Pa., 1960.
12. Snavely, B. L., and G. S. Bennett: Static Calibrations of Acoustic System MK1 Hydrophones, *Naval Ordnance Lab., White Oak, Md.*, NOLM 8698, Aug. 19, 1946.
13. Raymond, F. W.: Nonacoustical Method for Measuring Sensitivity of a Piezoelectric Hydrophone, *J. Acoust. Soc. Am.*, **35**:70 (January, 1963).
14. Bobber, R. J.: The Calibration of a Low-frequency Calibrating System, *Navy Underwater Sound Ref. Lab., Rept.* 23, Orlando, Fla.
15. Trott, W. J., and E. N. Lide: Two-projector Null Method for Calibration of Hydrophones at Low Audio and Infrasonic Frequencies, *J. Acoust. Soc. Am.*, **27**:951 (September, 1955).
16. Horton, C. W., and G. S. Innis: The Computation of Far-field Radiation Patterns from Measurements Made near the Source, *J. Acoust. Soc. Am.*, **33**:877 (1961).
17. Baker, D. D.: Determination of Far-field Characteristics of Large Underwater Sound Transducers from Near-field Measurements, *J. Acoust. Soc. Am.*, **34**:1737 (1962).
18. Baker, D. D., and K. McCormack: Computation of Far-field Characteristics of a Transducer from Near-field Measurements Made in a Reflective Tank, *J. Acoust. Soc. Am.*, **35**:736 (1963).
19. Ginty, G. F.: Transducer Test and Calibration Equipment, 1954, Navy Underwater Sound Laboratory, Fort Trumbull, New London, Conn. (In its *Res. Develop. Rept.* 260, Quarterly Report, July 1–Sept. 30, 1954, pp. 49–52.)
20. Wallace, J. D., and E. W. McMorrow: Sonar Transducer Pulse Calibration System, *J. Acoust. Soc. Am.*, **33**:75 (January, 1961).
21. Terry, R. L., and R. B. Watson: Pulse Technique for the Reciprocity Calibration of Microphones, *J. Acoust. Soc. Am.*, **23**:684 (1951).
22. McMorrow, E. W., J. D. Wallace, and J. J. Coop: Technique for Rapid Calibration of Hydrophones Using Random Noise Fields, *Tech. Mem.* ADC-EL41-EWM:JDW:JJC, *Naval Air Development Center, Johnsville, Pa.*, 1953.
23. Sabin, G. A.: Transducer Calibration by Impedance Measurements, *J. Acoust. Soc. Am.*, **28**:705 (1956).
24. Carter, J. L., and M. F. M. Osborne: Hydrophone Calibration by Explosion Waves, *Naval Res. Lab. Rept.* S-2179, Washington, D.C., 1944.
25. Schloss, F., and M. Strasberg: Hydrophone Calibrations in a Vibrating Column of Liquid, *J. Acoust. Soc. Am.*, **34**:958 (1962).
26. Beatty, L. G., R. J. Bobber, and D. L. Phillips: Sonar Transducer Calibration in a High-Pressure Tube, *J. Acoust. Soc. Am.*, **39**:48 (1966).

Carstensen (5) to a self-reciprocity method involving reflection from a reflecting surface. The reciprocity method is based on the *electroacoustic reciprocity principle*, similar to the well-known electrical reciprocity principle for passive bilateral networks. This principle is sometimes expressed by saying that in any network composed of linear bilateral elements excited by a zero-impedance generator, the reading of a zero-impedance ammeter is unchanged when the ammeter and generator are interchanged. The corresponding theorem for electroacoustic networks can be shown to relate current and open-circuit voltage on the electrical side to the diaphragm velocity and impressed force on the rigid, or blocked, diaphragm on the acoustical side (Fig. 3.2). The relationship is

$$\frac{|E|}{|v|} = \pm \frac{|F|}{|I|} \qquad \text{or} \qquad \frac{|E|}{|F|} = \pm \frac{|v|}{|I|}$$

where $|E|$ = absolute magnitude of electric voltage
$|I|$ = absolute magnitude of electric current
$|v|$ = magnitude of velocity on diaphragm
$|F|$ = magnitude of force on diaphragm

The $\pm$ sign arises because of the different kinds of electromechanical coupling in different types of transducers (the $+$ sign applies for piezoelectric, the $-$ sign for magnetostrictive and electromagnetic transducers). The above expressions can be applied to the problem of transducer calibration by relating the diaphragm velocity v to the pressure produced at a distance when the transducer is acting as a projector, and by relating the force F to the incident sound wave when the transducer is acting as a hydrophone. These relationships depend on the kind of sound field being considered, that is, whether the transducer generates spherical, cylindrical, or plane waves.

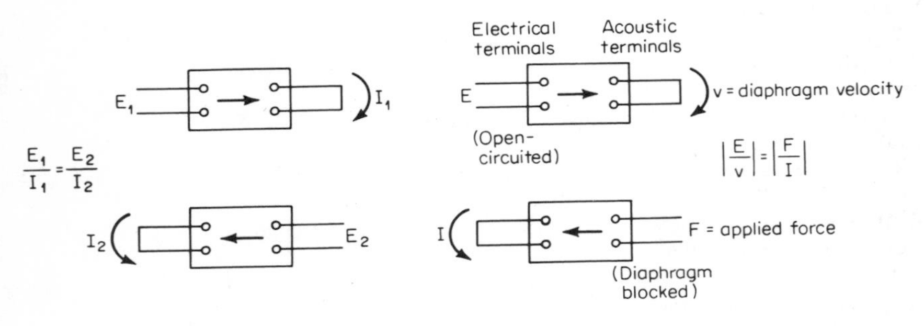

FIG. 3.2. Reciprocity in electric networks and electroacoustic transducers.

Spherical waves: If the transducer is considered to be a pulsating sphere and to generate spherical waves, the pressure at a large distance can be shown from acoustic theory (6) to be

$$|P|_r = \frac{f\rho}{2r}|Q|$$

where $|P|_r$ = absolute magnitude of pressure at distance r
$|Q|$ = absolute magnitude of volume velocity of source
ρ = fluid density

For a spherical source, the volume velocity $|Q|$ is the product of area A and surface velocity $|v|$ so that

$$|P|_r = \frac{f\rho}{2r}A|v|$$

When the transducer becomes a hydrophone, the force F of the incident sound field upon it is equal to its area multiplied by the incident sound pressure,* or

$$F = P_i A$$

Dropping the absolute magnitude signs from here on, and substituting for v and F in the reciprocity relationship, we obtain

$$\frac{E}{P_i A} = \frac{P_r}{I}\frac{2r}{\rho f A}$$

But the ratios E/P_i and P_r/I are the receiving and transmitting responses (for $r = 1$) as previously defined. Call these ratios M, for "microphone," and S, for "speaker." Then

$$M = S\frac{2r}{\rho f}$$

or

$$\frac{M}{S} = \frac{2r}{\rho f} \equiv J_s$$

The quantity J_s is called the *spherical-wave-reciprocity parameter* and is the ratio of the response of a reciprocal transducer acting as a projector to its response as a hydrophone when the transducer is regarded as a source of spherical waves.

Cylindrical waves: Now let the transducer be a pulsating cylinder generating cylindrical waves. If the surface velocity of the cylinder is v, the acoustic pressure at distance r can be shown from basic theory (7)

* This is true only at low frequencies where diffraction effects are absent. However, a rigorous derivation of reciprocity, based on the electroacoustic equations involved in Fig. 3.2, is not limited by this restriction.

to be

$$P_r = \pi \rho a v \left(\frac{cf}{r}\right)^{1/2}$$

where a = radius of cylinder
c = velocity of sound
If, as before, we change the transducer to a hydrophone, the force exerted upon it by an incident sound field of pressure P_i will be

$$F = AP_i = 2\pi aLP_i$$

where A = face area of cylindrical hydrophone
L = length of cylindrical hydrophone
Inserting in the reciprocity relationship we obtain

$$\frac{E}{P_i 2\pi aL} = \frac{P_r}{I}\left(\frac{r}{cf}\right)^{1/2}\frac{1}{\pi \rho a}$$

Writing, as before, $E/P_i \equiv M$, $P_r/I \equiv S$, we obtain

$$\frac{M}{S} = \frac{2L}{\rho c}\left(\frac{cr}{f}\right)^{1/2} \equiv J_c$$

The ratio J_c is called the *cylindrical-wave-reciprocity parameter* and was first derived by Bobber and Sabin (8). It is the parameter appropriate for transducers generating cylindrical waves.

Plane waves: When the transducer, acting as projector, is imagined to produce plane waves, the pressure produced at distance r when its diaphragm has the velocity v is simply

$$P_r = \rho c v$$

When the transducer is acting as a receiver of plane waves of pressure P_i, the force upon its surface of area A is

$$F = 2AP_i$$

where the factor 2 accounts for the back side of the transducer. Inserting in the reciprocity relationship, we obtain

$$\frac{E}{2AP_i} = \frac{P_r}{\rho c I}$$

so that

$$\frac{M}{S} = \frac{E/P_i}{P_r/I} = \frac{2A}{\rho c} = J_P$$

This is the *plane-wave-reciprocity parameter* originally derived by Simmons and Urick (9) and is the appropriate parameter for transducers generating plane waves.

Summary: The three reciprocity parameters can be written synoptically as follows:

$$\textit{Plane: } J_p = \frac{2}{\rho c}(\lambda r)^0 A$$

$$\textit{Cylindrical: } J_c = \frac{2L}{\rho c}(\lambda r)^{1/2} L$$

$$\textit{Spherical: } J_s = \frac{2}{\rho c}(\lambda r)$$

where the exponent of the term λr indicates the spreading law (Sec. 5.2) for the three kinds of acoustic radiation.

For a given transducer, the choice of parameters for a reciprocity calibration depends upon the geometry of the transducer and the distance at which the calibration is done. A long cylindrical transducer to be calibrated at a short distance $r \leq L^2/\lambda$ would require the cylindrical parameter, whereas a flat-faced pistonlike transducer calibrated at a distance $r \leq A/\lambda$ would require the plane-wave parameter. When large projectors must be calibrated at close distances, however, special techniques are needed (10, 11) in order to appropriately sample the complex sound field at distances comparable to the dimensions of the radiating faces of the projectors.

3.3 Reciprocity Calibration

Electroacoustic reciprocity makes possible the calibration of transducers without reference to a transducer of known response. A reciprocity calibration requires three transducers. One of the transducers must be reciprocal (in the sense of obeying the basic reciprocity condition); another must be a projector; and the third must be a hydrophone. The response of either the projector or the hydrophone is desired. The transducer being calibrated need not be reciprocal.

Let the three units be nondirectional and placed at equal distances r from each other, as shown in Fig. 3.3. The method in its simplest form involves two steps. In step 1,

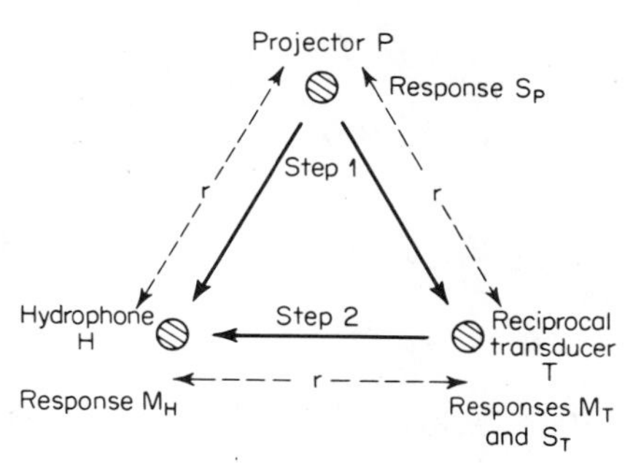

FIG. 3.3. Basic geometry for a reciprocity calibration.

the projector P is driven with current i and the voltages appearing at the output terminals of the transducer T (acting as hydrophone) and the hydrophone H are measured. In step 2, the transducer T acts as projector and is driven with the same current i as before. The voltage v'_H generated by the hydrophone H is measured. No other measurements are required.

Let the desired quantity be the receiving response M_H of the hydrophone H. Then from step 1

$$\frac{M_H}{M_T} = \frac{v_H}{v_T} \tag{1}$$

where M_T is the unknown receiving response of T. If the projector P, observed at distance r, is assumed to radiate spherical waves, then, from the definitions of transmitting and receiving responses, we have

$$S_P = \frac{P_r r}{i}$$

and

$$M_H = \frac{v_H}{P_r}$$

where P_r is the acoustic pressure produced by P at distance r. Multiplying these two equations together, we obtain

$$S_P M_H = \frac{v_H r}{i} \tag{2}$$

From the measurement of step 2

$$\frac{S_P}{S_T} = \frac{v_H}{v'_H} \tag{3}$$

Solving (1) for M_H, solving (3) for S_P, and inserting into (2) we obtain

$$S_T M_T = \frac{v_T v'_H}{v_H} \frac{r}{i}$$

But by reciprocity, the two responses of the transducer T are related by

$$M_T = J_S S_T$$

Proceeding to eliminate S_T, we find

$$M_T = \left(J_S \frac{v_T v'_H}{v_H} \frac{r}{i} \right)^{1/2}$$

and on substituting in (1), we obtain the desired receiving response of the hydrophone H:

$$M_H = \left(J_S \frac{v_H v'_H}{v_T} \frac{r}{i}\right)^{1/2}$$

The calibration is therefore obtained by three measured output voltages—the distance, the driving current, and the appropriate reciprocity parameter (in this case J_S). The other responses M_T, S_T, and S_P can be computed, if desired, without additional measurements. Standard procedures for making a reciprocity calibration are available (12).

3.4 Beam Patterns

The response of a transducer array varies with direction relative to the array. This property of array *directionality* is highly desirable, for it enables the direction of arrival of a signal to be determined and enables closely adjacent signals to be resolved. At the same time, directionality reduces noise, relative to the signal, arriving in other directions. In projector arrays, directionality serves to concentrate the emitted sound in a desired direction.

Directionality in hydrophone arrays results from the fact that sinusoidal signals arriving in some one direction tend to be in phase at all the array elements, whereas the noise background is out of phase. Similarly, broadband signals arriving in this direction tend to be correlated between pairs of array elements, whereas the noise is not. This direction of maximum inphase condition or maximum correlation is that of the *acoustic axis* of the array and is the direction of maximum sensitivity.

The response of an array varies with direction in a manner specified by the beam pattern of the array. If the response is written $R(\theta,\varphi)$ to indicate that it is a function of the angles θ and φ in polar coordinates, then we may write

$$R(\theta,\varphi) = R(0,0)v(\theta,\varphi)$$

where $R(0,0)$ = response in direction $\theta = 0$, $\varphi = 0$

$v(\theta,\varphi)$ = a response function normalized so that $v(0,0) = 1$

The beam pattern, or pattern function, is the square of v or

$$b(\theta,\varphi) = v^2(\theta,\varphi)$$

For hydrophones, $b(\theta,\varphi)$ is the mean-square voltage produced by an array of unit response when sound of unit pressure is incident on it in the direction θ, φ. For projectors, $b(\theta,\varphi)$ is the mean-square pressure produced at unit distance when unit current is fed into the projector. The direction (0,0) is arbitrary, but is ordinarily taken to be the direction of maximum response. In regular arrays, the beam pattern is symmetrical, and the direction (0,0) to which the axial response $R(0,0)$ refers is the acoustic axis of the array. Beam patterns, like responses, are commonly expressed in decibel units.

Arrays can normally be *steered*, either mechanically by physically rotating the array, or electrically by inserting in series or in parallel with each array element appropriate phasing networks (for narrow-band arrays) or time-delay networks (for broadband arrays) that effectively rotate the direction of maximum response into some desired direction. For example, the direction of maximum sensitivity of a plane array of elements located at the positions of the open circles in Fig. 3.4 can be rotated into a direction lying at angle θ_0 to a reference direction by delaying the output of each element by the time for sound to travel the distances l. In this way, the array is effectively converted into a line array along the line AB. The output voltage of an array steered in the (θ,φ) direction can be written

$$v(\theta,\theta_0, \varphi,\varphi_0) = R(\theta_0,\varphi_0)v(\theta_0 - \theta, \varphi_0 - \varphi)$$

where $R(\theta_0,\varphi_0)$ is the response when the array is steered along (θ_0,φ_0) direction and $v(\theta_0 - \theta, \varphi_0 - \varphi)$ is the square root of the beam pattern referred to this direction and again normalized so that $v(0,0) = 1$. For steered arrays, the beam pattern is not usually symmetrical, so that $b(\theta_0 - \theta, \varphi_0 - \varphi) \neq b(\theta - \theta_0, \varphi - \varphi_0)$.

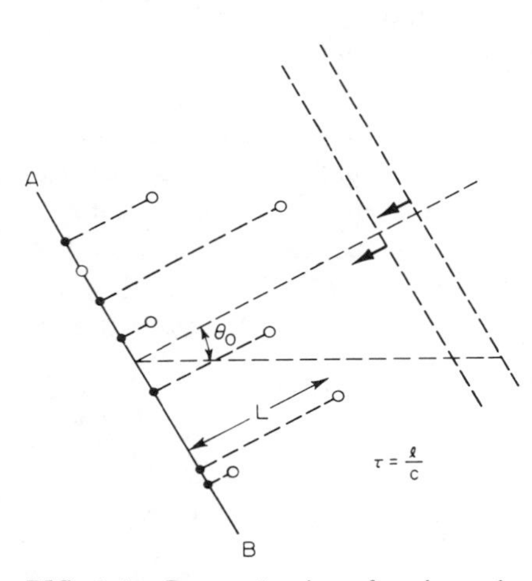

FIG. 3.4. Beam steering of an irregular plane array. Electrical time delays equal to l/c are inserted in series with each array element, where l is the perpendicular distance from the reference line AB and c is the velocity of sound. By so doing, the elements are brought to lie, in effect, along the line AB.

In their simplest form, many arrays are arranged with elements along a line or distributed along a plane. The acoustic axis of such *line* or *plane* arrays, when unsteered, lies at right angles to the line or plane. The beam pattern of a line array may be visualized as a doughnut-shaped figure having supernumerary attached doughnuts formed by the side lobes of the pattern. The three-dimensional pattern of a plane array is a searchlight type of figure with rotational symmetry about the perpendicular to the plane plus side lobes. Three-dimensional views of such patterns are drawn in Fig. 3.5.

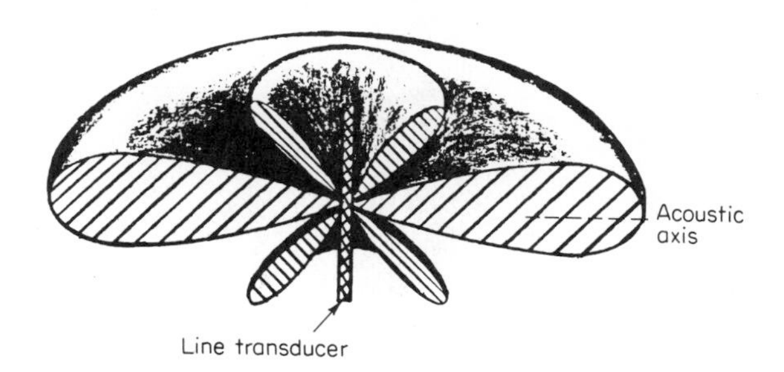

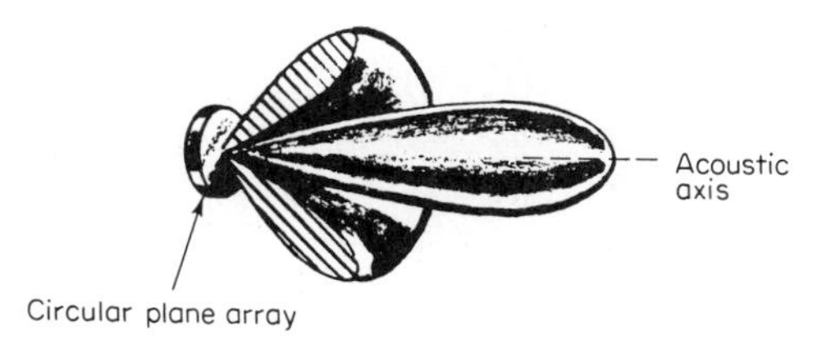

FIG. 3.5. Three-dimensional views of the beam pattern of a line and a circular-plane array.

Line of Equally Spaced Elements The beam pattern of a line of equally spaced, equally phased (i.e. unsteered) elements can be derived as follows. Let a plane sinusoidal sound wave of unit pressure be incident at an angle θ to a line of n such elements. As seen in Fig. 3.6, the output of the mth element relative to that of the zeroth element will be delayed by the amount of time necessary for sound to travel the distance $l_m = md \sin\theta$; the corresponding phase delay for sound of wavelength λ will be, at frequency $\omega = 2\pi f$,

$$u_m = \frac{2\pi}{\lambda} l_k = mu$$

where the phase delay between adjacent elements is in radians

$$u = \frac{2\pi d}{\lambda} \sin\theta$$

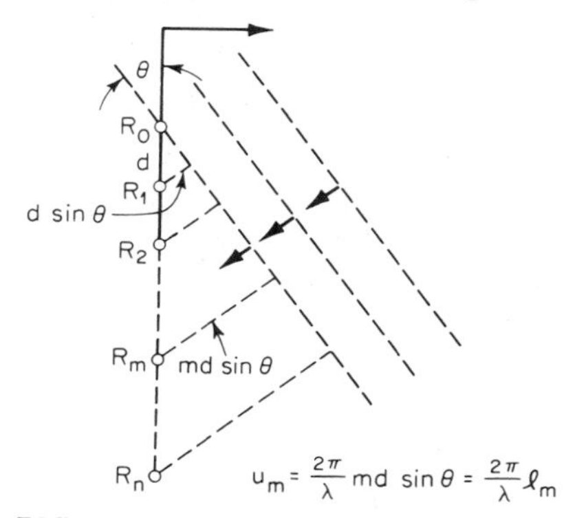

FIG. 3.6. Line array of equally spaced elements.

The output voltage of the mth element of voltage response R_m will be

$$v_m = R_m \cos(\omega t + mu)$$

and the array voltage will be the sum of such terms:

$$V = R_0 \cos\omega t + R_1 \cos(\omega t + u) + \cdots + R_m \cos(\omega t + mu) + \cdots + R_n \cos(\omega t + nu)$$

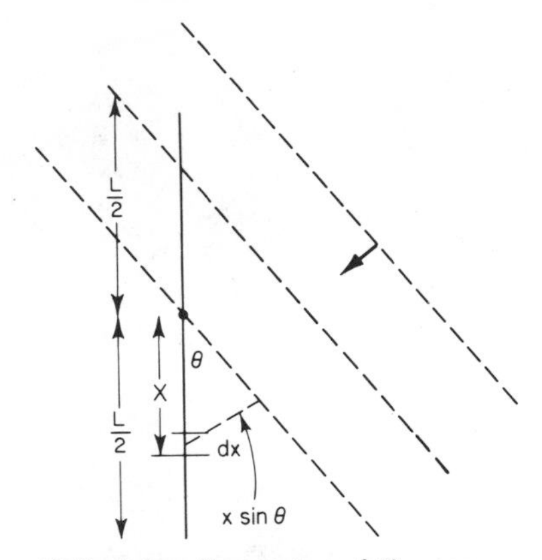

FIG. 3.7. Geometry of the continuous line array.

In the complex notation, the array voltage will be

$$V = (R_0 + R_1 e^{iu} + R_2 e^{2iu} + \cdots + R_n e^{(n-1)iu})e^{i\omega t}$$

If the array elements are all of unit response, we can drop the R's and write

$$V = (1 + e^{iu} + e^{2iu} + \cdots + e^{(n-1)iu})e^{i\omega t}$$

Multiplying by e^{iu} and subtracting, we find

$$V = \left(\frac{e^{inu} - 1}{e^{iu} - 1}\right) e^{i\omega t}$$

which, after manipulating, and neglecting the time dependence, becomes

$$V = \frac{\sin\,(nu/2)}{\sin\,(u/2)}$$

Finally, expressing u in terms of θ, we obtain the desired beam pattern, which is the square of this function normalized to unity at $\theta = 0$:

$$b(\theta) = \left(\frac{V}{n}\right)^2 = \left[\frac{\sin\,(n\pi d \sin\theta/\lambda)}{n \sin\,(\pi d \sin\theta/\lambda)}\right]^2$$

Continuous-line and Plane Circular Arrays When the array elements are so close together that they may be regarded as adjacent, the array becomes a continuous-line transducer and the beam pattern can be found by integration rather than by summation. For this case, let the line transducer be of length L and have a response per unit length of R/L. The contribution to the total voltage output produced by a small element of line length dx located a distance x from the center (Fig. 3.7) will be (neglecting the time dependence)

$$dv = \frac{R}{L} e^{(i2\pi/\lambda)x \sin\theta}\, dx$$

and the total voltage will be

$$\begin{aligned} V &= \frac{R}{L}\int_{-L/2}^{L/2} e^{(i2\pi/\lambda)x\sin\theta}\, dx \\ &= R\,\frac{e^{(i\pi L/\lambda)\sin\theta} - e^{-(i\pi L/\lambda)\sin\theta}}{(i2\pi L/\lambda)\sin\theta} \\ &= R\,\frac{\sin\,[(\pi L/\lambda)\sin\theta]}{(\pi L/\lambda)\sin\theta} \end{aligned}$$

The beam pattern will be the square of V normalized so that $b(\theta) = 1$, or

$$b(\theta) = \left(\frac{V}{R}\right)^2 = \left[\frac{\sin\,[(\pi L/\lambda)\sin\theta]}{(\pi L/\lambda)\sin\theta}\right]^2$$

In a similar manner, the beam pattern of a circular plane array of diameter D of closely spaced elements can be shown to be

$$b(\theta) = \left[\frac{2J_1[(\pi D/\lambda)\sin\theta]}{(\pi D/\lambda)\sin\theta}\right]^2$$

where $J_1[(\pi D/\lambda)\sin\theta]$ is the first-order Bessel function of argument $(\pi D/\lambda)\sin\theta$.

Generalized beam patterns for continuous-line and circular plane arrays are drawn in Fig. 3.8 in terms of the quantities $(L/\lambda)\sin\theta$ and $(D/\lambda)\sin\theta$.

Figure 3.9 is a nomogram for finding the angular width, between the axis and the -3 db down and -10 db down points, of the beam pattern of continuous-line and circular plane arrays. The dashed lines indicate how the nomogram is to be used. Thus, a circular plane array of 20 in. diameter at a wavelength of 4 in. (corresponding to a frequency of 15 kHz at a sound velocity of 5,000 fps) has a beam pattern 6° wide between

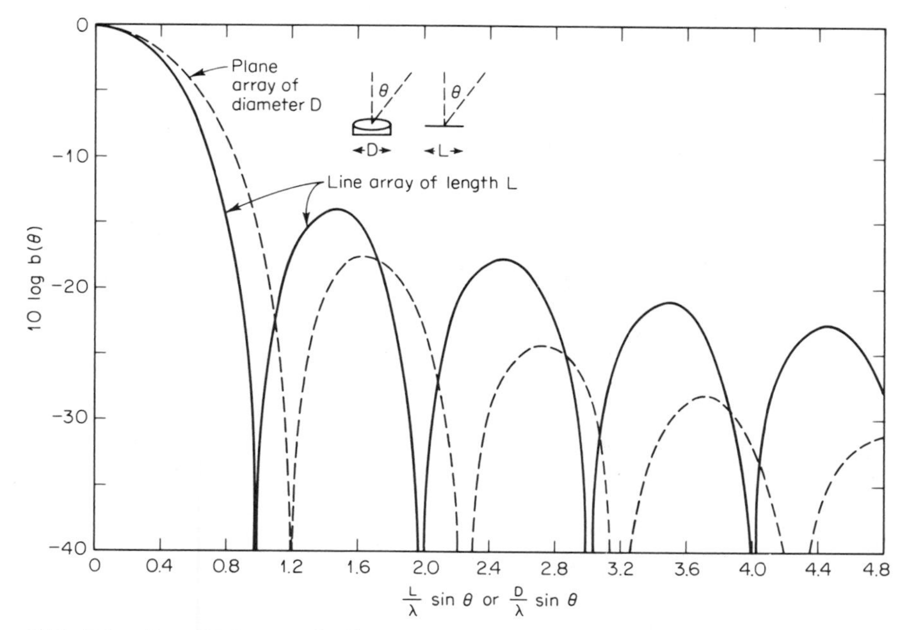

FIG. 3.8. Beam patterns of a line array of length L and of a circular plane of diameter D in terms of the dimensionless parameters $L/\lambda\sin\theta$ *and* $D/\lambda\sin\theta$.

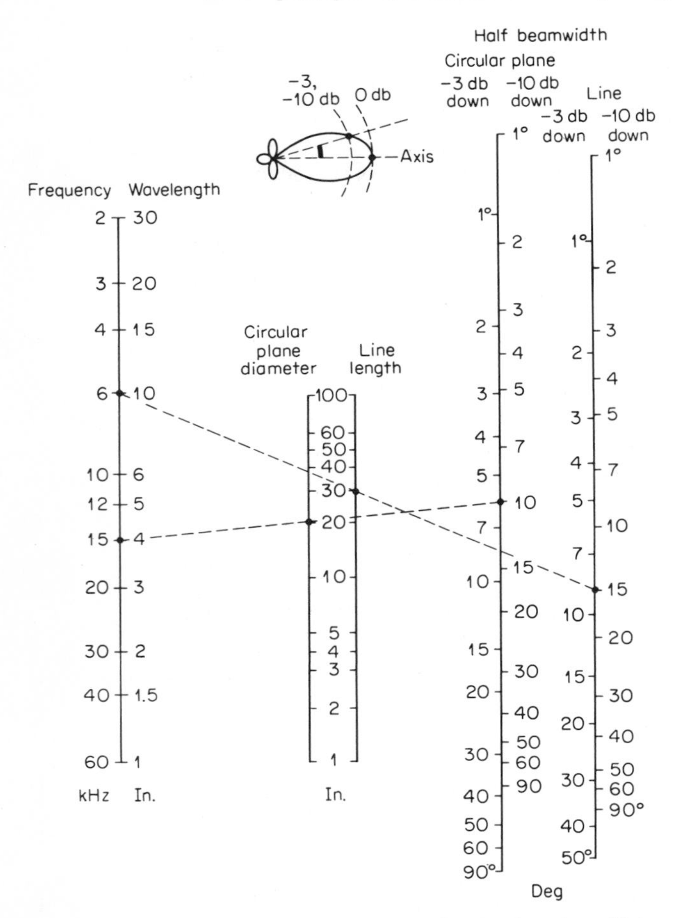

FIG. 3.9. Nomogram for finding the width of the beam pattern of circular-plane and line transducers.

the axis of the pattern and the −3 db down point and 10° wide between the axis and the −10 db down point. For a line array 30 in. long at a wavelength of 10 in. (frequency 6 kHz), the corresponding angles are 8 and 15°.

3.5 Product Theorem

So far, each element of an array of hydrophones has been nondirectional. When the array comprises identical hydrophones that are themselves directional, the beam pattern is given by the *product theorem*, which states that the beam pattern of an array of identical, equally spaced directional hydrophones is the product of (1) the pattern of each hydrophone alone and (2) the pattern of an identical array of nondirectional hydrophones.

The theorem is almost self-evident. Consider a line array of equally spaced elements with a separation d, all of unit response. If they are nondirectional, the array voltage produced by a unit sound pressure will be, as before,

$$V_{Nond}(\theta) = 1 + e^{iu} + e^{2iu} + \cdots$$

where u is the angular function $u = (2\pi d/\lambda) \sin \theta$. If each element itself has a response that varies with direction according to the function $v(\theta,\varphi)$, the array output will be

$$\begin{aligned} V_D(\theta,\varphi) &= v(\theta,\varphi) + v(\theta,\varphi)e^{iu} + v(\theta,\varphi)e^{2iu} + \cdots \\ &= v(\theta,\varphi)V_{Nond}(\theta) \end{aligned}$$

Hence

$$b_D(\theta,\varphi) = V_D^2(\theta,\varphi) = b(\theta,\varphi)b_{Nond}(\theta)$$

The beam pattern of the array of directional elements $b_D(\theta,\varphi)$ will therefore be the product of the beam pattern $b_{Nond}(\theta)$ of an identical array of nondirectional elements multiplied by the beam pattern $b(\theta,\varphi)$ of each element alone.

The theorem provides a handy shortcut in array design involving directional hydrophone elements. It applies only for additive arrays of elements of the same directivity, though not necessarily of the same sensitivity.

3.6 Shading and Superdirectivity

Shading is a method by which some degree of control can be exercised over the pattern of an array having some particular geometry. Amplitude shading involves adjusting the responses of array elements to provide the most desirable pattern for some particular purpose. Nearly always, an array is shaded with maximum response at the center and least response at the ends or sides of the array so that the sensitivity is *tapered* from a high value on the inside toward a lower value on the outside.

The effects of adjusting the response of array elements may be appreciated by comparing the beam patterns of a six-element line array shaded in different ways, as shown in Fig. 3.10. A convenient starting point is an unshaded array, whose shading formula, expressing the responses of the array elements, may be written 1, 1, 1, 1, 1, 1. The beam pattern of an unshaded line array has been derived above. When the sensitivity

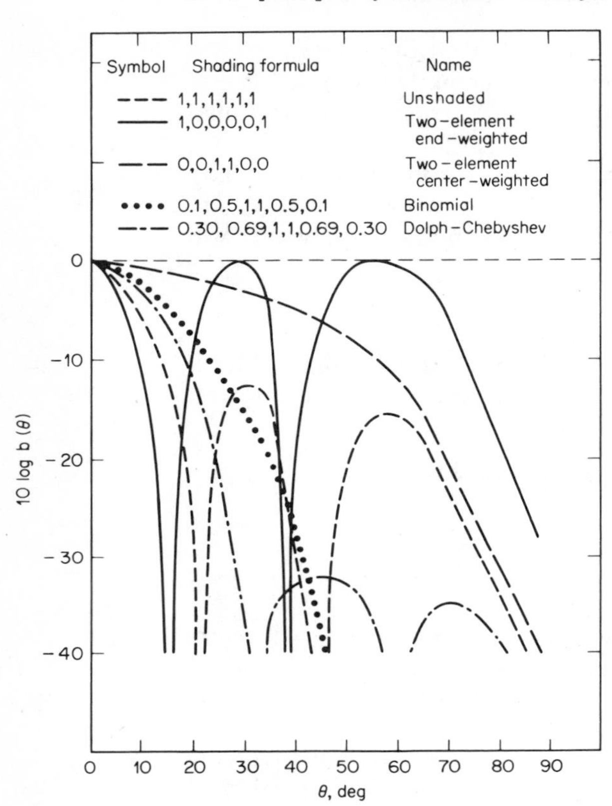

FIG. 3.10. Beam patterns of a six-element line array of half-wave spacing between elements for different shading formulas.

is all at the ends, with formula 1, 0, 0, 0, 0, 1, the pattern consists of a series of narrow lobes all of equal amplitude. When, on the other hand, the array is shaded strongly toward the center, with formula 0, 0, 1, 1, 0, 0, the side lobes disappear, leaving a single, broad main lobe. Thus the effect of tapering from center outward (0, 0, 1, 1, 0, 0) is widening of the main beam and reduction of the side lobes; the effect of reverse tapering (1, 0, 0, 0, 0, 1) is narrowing of the main beam at the expense of increased side lobes. *Binomial* shading, with sensitivities proportional to the coefficients of a binomial expansion (0.1, 0.5, 1, 1, 0.5, 0.1), yields the narrowest main lobe, with a total absence of side lobes; *Dolph-Chebyshev* shading (0.30, 0.69, 1, 1, 0.69, 0.30) gives the narrowest main lobe, with side lobes of a preassigned level. The procedures involved in a Dolph-Chebyshev design may be found in a paper by Davids, Thurston, and Meuser (13) and in a book by Albers (14). In general, the selection of shading factors to achieve a desired beam pattern can be done by the method of Lagrangian multipliers, as pointed out originally in a paper by Pritchard (15).

An extreme form of shading is called *superdirectivity*, in which narrow beams may be obtained with arrays of limited size. In a superdirective array, the elements are spaced less than one-fourth wavelength apart with the signs, or polarities, of adjacent elements reversed. The properties of superdirective arrays were pointed out by Pritchard (16). For example, the directivity index of an array of five elements one-eighth wavelength apart—and thus only one-half wavelength long—and having the shading factors $+1$, -4, $+6$, -4, $+1$ was computed to be 5.6 db. The price to be paid for superdirectivity is a low array sensitivity due to the phase reversals of the elements, together with relatively high side lobes—a price usually greater than the gain in directivity index that it buys.

Phase shading may also be employed in arrays. In this kind of shading, the spacing of the array elements is varied to obtain the desired pattern. This method has received considerable attention in the design of radio antennas (17–19) but has not been extensively employed in underwater sound.

3.7 Receiving Directivity Index

The parameter *directivity index* for sound reception is a measure of the amount by which a hydrophone array, through its beam pattern, discriminates against noise in favor of the signal. It is the increase in decibels in signal-to-noise ratio at the terminals of the directional hydrophone over the signal-to-noise ratio that would be observed with a nondirectional hydrophone. This increase depends on the directional characteristics of the signal and noise field. By convention, directivity index refers to a *plane-wave* (*unidirectional*) *signal* in *isotropic noise*. For signal and noise fields having other directional characteristics, the term *array gain* will be used to indicate the signal-to-noise ratio of a hydrophone array, relative to the signal-to-noise ratio of a nondirectional hydrophone.

Directivity index may best be understood by comparing the output of a directional hydrophone array with that of a nondirectional hydrophone of the same response when both are placed in the same isotropic noise field. Let the noise power per unit solid angle of the noise field be I_i, and let the axial sensitivity of the directional hydrophone and the sensitivity of the nondirectional hydrophone be unity. Then the mean-squared voltage produced by the nondirectional unit will be

$$R_{Nond} = m \int_{4\pi} I_i \, d\Omega = 4\pi m I_i$$

where $d\Omega$ = elemental solid angle

m = trivial proportionality constant

The mean-square voltage produced by the directional hydrophone will be

$$R_D = m \int_{4\pi} I_i b(\theta,\varphi)\, d\Omega = I_i m \int_{4\pi} b(\theta,\varphi)\, d\Omega$$

where $b(\theta,\varphi)$ is the normalized beam-pattern function. Then the directivity index is defined as

$$\text{DI} = 10 \log \frac{R_{Nond}}{R_D} = 10 \log \frac{4\pi}{\int_{4\pi} b(\theta,\varphi)\, d\Omega}$$

The DI of a directional array is therefore 10 times the logarithm of the ratio of the mean-squared output voltage of a nondirectional hydrophone in an isotropic noise field to that of the directional array of the same axial sensitivity placed in the same noise field.

If θ and φ are angles in polar coordinates, the expression becomes, on expanding $d\Omega$,

$$\text{DI} = 10 \log \frac{4\pi}{\int_0^{2\pi} \int_{-\pi/2}^{\pi/2} b(\theta,\varphi) \sin\theta\, d\theta\, d\varphi}$$

If, further, the beam pattern has rotational symmetry and is nondirectional in the plane in which φ is reckoned, then the above expression simplifies to

$$\text{DI} = 10 \log \frac{4\pi}{2\pi \int_{-\pi/2}^{\pi/2} b(\theta) \sin\theta\, d\theta}$$

DI is always a positive number of decibels.*

* Prior to about 1948, directivity index was regarded as a negative rather than a positive quantity.

TABLE 3.2 Directivity of Simple Transducers

Type	Pattern function	DI = 10 log
Continuous line of length L $L \gg \lambda$	$\left[\dfrac{\sin(\pi L/\lambda)\sin\theta}{(\pi L/\lambda)\sin\theta}\right]^2$	$\dfrac{2L}{\lambda}$
Piston of diameter D in an infinite baffle $D \gg \lambda$	$\left(\dfrac{2J_1[(\pi D/\lambda)\sin\theta]}{(\pi D/\lambda)\sin\theta}\right)^2$	$\left(\dfrac{\pi D}{\lambda}\right)^2$
Line of n elements of equal spacing d	$\left[\dfrac{\sin(n\pi d\sin\theta/\lambda)}{n\sin[(\pi d/\lambda)\sin\theta]}\right]^2$	$\dfrac{n}{1 + \dfrac{2}{n}\displaystyle\sum_{\rho=1}^{n-1} \dfrac{(n-\rho)\sin(2\rho\pi d/\lambda)}{2\rho\pi d/\lambda}}$
Two-element array; as above but $n = 2$	$\left[\dfrac{\sin(2\pi d\sin\theta/\lambda)}{2\sin[(\pi d/\lambda)\sin\theta]}\right]^2$	$\dfrac{2}{1 + \left[\dfrac{\sin(2\pi d/\lambda)}{2\pi d/\lambda}\right]}$

For simple arrays, such as the line and circular plane, DI can be evaluated mathematically. Expressions for the DI of some simple transducer arrays are given in Table 3.2 in terms of the array dimensions. For arrays which cannot be approximated by these simple forms, the DI can be found by integration of the beam pattern, for which some graphical aids have been developed (20).

A nomogram for finding the directivity index of line and circular-piston arrays in terms of their dimensions and the frequency is given in Fig. 3.11. The DI is the point of intersection of the line joining the dimension and frequency scales with the center scale of the diagram. For example, the DI of a circular-piston array of 5 in. diameter at a frequency of 30 kHz is 18 db.

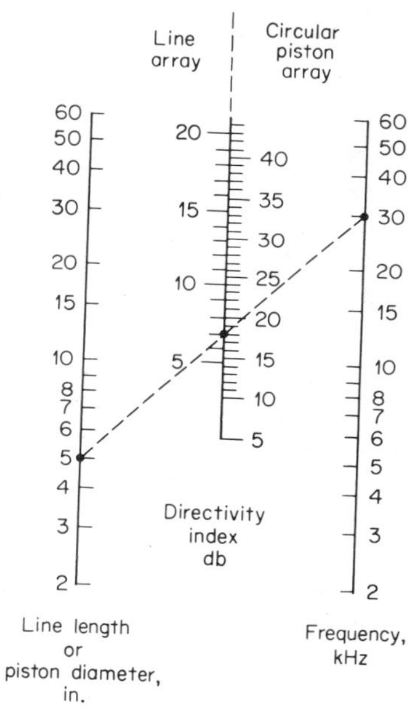

FIG. 3.11. Directivity-index nomogram for line and circular-piston arrays in an infinite baffle.

3.8 Array Gain

The concept of directivity index may be extended to general signal and noise fields characterized by directional functions $N_S(\theta,\varphi)$ and $N_N(\theta,\varphi)$ representing the signal and noise power per unit solid angle, respectively, incident on the hydrophone array. Let the signal and noise fields be of equal magnitude when measured with a nondirectional hydrophone, so that

$$\int N_S(\theta,\varphi)\, d\Omega = \int N_N(\theta,\varphi)\, d\Omega$$

Then the *array gain* of a directional array having beam pattern $b(\theta,\varphi)$ when placed in this signal and noise field will be

$$\text{AG} = 10 \log \frac{\int N_S(\theta,\varphi) b(\theta,\varphi)\, d\Omega}{\int N_N(\theta,\varphi) b(\theta,\varphi)\, d\Omega}$$

For a unidirectional signal in isotropic noise, this expression reduces to that for directivity index given above.

For deep-sea ambient noise at a bottomed location, observed values of N_N for different frequencies and wind speeds may be found in Sec. 7.9.

An alternate and more useful approach to array gain involves the *coherence* of signal and noise across the dimensions of the array. By coherence is meant the degree of similarity of waveform of signal and noise between any two elements of the array. Coherence is measured by the *crosscorrelation coefficient* of the outputs of different elements of the array. If $v_1(t)$ and $v_2(t)$ are voltages generated by two array elements as functions of time, then the crosscorrelation coefficient between them is defined as

$$\rho_{12} = \frac{\overline{v_1(t)v_2(t)}}{[\overline{(v_1)^2}\,\overline{(v_2)^2}]^{1/2}}$$

where the bars indicate time averages and the denominator serves as a normalization factor. The crosscorrelation coefficient of two functions of time is therefore the normalized time-averaged product of the two functions.

The array gain depends on the crosscorrelation coefficients of signal and noise between the elements of the array. Consider a linear additive array of n elements of equal sensitivity. Let the individual output voltages, as functions of time, including any phase shifts or delays incorporated for steering, be denoted by $s_1(t), s_2(t), \ldots, s_n(t)$. If the array elements are in series, the average voltage appearing across the array terminals will be $\overline{s_1(t) + s_2(t) + \cdots + s_n(t)}$. The average signal power into the load across the array terminals will be

$$\overline{S^2} = m\overline{[s_1(t) + s_2(t) + \cdots + s_n(t)]^2}$$

where m is a proportionality factor. Similarly, the noise power will be

$$\overline{N^2} = m\overline{[n_1(t) + n_2(t) + \cdots + n_n(t)]^2}$$

where the n's are the outputs of the array elements produced by noise. The average array signal-to-noise ratio will therefore be

$$\frac{\overline{S^2}}{\overline{N^2}} = \frac{\overline{[s_1(t) + s_2(t) + \cdots + s_n(t)]^2}}{\overline{[n_1(t) + n_2(t) + \cdots + n_n(t)]^2}}$$

Expanding and dropping the t's, one obtains

$$\frac{\overline{S^2}}{\overline{N^2}} = \frac{(\overline{s_1s_1} + \overline{s_1s_2} + \cdots + \overline{s_1s_n}) + (\overline{s_2s_1} + \overline{s_2s_2} + \cdots + \overline{s_2s_n}) + \cdots}{(\overline{n_1n_1} + \overline{n_1n_2} + \cdots + \overline{n_1n_n}) + (\overline{n_2n_1} + \overline{n_2n_2} + \cdots + \overline{n_2n_n}) + \cdots}$$

For simplicity, let

$$\overline{s_1^2} = \overline{s_2^2} \cdots = \overline{s_n^2} = \overline{s^2}$$

and similarly let

$$\overline{n_1^2} = \overline{n_2^2} \cdots = \overline{n_n^2} = \overline{n^2}$$

Then by the definition of the crosscorrelation coefficient, the array signal-to-noise power becomes

$$\frac{\overline{S^2}}{\overline{N^2}} = \frac{\overline{s^2}}{\overline{n^2}} \frac{[(\rho_s)_{11} + (\rho_s)_{12} + \cdots + (\rho_s)_{1n}] + [(\rho_s)_{21} + (\rho_s)_{22} + \cdots + (\rho_s)_{2n}] + \cdots}{[(\rho_n)_{11} + (\rho_n)_{12} + \cdots + (\rho_n)_{1n}] + [(\rho_n)_{21} + (\rho_n)_{22} + \cdots + (\rho_n)_{2n}] + \cdots}$$

$$= \frac{\overline{s^2}}{\overline{n^2}} \frac{\sum_j \sum_i (\rho_s)_{ij}}{\sum_j \sum_i (\rho_n)_{ij}}$$

Here $(\rho_s)_{ij}$ and $(\rho_n)_{ij}$ are the crosscorrelation coefficients between the ith element and the jth element of the signal and of the noise, respectively. The *array gain* is by definition the ratio, in decibel units, of the signal to noise of the array to the signal to noise of a single element, so that

$$\mathrm{AG} = 10 \log \frac{\overline{S^2}/\overline{N^2}}{\overline{s^2}/\overline{n^2}} = 10 \log \frac{\sum_i \sum_j (\rho_s)_{ij}}{\sum_i \sum_j (\rho_n)_{ij}}$$

The gain of the array therefore depends on the sum of the crosscorrelation coefficients between all pairs of elements of the array, for both noise and signal. When the mean-square outputs of the individual array elements are not all the same, as when amplitude shading is used, the array gain becomes

$$\mathrm{AG} = 10 \log \sum_i \sum_j \frac{a_i a_j (\rho_s)_{ij}}{a_i a_j (\rho_n)_{ij}}$$

where a_i is the rms voltage produced by the ith element due to the signal or the noise.

The quantities ρ_s and ρ_n are basic properties of the signal and noise sound field in which the array is placed. Accordingly, the same array will have a different array gain in different signal and noise fields.

Both ρ_s and ρ_n will depend on the electrical time delays that may be introduced into the array for steering; indeed, the purpose of these electrical delays is to maximize ρ_s when the array is steered in the direction from which the signal is coming.

Table 3.3 gives expressions for ρ_s and ρ_n for a unidirectional signal and for isotropic noise, respectively, for a single frequency and for a frequency band, with and without electrical delays. Computed curves of the correlation of isotropic noise as a function of electrical delay and bandwidth may be found in a report by Jacobson (21). For deep-water ambient noise originating from random sources distributed over the sea surface,

TABLE 3.3 Crosscorrelation Coefficients of Signals and Noise

	Unidirectional signal ρ_s	Isotropic noise ρ_n
Single frequency, zero time delay	$\cos \omega\tau_w$	$\dfrac{\sin(\omega\, d/c)}{\omega\, d/c}$
Single frequency, time delay	$\cos \omega(\tau_w + \tau_e)$	$\dfrac{\sin(\omega\, d/c)}{\omega\, d/c} \cos \omega\tau_e$
Flat bandwidth, zero time delay	$\dfrac{\sin[\frac{1}{2}(\omega_2 - \omega_1)\tau_w]}{\frac{1}{2}(\omega_2 - \omega_1)\tau_w} \cos \dfrac{(\omega_2 + \omega_1)}{2} \tau_w$	$\dfrac{1}{(\omega_2 - \omega_1)d/c}\left[\mathrm{Si}\left(\omega_2 \dfrac{d}{c}\right) - \mathrm{Si}\left(\omega_1 \dfrac{d}{c}\right)\right]$
Flat bandwidth, time delay	$\dfrac{\sin[\frac{1}{2}(\omega_2 - \omega_1)(\tau_w + \tau_e)]}{\frac{1}{2}(\omega_2 - \omega_1)(\tau_w + \tau_e)} \cdot \cos \dfrac{(\omega_2 + \omega_1)}{2} (\tau_w + \tau_e)$	$\dfrac{1}{2(\omega_2 - \omega_1)d/c}\left[\mathrm{Si}\left(\omega_2\left(\dfrac{d}{c} + \tau_e\right)\right) - \mathrm{Si}\left(\omega_1\left(\dfrac{d}{c} + \tau_e\right)\right) - \mathrm{Si}\left(\omega_2\left(\dfrac{d}{c} - \tau_e\right)\right) + \mathrm{Si}\left(\omega_1\left(\dfrac{d}{c} - \tau_e\right)\right)\right]$

Symbols for table:

$\omega = 2\pi$ (frequency)

τ_w = travel time of signal between array elements $= (d/c) \cos \theta$

d = separation of array elements

θ = angle to line joining the two elements

c = velocity of sound

τ_e = electrical (steering) delay

$\omega_2 = 2\pi$(upper frequency of band)

$\omega_1 = 2\pi$(lower frequency of band)

$$\mathrm{Si}(x) = \int_0^x \frac{\sin u}{u}\, du$$

expressions for the crosscorrelation coefficient between separated receivers have been derived by Cron and Sherman (22).

It is instructive to compute the array gain for some simple situations. When both the signal and noise are either completely coherent or completely incoherent across the array, the array gain becomes zero db; the array is unable to distinguish signal from noise. On the other hand, for a *perfectly coherent* signal in *incoherent* noise, such as when

$$(\rho_n)_{ij} = 0 \qquad i \neq j$$
$$(\rho_n)_{ij} = 1 \qquad i = j$$

it is easy to show, by expansion of the double summations of the array-gain expression, that *an array of n elements has a gain of 10 log n.* However, with a perfectly coherent signal in a background of only partly coherent noise, such that

$$(\rho_n)_{ij} = \rho \qquad i \neq j$$
$$(\rho_n)_{ij} = 1 \qquad i = j$$

then, by expansion, the array gain becomes

$$10 \log \frac{n}{1 + (n - 1)\rho}$$

and the array gain is less than 10 log n. It is obvious therefore that, in general, the gain of an array degrades as the signal coherence decreases and as the noise coherence increases. When, as in the actual ocean, the medium in which the array operates is not statistically time-stationary, but causes correlated amplitude and phase fluctuations in the received signal, the array performance is further degraded (23–26). The gain of an array, in short, depends on the statistics of the desired and undesired portions of the sound field in which the array operates.

As a simple example of a computation, we compute the gain of a three-element, equally spaced, unshaded line array. The coherence of signal and noise, plotted as a continuous function of separation d in the direction of the line, is shown in Fig. 3.12. The values of ρ_s and ρ_n corresponding

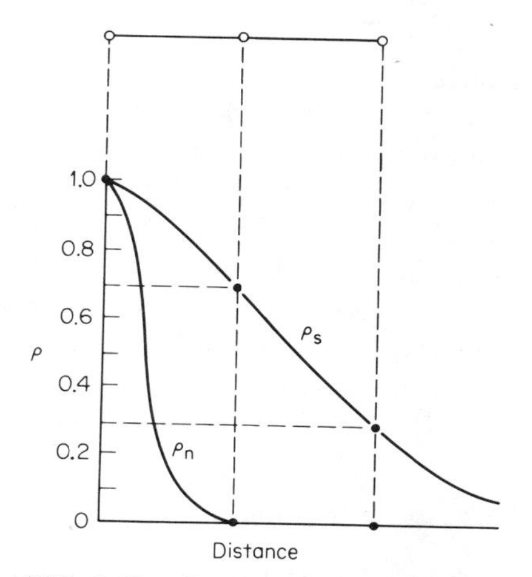

FIG. 3.12. Example of signal and noise coherence along a three-element array.

to the spacings of the elements are those of the following matrixes:

ρ_s

j \ i	1	2	3
1	1	0.7	0.3
2	0.7	1	0.7
3	0.3	0.7	1

ρ_n

j \ i	1	2	3
1	1	0	0
2	0	1	0
3	0	0	1

The array gain is 10 times the logarithm of the ratio of the sums of the matrix elements, or

$$\begin{aligned} \text{AG} &= 10 \log \frac{(1 + 0.7 + 0.3) + (0.7 + 1 + 0.7) + (0.3 + 0.7 + 1)}{(1 + 0 + 0) + (0 + 1 + 0) + (0 + 0 + 1)} \\ &= 10 \log \frac{6.4}{3} \\ &= 3.3 \text{ db} \end{aligned}$$

The effect of the degradation of signal coherence is, in this example, to degrade the array gain from 10 log 3 = 4.8 db, when all the elements of ρ_s are unity, to 3.3 db. On the other hand, occasionally very large array gains can be had by tailoring the array to the prevailing signal and noise fields. The simplest example is a vertical array used to discriminate against undesired sound arriving vertically (such as the bottom return, at times corresponding to normal incidence) in favor of a signal arriving horizontally. If a narrow frequency band can be tolerated, a half-wavelength spacing between elements will markedly reduce the denominator in the array gain expression, while leaving the numerator unaffected, and so yield a large array gain. The array, in a more customary view, has cancelled out the noise. In general, array design for maximum gain in a particular signal and noise field is best done by evaluating on a digital computer the gain of a number of selected arrays and selecting the best. However, it should be noted that in some applications beam-width and side-lobe level will be more important than array gain and will govern the acoustical design.

3.9 Multiplicative Arrays

Thus far we have been concerned with arrays that are both *linear* and *additive*—that is, they comprise elements that have an output linearly proportional to the acoustic pressure and which are simply added together, or summed, to give the array output. Instead of being added, the element outputs may be *multiplied* together so as to form a multiplicative or correlative array. Such arrays have properties different from those of linear arrays.

Figure 3.13*a* shows a four-element linear array in which the elements are simply added together. In Fig. 3.13*b*, the four elements are multiplied together and the output of the multiplier is averaged or smoothed. This kind of multiplicative array, called an intraclass correlator, was originally studied by Faran and Hills (27) and more recently by Fakley (28). When the outputs of various multipliers are themselves multiplied together, a time-average-product (TAP) array (so called by Berman and Clay (29)) is obtained. One arrangement of a four-element TAP array is shown in Fig. 3.13*c*.

A multiplicative array of only two elements can be made to have the same beam pattern as a linear array of an arbitrary number of equally spaced elements. To see this, consider a linear array of an odd $(2n - 1)$ number of elements spaced a distance d apart. If the array is assumed to be amplitude-shaded symmetrically about the center element, the

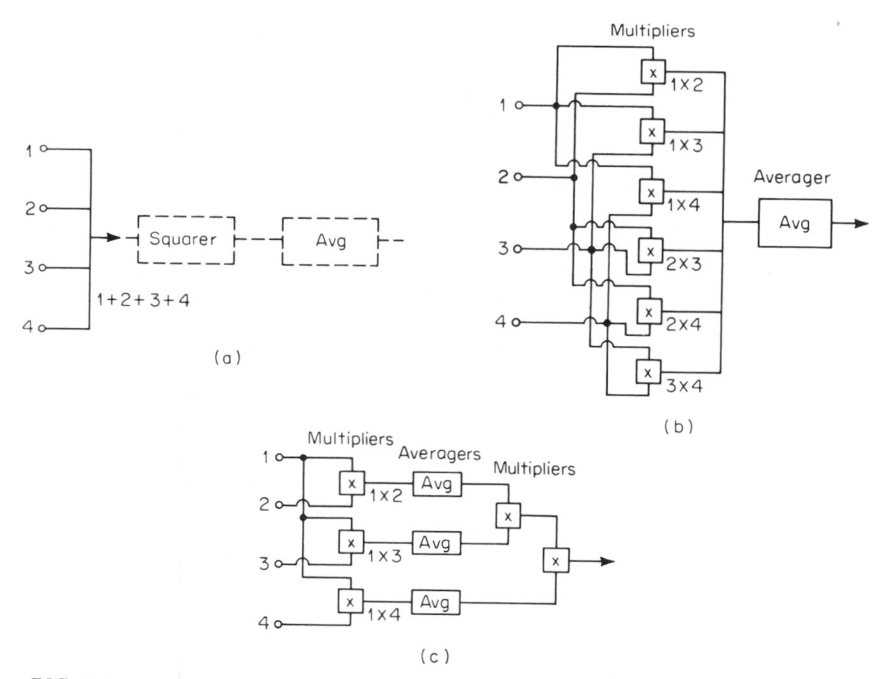

FIG. 3.13. Schematic diagrams of (a) additive array, (b) intraclass correlator, and (c) TAP array (Class I of Berman and Clay).

array output can be written

$$v(u) = \sum_{k=1}^{n} a_k \cos 2ku$$

where a_k is the common shading factor of the two elements a distance kd away from the center element, and $u = (\pi d/\lambda) \sin \theta$. Similarly, for an even number of elements

$$v(u) = \sum_{k=1}^{n} a_k \cos (2k - 1)u$$

where a_k is the common shading factor for elements distant $(k - \frac{1}{2})d$ from the center. In these expressions, each of the $\cos 2ku$ or $\cos (2k - 1)u$ terms can be expanded in a power series of $\cos u$, thus:

$$\begin{aligned} \cos u &= \cos u \\ \cos 2u &= 2 \cos^2 u - 1 \\ \cos 3u &= 4 \cos^3 u - 3 \cos u \\ \cos 4u &= 8 \cos^4 u - 8 \cos^2 u + 1 \\ &\cdots\cdots\cdots \end{aligned}$$

These polynomial expansions of $\cos mu$ in powers of $\cos u$ are, in the general case, the Chebyshev polynomials, written

$$T_m(\cos u) = \cos mu$$

These were used originally by Dolph (30) in antenna design, as mentioned above. The array output can therefore be written as a sum of terms like $\cos^m u$, appropriately weighted by the shading factors of the particular array. This means that the beam pattern of an array of any number of elements can be synthesized by an array of only two elements working into appropriate power-law amplifiers and combined in accordance with the shading factors and the terms of the $\cos mu$ expansions. An example, taken from Brown and Rowlands (31), of how the beam pattern of a linear array of $2n + 1$ elements can be achieved, in principle, with only two elements plus nonlinear amplifiers is shown in Fig. 3.14.

This technique is appealing in applications where a reduction in the number of hydrophone elements is necessary. Here one may trade physical size, or number of elements, or both for complexity in processing. The method is valid only at high signal-to-noise ratios, since the effect of noise (so far neglected) is to destroy the beam pattern unless a long averaging time is used.

The saving in number of array elements is coupled with a degradation of the signal-to-noise ratio when the signal-to-noise ratio is less than unity. Using communication theory, Brown and Rowlands (31) showed

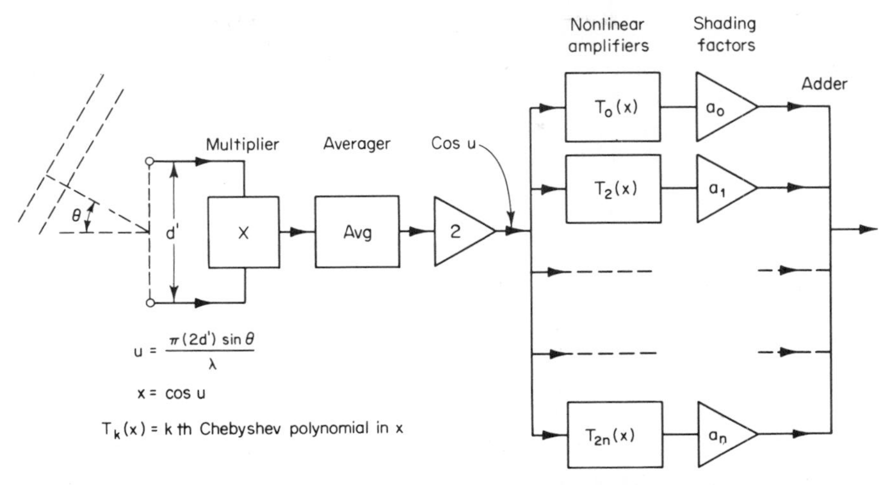

FIG. 3.14. Synthesis of the beam pattern of a linear array of an odd number of equally spaced elements by two elements plus amplifiers generating Chebyshev polynomials of cos u. ***(After Ref. 31.)***

that for improving the signal-to-noise ratio of small signals buried in noise, one cannot do better with n elements than to form an additive array of the n elements. If the n elements are split into two additive arrays of $n/2$ elements each and the outputs are multiplied together, it was shown by Faran and Hills (27) and Jacobson (32) that the array gain in uncorrelated noise is 10 log $(n/2)$, or a loss of 3 db over simple addition of the elements. The advantages of multiplicative processing lie in narrower beams and in better angular resolution of closely spaced sources. In one study (33), for example, it was shown that a four-element array two wavelengths in overall length and having a bandwidth-time product of 9 has the same resolution as a linear array approximately five wavelengths long. Another benefit lies in the flexibility in design that multiplicative processing provides. Some examples of possible nonlinear arrays are described in an article by Tucker (34).

A number of different beams may be obtained with a single array of hydrophones by means of *dimus* processing of the outputs of the array elements. In the dimus (*di*gital *mu*ltibeam *s*teering) technique (35, 36), the time delays needed to form different beams are obtained with digital shift registers using successive infinitely clipped samples of the output of the elements. The advantage of dimus is that the various directional beams are formed simultaneously; the array can "look" acoustically in different directions at the same time, though with a loss of about 1 db in signal-to-noise ratio.

With linear arrays, the beam width, array gain, and resolution are all interconnected. With multiplicative arrays there is no longer any direct relation between these array properties. They have to be determined mathematically for particular configurations of multipliers and averagers. Indeed, with multiplicative arrays, the terms DI and DT in the sonar equations must be considered together because it is impossible to separate clearly the array from the signal processing which accompanies it.

In summary, it may be said that multiplicative arrays find applications in conditions of high signal-to-noise ratio where narrow beams or high resolution are desirable or where a reduction in size or number of elements over a corresponding linear array is mandatory. In these applications, the tradeoffs in processing complexity and signal-to-noise ratio must be seriously considered.

REFERENCES

1. Hunt, F. V.: "Electroacoustics," John Wiley & Sons, Inc., New York, 1954.
2. Design and Construction of Crystal Transducers, Design and Construction of Magnetostriction Transducers, *Natl. Defense Res. Comm. Div. 6 Sum. Tech. Repts.* 12 and 13, 1946.
3. MacLean, W. R.: Absolute Measurement of Sound without a Primary Standard, *J. Acoust. Soc. Am.*, **12**:140–146 (1940).
4. Ebaugh, P. E., and R. Meuser: Practical Application of the Reciprocity Theorem in the Calibration of Underwater Sound Transducers, *J. Acoust. Soc. Am.*, **19**:695–700 (1947).
5. Carstensen, E. L.: Self-reciprocity Calibration of Electroacoustic Transducers, *J. Acoust. Soc. Am.*, **19**:961–965, 1947.
6. Morse, P. M.: *Vibration and Sound,* 2d ed., eq. 27.4, McGraw-Hill Book Company, New York, 1948.
7. Morse, P. M.: "Vibration and Sound," 2d ed., eq. 26.2, McGraw-Hill Book Company, New York, 1948.
8. Bobber, R. J., and G. A. Sabin: Cylindrical Wave Reciprocity Parameter, *J. Acoust. Soc. Am.*, **33**:446 (1961).
9. Simmons, B. D., and R. J. Urick: The Plane Wave Reciprocity Parameter and Its Applications to the Calibration of Electroacoustic Transducers at Close Distances, *J. Acoust. Soc. Am.*, **21**:633 (1949).
10. Horton, C. W., and G. S. Innis: The Computation of Far-field Radiation Patterns from Measurements Made near the Source, *J. Acoust. Soc. Am.*, **33**:877 (1961).
11. Trott, W. J.: Underwater Sound Transducer Calibration from Nearfield Data, *J. Acoust. Soc. Am.*, **36**:1557 (1964).
12. Procedures for Calibration of Electroacoustic Transducers, Z24.24-1957, USA Standards, New York, 1957.
13. Davids, N., E. G. Thurston, and R. E. Meuser: The Design of Optimum Directional Acoustic Arrays, *J. Acoust. Soc. Am.*, **24**:50 (1952).
14. Albers, V. O.: "Underwater Acoustics Handbook," chap. 11, The Pennsylvania State University Press, University Park, Pa., 1960.
15. Pritchard, R. L.: Optimum Directivity Patterns for Linear Point Arrays, *J. Acoust. Soc. Am.*, **25**:879 (1953).

16. Pritchard, R. L.: Maximum Directivity Index of a Linear Point Array, *J. Acoust. Soc. Am.*, **26**:1034 (1954).
17. Sandler, S. S.: Some Equivalence between Equally and Unequally Spaced Arrays, *Inst. Radio Engrs. Trans. Antennas Propagation*, **8**:496 (1960).
18. Harrington, R. F.: Sidelobe Reduction by Non-uniform Element Spacing, *Inst. Radio Engrs. Trans. Antennas Propagation*, **9**:187 (1961).
19. Ishimaru, A., and Y. S. Chen: Thinning and Broadbanding Antenna Arrays by Unequal Spacings, *Inst. Radio Engrs. Trans. Antennas Propagation*, **13**:34 (1965).
20. Basic Methods for the Calibration of Sonar Equipment, *Nat. Defense Res. Comm. Div. 6 Sum. Tech. Rept.* 10, pp. 21–26, 1946.
21. Jacobson, M. J.: Space-Time Correlation in Spherical and Circular Noise Fields, *J. Acoust. Soc. Am.*, **34**:971 (1962).
22. Cron, B. F., and C. H. Sherman: Spatial Correlation Functions for Various Noise Models, *J. Acoust. Soc. Am.*, **34**:1732 (1962).
23. Bourret, R. C.: Directivity of a Linear Array in a Random Transmission Medium, *J. Acoust. Soc. Am.*, **33**:1793 (1961).
24. Berman, H. G., and A. Berman: Effect of Correlated Phase Fluctuation on Array Performance, *J. Acoust. Soc. Am.*, **34**:555 (1962).
25. Brown, J. L.: Variation of Array Performance with Respect to Statistical Phase Fluctuations, *J. Acoust. Soc. Am.*, **34**:1927 (1962).
26. Lord, G. E., and S. R. Murphy: Reception Characteristics of a Linear Array in a Random Transmission Medium, *J. Acoust. Soc. Am.*, **36**:850 (1964).
27. Faran, J. J., and R. Hills: Application of Correlation Techniques to Acoustic Receiving Systems, *Harvard Univ. Acoust. Res. Lab. Tech. Mem.* 27, September, 1952. Also, *Tech. Mem.* 28, November, 1952.
28. Fakley, D. C.: Time Averaged Product Array, *J. Acoust. Soc. Am.*, **31**:1307 (1959).
29. Berman, A., and C. S. Clay: Theory of Time-average Product Arrays, *J. Acoust. Soc. Am.*, **29**:805 (1957).
30. Dolph, C. L.: A Current Distribution of Broadside Arrays Which Optimizes the Relationship between Beam Width and Side-lobe Level, *Proc. Inst. Radio Engrs.*, **34**:335 (June, 1946).
31. Brown, J. L., and R. O. Rowlands: Design of Directional Arrays, *J. Acoust. Soc. Am.*, **31**:1638 (1959).
32. Jacobson, M. J.: Analysis of a Multiple Receiver Correlation System, *J. Acoust. Soc. Am.*, **29**:1342 (1957).
33. Linder, I. W.: Resolution Characteristics of Correlation Arrays, *J. Res. Natl. Bur. Stds.*, **6JD**(3):245 (1961).
34. Tucker, D. G.: Sonar Arrays, Systems and Displays, in V. M. Albers (ed.), "Underwater Acoustics," Plenum Press, New York, 1963.
35. Anderson, V. C.: Digital Array Phasing, *J. Acoust. Soc. Am.*, **32**:867 (1960).
36. Rudnick, P.: Small Signal Detection in the DIMUS Array, *J. Acoust. Soc. Am.*, **32**:871 (1960).
37. Bazhenov, V. A.: "Piezoelectric Properties of Wood" (translation), Consultants Bureau, New York, 1961.

FOUR

Generation of Underwater Sound: Projector Source Level

All active sonars utilize some form of *projector* to generate acoustic energy. Sonar projectors normally, but not always, generate sound through a process of converting electric energy. They usually consist of arrays of individual elements, by means of which a directional beam is formed to send the generated energy into directions where it is wanted. In many sonars, the projector is also used as a hydrophone in order to save cost, weight, and space.

In the sonar equations, the parameter *source level* specifies the amount of sound radiated by a projector. It is defined as the intensity of the radiated sound in decibels relative to the intensity of a plane wave of rms pressure 1 dyne/cm^2, referred to a point 1 yd from the acoustic center of the projector in the direction of the target. Because a directional projector ordinarily "points" in the direction of the target, the reference point for source level lies along the axis of the projector's beam pattern.

The *transmitting directivity index* of a projector is the difference, measured at a point on the axis of the beam pattern, between the level of the sound generated by the projector and the level that would be produced by a nondirectional projector radiating the same total amount of acoustic power. Figure 4.1 shows the beam pattern of a directional projector and of the equivalent nondirectional projector radiating the same total acoustic power. If the intensity represented by the directional pattern along its axis is I_D, and if the intensity represented by the nondirectional pattern is I_{Nond}, then the transmitting directivity index is

$$\mathrm{DI}_T = 10 \log \frac{I_D}{I_{Nond}}$$

where the subscript T emphasizes that the transmitting, rather than the receiving, directivity index is referred to. In terms of the geometry of the projector array and the wavelength, the same expressions apply for both the transmitting and the receiving directivity indexes.

When an electric voltage is impressed upon a material having the property of being piezoelectric or magnetostrictive, the material is caused to change its shape and thereby creates sound in the medium in which it is placed. This process of converting electric energy into sound, and, also, the design and construction of electroacoustic projectors, involves specialized theory and engineering practice. Much useful material on these subjects may be found in the World War II Summary Technical Reports of Division 6 of the NDRC on crystal and magnetostrictive transducers (1). Figure 4.2 shows the construction and beam pattern of a 24-kHz circular projector 19 in. in diameter, as constructed and measured during the World War II years. It is a mosaic of 728 ADP (ammonium dihydrogen phosphate) crystals cemented to a backing plate with an air cavity behind it. This unit was estimated to have an efficiency of 50 percent (ratio of acoustic-power output to electric-power input).

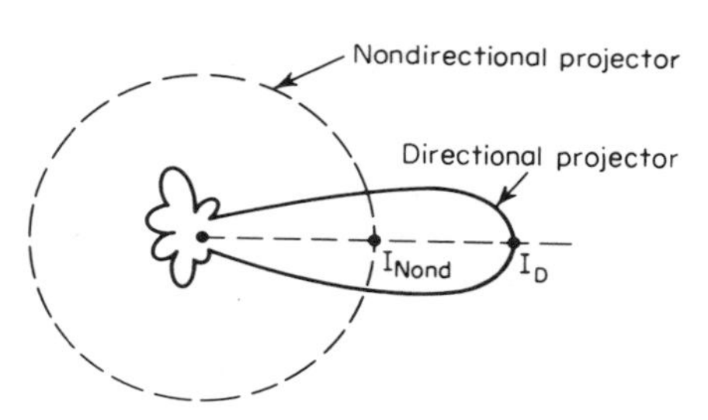

FIG. 4.1. Beam patterns of a directional projector and the equivalent nondirectional projector.

Not all projectors are electroacoustic. An explosive charge detonated in water converts chemical energy into sound and has interesting acoustical properties, which will be described in Sec. 4.3.

4.1 Relation between Source Level and Radiated Acoustic Power

The source level of a projector is related in a simple way to the acoustic power it radiates and to its directivity index. Let a nondirectional projector be located in a homogeneous, absorption-free medium. At a large distance r, let the intensity of the sound emitted by the projector be I_r. If r is taken to be very large, the intensity is related to the rms pressure p_r in dynes per square centimeter by the plane-wave expression

$$I_r = \frac{p_r^2}{\rho c} \times 10^{-7} \qquad \text{watt/cm}^2$$

where ρ = density, g/cm³
c = velocity of sound, cm/sec

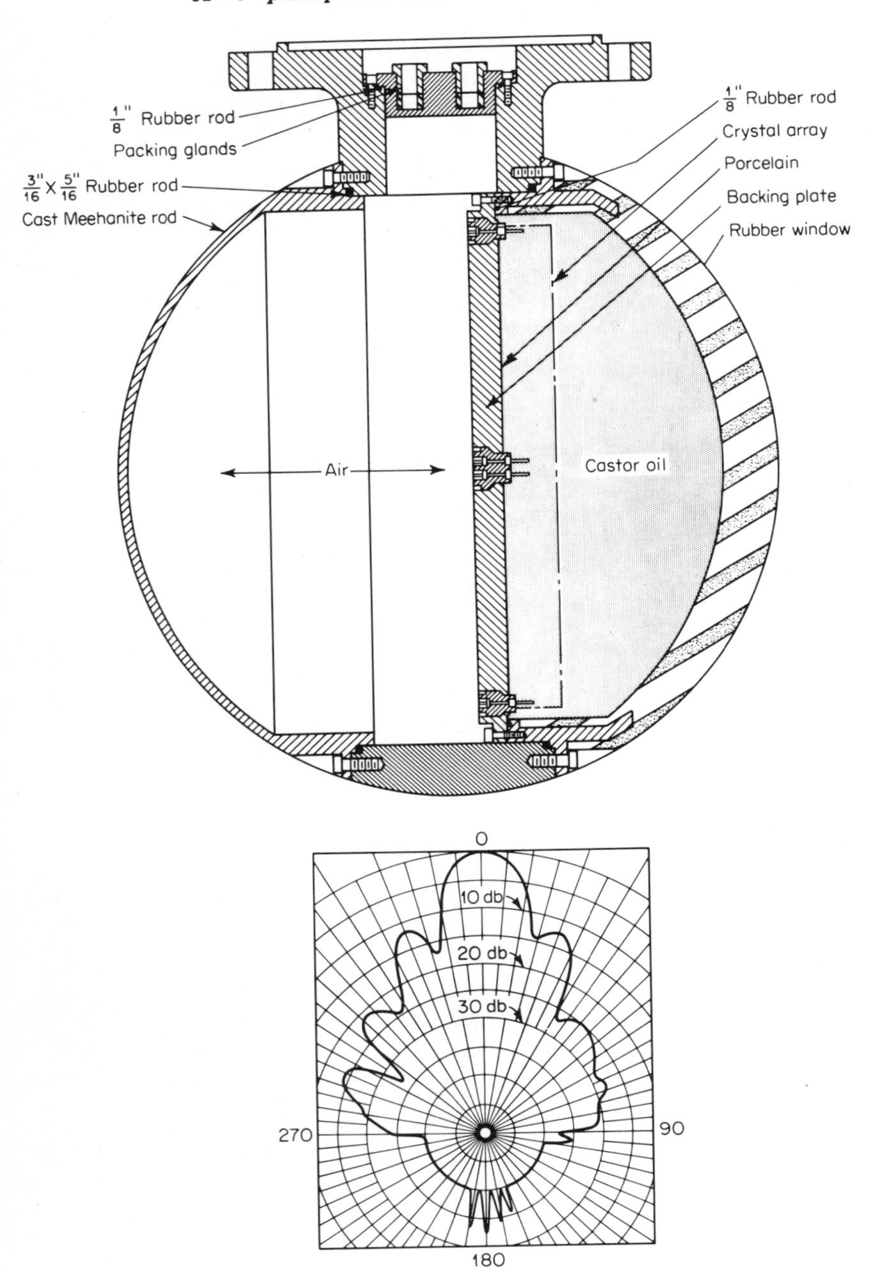

FIG. 4.2. Construction and beam pattern of a piezoelectric projector comprising a crystal array cemented to a backing plate. (*After Ref. 1, Report 12, pp. 243, 246.*)

Substituting $\rho = 1$ g/cm^3 and $c = 1.5 \times 10^5$ cm/sec = 4,920 ft/sec and converting from centimeters to yards, we find

$$I_r = 5.58 \times 10^{-9}\, p_r^2 \qquad \text{watt/yd}^2$$

For a nondirectional projector, this intensity corresponds to a radiated power output of

$$P_r = 4\pi r^2 I_r = 70.0 \times 10^{-9}\, p_r^2 r^2 \qquad \text{watt}$$

At a distance of 1 yd, the power is

$$P = 70.0 \times 10^{-9}\, p_1^2$$

where p_1 is the rms pressure at 1 yd in dynes per square centimeter. Converting to decibels and remembering that $10 \log p_1^2$ is the source level SL, we have

$$10 \log p = -71.5 + \text{SL}$$

or

$$\text{SL} = 71.5 + 10 \log P$$

If the projector is directional with transmitting directivity index DI_T, we obtain, by the definition of DI_T,

$$\text{SL} = 71.5 + 10 \log P + \text{DI}_T$$

This function is graphed for convenience in Fig. 4.3.

It should be noted that the value of SL computed above is not neces-

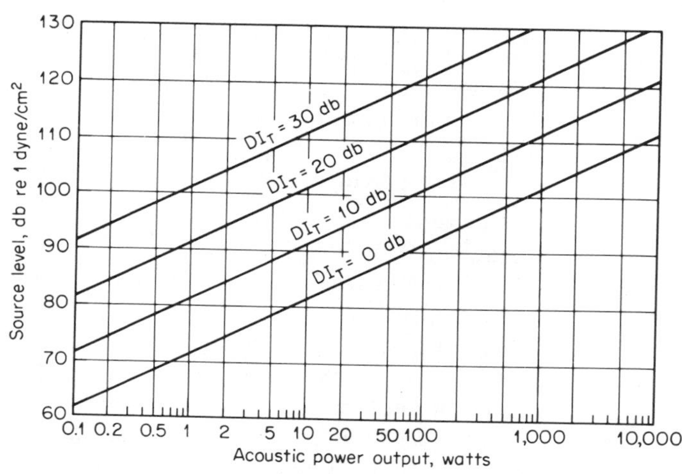

FIG. 4-3. Plots of the function $\text{SL} = 71.5 + 10 \log P + \text{DI}_T$ *relating source level, power output, and transmitting directivity index.*

sarily the value that would be measured by a calibrated hydrophone placed 1 yd from the projector. As in the case of the source level of the radiated noise of ships (Chap. 10), SL is a *reduced* quantity obtained through measurements made at large distances, with 1 yd merely an arbitrary standard distance to which the sonar parameters are referred.

The quantity P is the total *acoustic* power radiated by the projector. For an electroacoustic projector, P is less than the *electric* power fed into the projector. If the electric power is P_e, the ratio of these powers is the efficiency E of the projector, or

$$E = \frac{P_e}{P}$$

The efficiencies of most sonar projectors range from 20 to 70 percent and depend upon the bandwidth employed; sharply tuned high-Q projectors have higher efficiencies than broadband projectors.

The radiated acoustic powers of shipboard sonar equipments range from a few hundred watts to some tens of kilowatts and have transmitting directivity indexes lying between 10 and 30 db (2). It follows that the source levels of active shipboard sonars range from about 110 to about 140 db.

4.2 Limitations on Sonar Power

To achieve the maximum range with active sonars, it is desirable to generate the maximum amount of acoustic power—at least up to the point where the just-detectable echo occurs in a background of reverberation rather than noise. In attempting to do so, two peculiar limitations, other than electrical failures, are encountered. One, caused by *cavitation*, is primarily a property of the fluid medium into which the acoustic power is radiated; the other is caused by *interaction effects* between radiating elements of densely packed arrays of projector elements.

Cavitation Limitation When the power applied to a sonar projector is increased, it is found that cavitation bubbles begin to form on the face and just in front of the projector. These bubbles are a manifestation of the rupture of the water caused by the negative pressures of the generated sound field. These negative pressures tear the liquid apart, so to speak, when they exceed a certain value called the *cavitation threshold*. The cavitation threshold may be expressed as a peak pressure in atmospheres or as a plane-wave intensity in watts per square centimeter. Since 1 atm is almost equal to 10^6 dynes/cm^2, the relation between the two is

$$I_c = \frac{[(0.707)10^6 p_c]^2}{\rho c} \times 10^{-7} = 0.3 p_c^2 \qquad \text{watt/cm}^2$$

where I_c = cavitation threshold, watts/cm^2
p_c = peak pressure of sound wave causing cavitation, atm
$\rho = 1.0$ g/cm^3
$c = 1.5 \times 10^5$ cm/sec

A cavitation threshold of 1 atm is therefore equivalent to a plane-wave intensity of 0.3 watt/cm^2. When multiplied by the face area of the projector (in square centimeters), the cavitation threshold represents the maximum working value for the power output (in watts) of the projector. When this limit is exceeded by driving the projector harder, a number of deleterious effects begin to occur, such as erosion of the projector surface, a loss of acoustic power in absorption and scattering by the cavitation bubble cloud, a deterioration in the beam pattern of the projector, and a reduction in the acoustic impedance into which the projector must operate. In all these ways, the cavitation threshold represents the onset of a gradual deterioration of projector performance.

The physics of cavitation in liquids and the many phenomena accompanying it have received a great deal of attention in the literature. The reader interested in the processes of phenomena of cavitation is referred to a comprehensive review by Blake (3), to subsequent literature quoted by Strasberg (4), and to a summary article in a book edited by Mason (5).

The onset of cavitation is associated with the presence of cavitation nuclei in the liquid. As first postulated by Blake (6), these are believed to be extremely small microscopic bubbles of air that occur in cracks or cavities of small solid particles suspended in the liquid. Alternatively, cavitation nuclei may exist as free air bubbles surrounded by skins of organic impurities that hinder the escape of air into solution. In a sound field, these tiny bubbles form the nuclei into which dissolved air diffuses under the influence of the negative-going portions of the sound wave. When the negative pressure begins to exceed the cavitation threshold, the nuclei begin to grow in size by a process called (by Blake) *rectified diffusion*, by which more dissolved air diffuses into the bubble than out of it.

The occurrence of cavitation is accompanied by a variety of interesting physical and chemical phenomena, such as luminescence and the breakdown of chemical compounds. In laboratory experiments, cavitation is produced by focusing sound on a small region of the liquid under study, such as by exciting a spherical container into vibrational resonance. At the center of such a pulsating sphere, the occurrence of cavitation can be seen as the appearance of a bubble cloud or heard as the hiss of collapsing cavitation bubbles.

Practically speaking, the cavitation threshold may be said to occur at the point of departure from linearity of the input-output curve of the projector illustrated in Fig. 4.4. This point also marks the beginning of harmonic distortion in the output of a sinusoidal wave. The cavitation threshold of liquids is notably sensitive to a variety of factors, such as temperature, dissolved-air content, and previous pressure history, as well as to the particular criterion used for establishing the onset of cavitation.

The cavitation threshold of a projector may be raised, and more acoustic power may be radiated, by (1) increasing the frequency, (2) decreasing the pulse duration, and (3) increasing the depth (hydrostatic pressure) of the projector. Figure 4.5 shows the variation of cavitation threshold of fresh water with frequency as measured and compiled by Esche (7). The shaded area is the range of observed values, and the dashed curve is an estimated average curve. Below 10 kHz, the method of cavitation detection used by Esche led to thresholds lower than 1 atm, the value indicated by the dashed curve. A number of other measurements may be mentioned. At 25 kHz, Strasberg (4) found values ranging from 2.5 atm in tap water saturated with air at atmospheric pressure to 6.5 atm in water devoid of dissolved air. At 60 kHz, Blake (6) measured thresholds ranging from 3.5 to 4.6 atm for air-saturated and degassed water. Finally, at 550 kHz, Rosenberg (8) measured thresholds between 150 and 380 atm, a range of values lying close to the extension of the curve of Fig. 4.5.

Figure 4.6 shows the cavitation threshold of four full-scale sonar projectors measured by Liddiard (9) in open water as a function of pulse duration, using the criterion shown in Fig. 4.4. The projectors used were flat-faced piston magnetostrictive and crystal projectors resonant in the frequency range 14.2 to 31.6 kHz. For water, the cavitation limit is seen to increase with a shortening of the pulse duration below about 5 msec—in keeping with the idea that cavitation nuclei require a finite time interval to grow to a size where their effects are observable. As an interesting comparison, the results of laboratory measurements on castor oil by Briggs, Johnson, and Mason (10) are plotted. Castor oil, as well as other liquids of high viscosity and free of dissolved air, can tolerate much higher cavitation thresholds and is, accordingly, useful as filling liquid in transducer construction.

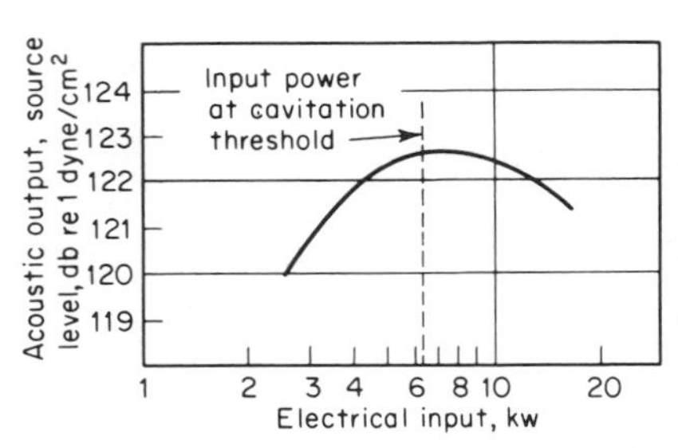

FIG. 4.4. Input versus output curve for a projector in the vicinity of the cavitation threshold. (After Ref. 9.)

The effect of an increased depth of operation is to increase the cavitation threshold by the amount of the hydrostatic pressure. Since each 33 ft of depth represents an increased pressure of 1 atm, the cavitation

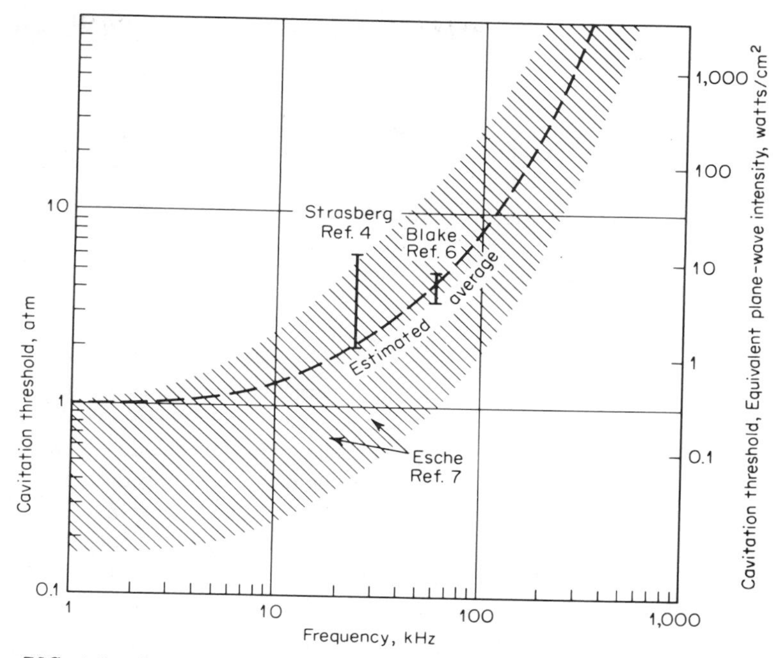

FIG. 4.5. Frequency dependence of the cavitation threshold. CW data on fresh water at atmospheric pressure.

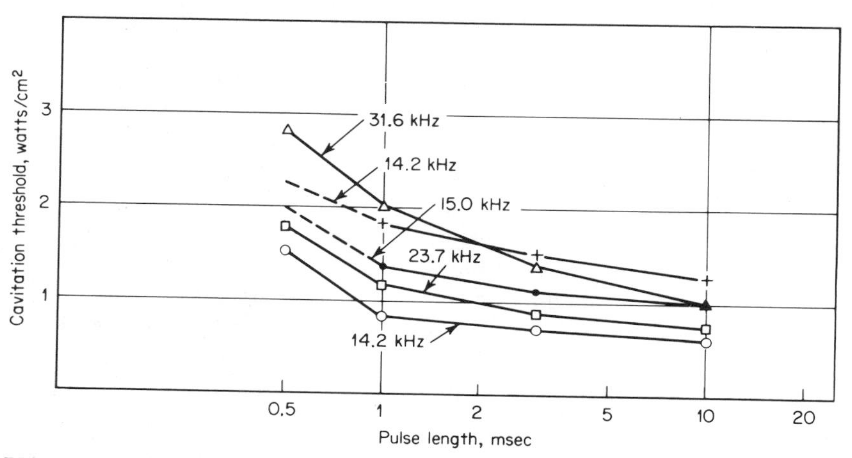

FIG. 4.6. Variation of cavitation threshold with pulse duration for five projectors. [After Liddiard (9).]

threshold at depth h (in feet) becomes (in atmospheres)

$$P_c(h) = P_c(0) + \frac{h}{33}$$

where $P_c(h)$ = threshold at depth h
$P_c(0)$ = threshold at zero depth

The amount by which the cavitation threshold exceeds the ambient hydrostatic pressure may be thought of as the *tensile strength* of the fluid. Since the pressure at depth h equals $1 + h/33$ atm, we may write

$$P_c(h) = 1 + \frac{h}{33} + T$$

where T is the tensile strength in atmospheres.

When actual sonar projector arrays are used, cavitation limits lower than those measured by other methods are found. These lower values are the result of "hot spots," or regions of high acoustic intensity, near the face of the projector. The effect of the nonuniform sound field in front of projectors on the cavitation limit has been investigated theoretically by Sherman (11), who defined a factor γ to express the near-field effect on the cavitation limit. The factor γ is the ratio of the cavitation limit for a particular transducer type to its value in a plane acoustic wave. For circular pistons, plane and hemispherical, γ was found to range between 0.3 and 0.6, with an average of about 0.5.

In summary, the cavitation threshold for long-pulse sonar projectors can be written

$$I_c = 0.3\gamma\left(P_c(0) + \frac{h}{33}\right)^2 \qquad \text{watts/cm}^2$$

$$= 0.15\left(P_c(0) + \frac{h}{33}\right)^2 \qquad \text{watts/cm}^2$$

for a value of $\gamma = 0.5$ and with h in feet. The threshold P_c approximates 1 atm at low frequencies and increases with frequency as indicated in Fig. 4.5. At a depth of 33 ft, the low-frequency cavitation limit is thus 0.6 watt/cm^2 for pulse durations in excess of about 5 msec. For shorter pulses, higher limits exist, as illustrated in Fig. 4.6.

It should be noted that the cavitation limit so predicted for sonar projectors is a conservative estimate since it corresponds (with $\gamma = 0.5$) to the *onset* of cavitation at a *single point* on the transducer surface. In practice, it is likely that considerably higher powers, higher, perhaps, by a factor of 2 or 3, can be usefully radiated without incurring excessively deleterious effects on the operation and performance of the projector.

Interaction Effects A near-field effect in many ways similar to that just described occurs in large arrays of closely spaced resonant projector elements. When such an array is driven electrically, it is found that the velocity of motion of each element is not uniform, but varies from element to element of the array in a complex manner. This erratic behavior is caused by acoustic interactions between one element and another. For example, of two elements of the array, one may act as a sink for the acoustic output of another. In doing so, it absorbs, rather than radiates, sound and may possibly be driven to destruction. Interaction effects, unless compensated for in design, reduce the power output of a projector array and deteriorate its beam pattern. They are important for arrays intended to radiate large amounts of acoustic power.

From a theoretical standpoint, interaction effects can be considered in terms of the mutual radiation impedance of a pair of sound sources, as has been done in a number of papers (12–19). Various control measures are available, as pointed out by Carson (20). First, and most simply, the mutual radiation impedance can be reduced by separating the elements of the array, though at a great cost in reduction of acoustic efficiency and power and in deterioration of the beam pattern. A second way of reducing the mutual impedance between elements is insertion of a reactance in series or in parallel with each element—a treatment effective at only a single frequency. A third method is to make the elements of the array individually large enough so that their self-radiation impedance overwhelms the mutual radiation impedance between elements—a "cure" which is costly from several engineering aspects. Finally, and most elegantly, individual amplifiers may be used to drive each array element in just the right amplitude and phase to yield the desired uniform inphase velocity of motion; a clever method of doing this, according to Carson, might be to place an accelerometer on the radiating face of each element to provide feedback to the driver amplifiers.

4.3 Explosions as Sources of Underwater Sound

A kind of underwater sound source that does not suffer from the difficulties described in the preceding section, but has other limitations, is the underwater explosion. Underwater explosive sound sources are commonly small charges of explosive material ranging from a few grains to a few pounds in weight. A photograph of a standard United States Navy explosive sound source is shown in Fig. 4.7. This is an air-dropped pressure-detonated "sound signal" Mk 64 containing a small charge of TNT and detonating at 800 ft. It is used primarily for explosive echo ranging against submarine targets. A variety of similar charges detonat-

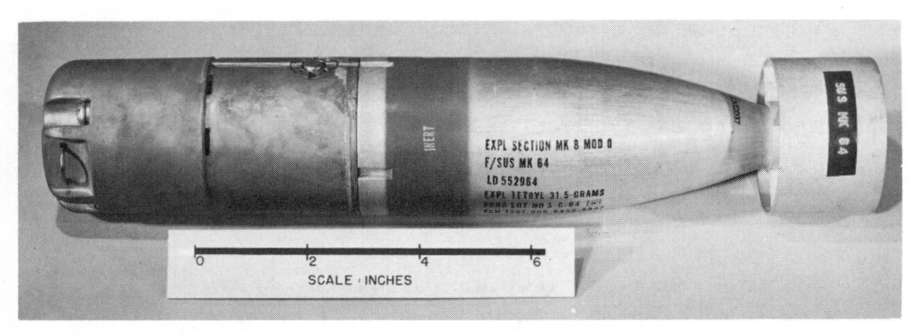

FIG. 4.7. A standard United States Navy explosive sound signal.

ing at other depths and containing other weights of explosive charge are in use in the United States Navy.

The basic phenomena of underwater detonation were established during World War II and are summarized in a book by Cole (21), as well as in a three-volume compendium of papers published by the Office of Naval Research (22).

The Pressure Signature When an explosion is initiated in a mass of explosive material, a pressure wave, initiated inside the material, propagates into the surrounding medium. This pressure wave is part of a complex series of phenomena occurring during the conversion of a solid explosive material into gaseous reaction products. The conversion process, and the accompanying pressure wave, proceeds in the material at a speed depending on the kind of material. In "high" explosives like TNT, the velocity of detonation lies between 15,000 and 30,000 ft/sec, and a *shock wave* that propagates in all directions is produced in the medium. In other materials, such as black powder, the explosion process is one of *deflagration*, or burning, rather than *detonation* and occurs at a velocity of only 0.1 to 1 ft/sec. The pressure pulses produced by these two processes are greatly different, as Fig. 4.8 illustrates. The detonating material creates a shock wave characterized by an infinitely steep front, a high peak pressure, and a rapid decay. The deflagrat-

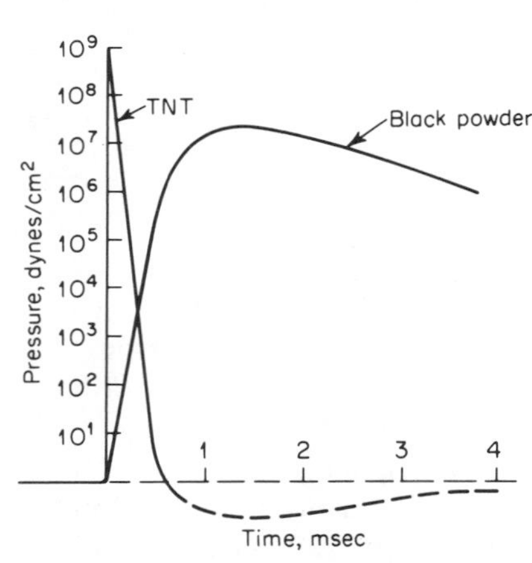

FIG. 4.8. Pressure pulses at 1 yd from a 1-lb charge of TNT and from a 1-lb charge of black powder in a heavy metal container.

ing material gives rise (23) to a relatively low, broad pressure blob. Only "high" TNT-like explosives are used as sources of underwater sound because they produce a much greater amount of acoustic energy at the higher frequencies useful in sonar.

The shock wave from a detonation is normally followed by a series of pressure pulses called *bubble pulses*. These are caused by successive oscillations of the globular mass of gaseous materials called the *gas globe* that remains after the detonation is completed. At the instant of minimum volume of the oscillating gas globe, a positive pressure pulse is generated; successive pulsations generate additional pulses, with each successive pulse being weaker than the preceding one. No bubble pulses occur at depths so shallow that the gas globe breaks the sea surface; neither do they occur when small charges are contained in a case that does not rupture when the charge is fired.

The pressure "signature" of a detonating explosion, as observed at short ranges, accordingly consists of the shock wave followed by a small number of bubble pulses, such as illustrated in Fig. 4.9. At longer ranges, the signature is complicated by refraction and multipath-propagation effects in the sea. These complications range from the simple addition of a negative-going reflected pulse from the sea surface, when a surface-reflected path exists between the explosion and a receiver, to the distorted pulses seen by Brockhurst, Bruce, and Arons (24) in shadow zones, to the long, drawn-out, complex sofar signature of shots fired at great ranges in the deep sea.

Of the total chemical energy of the explosive, about 40 percent is radiated as acoustic energy to distances beyond 1 yd for a 1-lb charge. Table 4.1 is a breakdown given by Arons and Yennie (25) of the dissipation of the chemical energy on the detonation of TNT. If the ratio

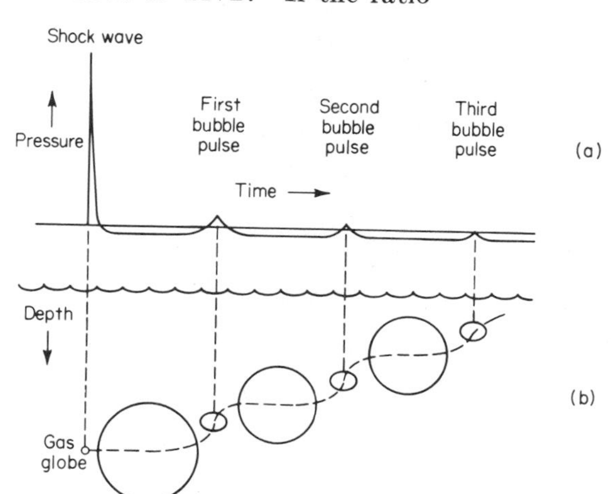

FIG. 4.9. (a) Shock wave and bubble pulses from an explosion. (b) Pulsations and migration in depth of the gas globe. (After Ref. 36.)

TABLE 4.1 Partition of the Energy of an Explosion of TNT (From Ref. 25)

Total acoustic radiation at $R = w^{1/3}/0.352$ ft (shock wave plus first and second bubble pulses)	410 cal/g
Shock-front dissipation at short ranges, up to $R = w^{1/3}/0.352$ ft	200
Energy remaining at third bubble maximum	95
Unaccounted for (turbulence, viscosity, heat conduction, chemical changes in the gaseous products)	345
Total	1,050 cal/g

of acoustic to chemical energy is taken as a measure of acoustic efficiency, explosives may be said to be comparable in efficiency to electroacoustic sonar projectors.

The Shock Wave The characteristics of the shock wave from explosions of various sizes over a wide range of distances in the sea were established by Arons and Yennie (26) by means of field measurements made in 1948 and earlier. Using charges ranging in size from ½ to 55 lb, they observed the amplitude and shape of the shock wave in 40 ft of water in Vineyard Sound, Massachusetts, at distances from a few feet to nearly 1 mile. It was found that the shock wave, expressed as a relationship between pressure and time, could be approximated by the function

$$p = p_0 e^{-t/t_0}$$

where p is the instantaneous pressure at time t after the onset of the shock front, p_0 is the peak pressure occurring at time $t = 0$, and t_0 is the time constant of the exponential pulse, or the time for the pressure to decay to $1/e = 0.368$ of its initial value p_0. Both p_0 and t_0 were found to be functions of the charge weight and range r. Arons, Yennie, and Cotter (27) found that for a charge of weight w pounds at a distance r feet, the peak pressure p_0 in lb/in.² could be written

$$p_0 = 2.16 \times 10^4 \left(\frac{w^{1/3}}{r}\right)^{1.13}$$

and the time constant in microseconds,

$$t_0 = 58w^{1/3} \left(\frac{w^{1/3}}{r}\right)^{-0.22}$$

Figures 4.10 and 4.11 show p_0 and t_0 as a function of range for a 1-lb charge ($w = 1$), using the more customary underwater sound units of pressure in dynes per square centimeter and range in yards. For other charge weights, the peak pressure at a given range r may be found by reading off its value from Fig. 4.10 at a "reduced" range $rw^{-1/3}$; the time

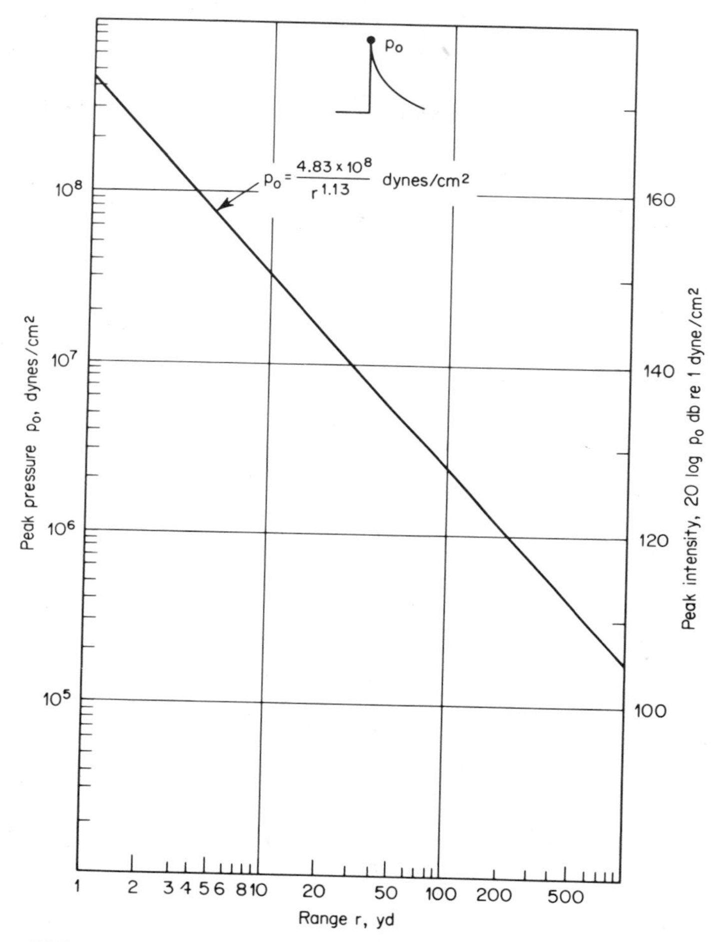

FIG. 4.10. Peak pressure of a 1-lb charge of TNT as a function of range. To find the peak pressure of a charge of weight w lb, multiply r by $w^{-1/3}$. (Adapted from Ref. 27.)

constant may be found by reading off a value from Fig. 4.11 at the "reduced" range $rw^{-1/3}$ and then multiplying by $w^{1/3}$.

The occurrence of the factor $w^{1/3}$ in the peculiar manner shown by the above expressions is a result of the *similarity principle* for shock waves (21, pp. 110–114), which states that the parameters of the shock wave are the same for all charge weights when the distances are reckoned in units of charge radii, or in units of $w^{1/3}$ for a spherical charge. The other peculiarity of the expressions for p_0 and t_0 is the occurrence of the expo-

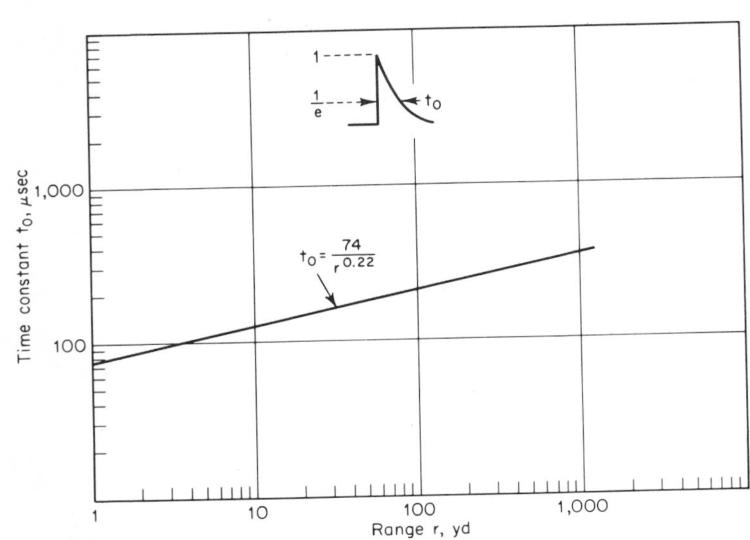

FIG. 4.11. Time constant of the shock wave for a 1-lb charge. To find t_0 for a charge of weight w lb, read off the value for a reduced range $rw^{-1/3}$ and multiply by $w^{1/3}$. (Adapted from Ref. 27.)

nents 1.13 and −0.22. In an absorption-free linear fluid one would expect the exponents to be unity and zero, respectively, corresponding to spherical spreading and a time constant independent of distance. The excess attenuation of the peak pressure with range, as well as the broadening of the shock-wave decay, is due to a combination of the effects of nonlinearity of the underwater medium and greater attenuation in the sea at higher frequencies, with the latter effect overwhelming the former at the ranges of interest to sonar. The effects of absorption and high amplitude on the form of the shock wave were first studied by Arons, Yennie, and Cotter (27) and Arons (28). In addition to the effects of range on p_0 and t_0, long-range shock pulses have a finite time-of-rise to the peak pressure p_0 (amounting to about 20 μ at 1 kyd)—an effect attributable to high-frequency absorption, but not observable with certainty with the instrumentation used to obtain the empirical expressions for the form of the shock wave. A more modern study of the effects peculiar to the propagation of high-amplitude waves to long ranges has been made by Marsh, Mellen and Konrad (29).

Bubble Pulses Following the shock wave occur a series of positive pressure pulses emitted by the pulsating gas globe at the successive instants of minimum volume. The amplitude of these pulses decreases progressively in size as the energy of the gas globe is dissipated. Measurements show that the peak pressure of the second bubble pulse is only about one-fifth that of the first; accordingly, only one or two bubble oscillations are practically significant. Similar bubble pulses are also radiated by the sudden collapse of a gaseous cavity under pressure, as when a glass bottle containing air breaks, or *implodes*, at a depth in the sea (30).

For explosives the time interval T, in seconds, between the shock wave and the first bubble pulse is found by theoretical analysis (31) to be

$$T = \frac{Kw^{1/3}}{(d + 33)^{5/6}}$$

where K = a proportionality constant
w = charge weight, lb
d = depth of detonation below sea surface, ft

K depends slightly on the type of explosive and is equal to 4.36 for TNT (21, p. 282). Figure 4.12 is a plot of the above relationship for a number of different weights of charge. The reciprocal of T is the frequency of the first bubble maximum in the acoustic-energy spectrum of the charge.

With increasing depth of detonation, the interval T decreases and the pressure amplitude of the bubble pulses relative to the shock wave

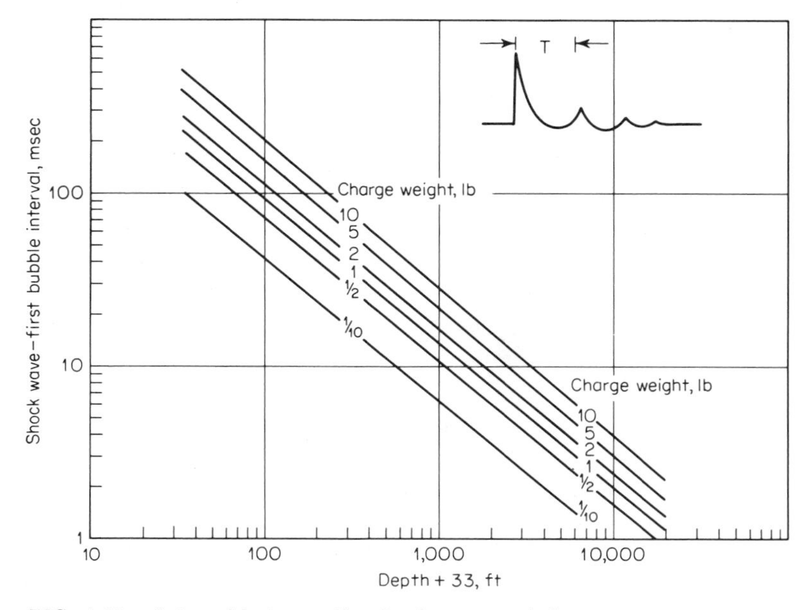

FIG. 4.12. Interval between the shock wave and the first bubble pulse as a function of the quantity (depth + 33) ft for various charge weights.

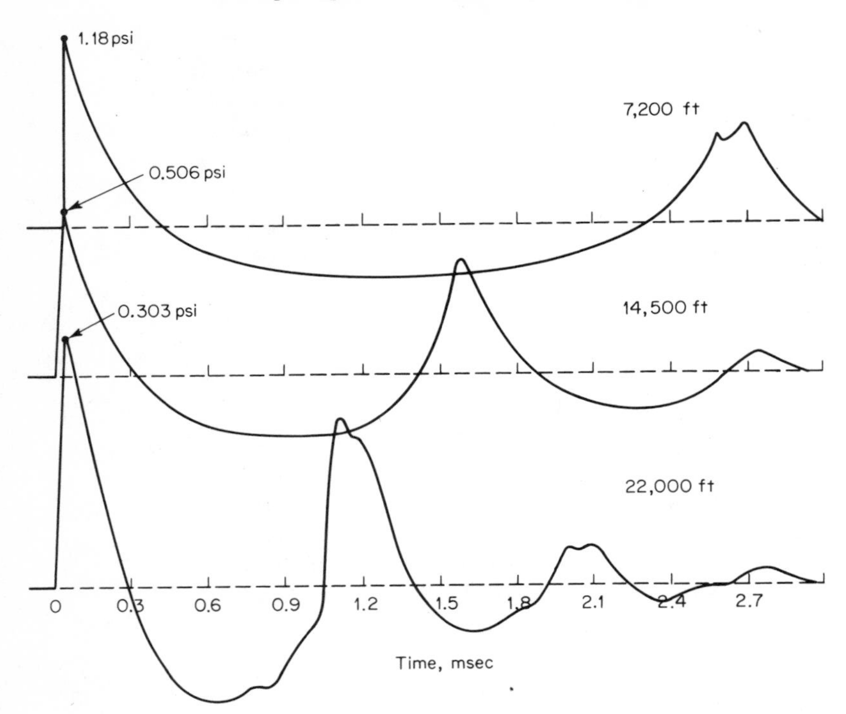

FIG. 4.13. Pressure signature of a 1-lb charge at different depths, as recorded near the sea surface above. (*After Ref. 37.*)

increases. These effects are illustrated by Fig. 4.13, which shows the pressure signature of a 1-lb charge fired at three depths and recorded near the sea surface overhead. Another effect of depth is a shortening of the apparent duration of the shock wave. This shortening is caused by the increased negative-going portions of the bubble oscillation at the deeper depths.

Acoustic-energy Flux Density and Energy Spectrum In order to avoid the complication caused by the propagation effects just mentioned, it is necessary to describe an explosion as a source of sound in terms of its acoustic-energy flux density and the distribution of the acoustic energy in frequency.

In units of ergs per square centimeter, the energy-density of a plane wave is (Secs. 1.3 and 2.5)

$$E = \frac{1}{\rho c} \int_0^\infty p^2(t)\, dt$$

where $p(t)$ = pressure signature, dynes/cm²
ρ = density of medium
c = sound velocity of medium

For sonar purposes, a more practical unit than ergs per square centimeter is (Sec. 1.3) the energy-density of a plane wave of rms pressure of 1 dyne/cm² integrated over a period of 1 sec. The two units differ by the factor $1/\rho c$, amounting to 52 db.

The total acoustic energy in the shock wave can be found by substituting the expression $p(t) = p_0 e^{-t/t_0}$ in the integral expression for E. The result is

$$E = \frac{p_0^2 t_0}{2\rho c} \qquad \text{ergs/cm}^2$$

which is the same as that of a square-topped sinusoidal pulse of peak pressure p_0 lasting for t_0 seconds. At 100 yd, we conclude that the shock wave from a 1-lb charge, using values for p_0 and t_0 from Figs. 4.10 and 4.11, has an energy flux density of +88 db (1 dyne/cm²)² sec. In comparison, a pinging sonar of source level 130 db and a pulse duration of ½ sec has an energy flux density at 100 yd of 130 − 40 − 3 = 87 db re (1 dyne/cm²)² sec.

Since broadband reception is nearly always employed with explosive sources, it is necessary to consider the frequency distribution of the acoustic flux energy-density of an explosive source. This subject has been considered theoretically and experimentally by Weston (32). For an exponential pulse of peak pressure p_0 and time constant t_0, the energy flux spectral density $E_0(f)$ can easily be found by Fourier analysis to be

$$E_0(f) = \frac{2p_0^2}{\rho c(1/t_0^2 + 4\pi f^2)} \qquad \text{ergs/(cm}^2\text{)(Hz)}$$

where f is the frequency. For a single simulated bubble pulse of peak pressure p_1 having an exponential rise and an exponential decay of time constant t_1, Weston showed that the energy flux spectral density is given by

$$E_0'(f) = 8\left[\frac{p_1/t_1}{1/t_1^2 + 4\pi^2 f^2}\right]^2 \qquad \text{ergs/(cm}^2\text{)(Hz)}$$

For values of p_0, t_0, p_1, and t_1 appropriate for a 1-lb charge, the functions $E_0(f)$ and $E_0'(f)$ are plotted, following Weston, as the dashed lines in Fig. 4.14. The solid curve is the combined spectrum of shock wave and first two bubble pulses. At the higher frequencies, the combined spectrum is given by simple energy addition and is determined principally by the energy of the shock wave; at low frequencies, the combined spectrum falls off with decreasing frequency due to phase cancellation of

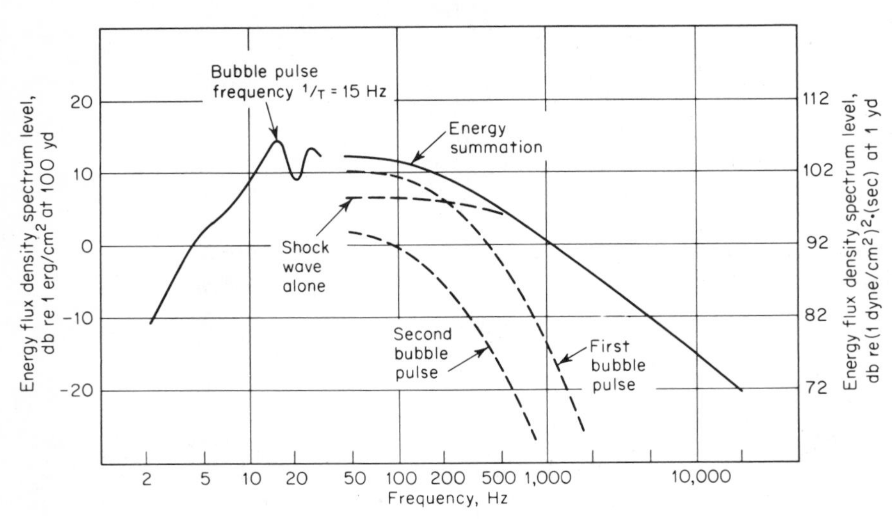

FIG. 4.14. Energy spectrum of a 1-lb charge at a depth of 20 fathoms, showing addition of shock-wave and bubble-pulse energies at frequencies greater than $1/T$. [After Weston (32).]

shock wave and bubble pulses; in between occur a number of oscillations representing interference between the two, with the principal peak occurring at a frequency $1/T$ equal to the reciprocal of the shock wave–first bubble interval. The number and amplitude of these oscillations depend upon the bandwidth being considered.

Figure 4.15 shows explosive spectra for a number of charge weights, all based on the measurements of Weston at a distance of 100 yd and a depth of 120 ft. Generally confirming measurements were subsequently made by Stockhausen (33). The energy spectrum of a charge of any weight w pounds at frequency f may be obtained from that of a 1-lb charge by entering the 1-lb spectrum at a frequency equal to $w^{1/3}f$ and multiplying the value as read by $w^{4/3} = 13.3 \log w$.

At depths greater than 120 ft the spectrum is modified by the changing characteristics of the bubble pulses. The characteristics of deep charges have been investigated in papers by Blaik and Christian (34) and Christian and Blaik (35). Figure 4.16 shows octave-band spectra compiled from data in these papers, along with the shallow-depth spectrum of a 1-lb charge from the preceding figures. Different and more complicated scaling laws must be employed to find the spectrum when both charge weight and depth are varied.

Advantages and Disadvantages of Explosive Sound Sources Explosives have some clear advantages, as well as some real disadvantages, in

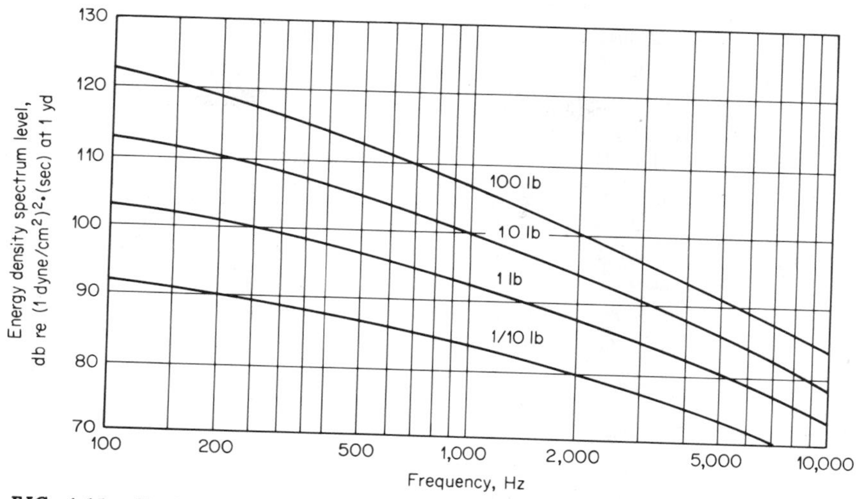

FIG. 4.15. Explosive source spectra for charges of different weights. [Based on data of Weston (32) at 100 yd and reduced to 1 yd by adding 40 db.]

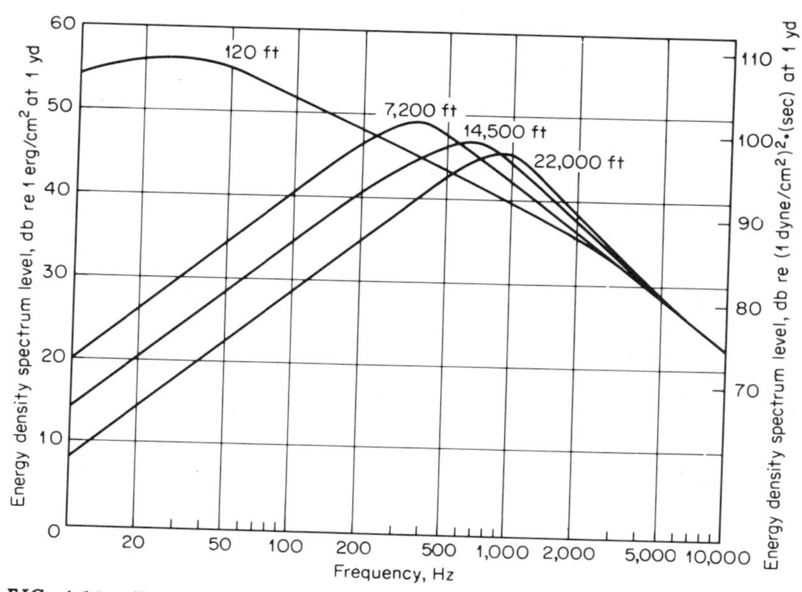

FIG. 4.16. Energy-density spectra of a 1-lb charge at different depths reduced from measurements made near the surface in octave frequency bands. [120-ft spectrum after Weston (32); deep spectra after Christian and Blaik (35).]

comparison with conventional-pulsed electroacoustic sonar sources. On the desirable side is the fact that they are mobile—in the sense of being free of connecting cables—and therefore can be easily launched and detonated at any depth. They yield a short high-power broadband pulse that is useful when range resolution is important. They are nondirectional and need not be beamed to send an adequate amount of acoustic energy in the direction of the target.

On the negative side are the disadvantages that explosives cannot be made repeatable so as to yield repeated "looks" at the target and that the short duration makes processing of the received signal difficult. Also, the combination of high power and nondirectionality yields a high background of reverberation that, in echo ranging, obscures the desired echo. Finally, the explosive signature cannot easily be coded, or its spectrum altered, so as to take advantage of propagation and target characteristics or to make signal processing easier.

For these reasons, explosive sources in sonar are restricted to research studies and to air-dropped applications where their mobility and depth flexibility are of paramount importance.

REFERENCES

1. Design and Construction of Crystal Transducers, Design and Construction of Magnetostriction Transducers, *Natl. Defense Res. Comm. Div. 6 Sum. Tech. Repts.* 12 and 13, 1946.
2. Sahlein, H. F.: Sonar Equipment Summary, *U.S. Navy Electron. Lab. Rept.* 379, 1953.
3. Blake, F. G.: The Tensile Strength of Liquids: A Review of the Literature, *Harvard Univ. Acoust. Res. Lab. Tech. Mem.* 9, 1949.
4. Strasberg, M.: Onset of Ultrasonic Cavitation in Tap Water, *J. Acoust. Soc. Am.*, **31**:163 (1959).
5. Mason, W. P. (ed.): Physics of Acoustic Cavitation in Liquids, in H. G. Flynn, "Physical Acoustics," vol. 1, part B, chap. 9, Academic Press, Inc., New York, 1964.
6. Blake, F. G.: The Onset of Cavitation in Liquids, *Harvard Univ. Acoust. Res. Lab. Tech. Mem.* 12, 1949.
7. Esche, R.: Schwingungskavitation in Flussigkeiten, *Akust. Beih.* (*Acustica*), **4**:AB208 (1952).
8. Rosenberg, L. D.: La Genation et l'etude des vibrations ultrasonares, *Acustica*, **12**:40 (1962).
9. Liddiard, G. E.: Short Pulse Cavitation Thresholds for Projectors in Open Water, *U.S. Navy Electron. Lab. Rept.* 376, 1953.
10. Briggs, H. B., J. B. Johnson, and W. P. Mason: Properties of Liquids at High Sound Pressures, *J. Acoust. Soc. Am.*, **19**: 664 (1947).
11. Sherman, C. H.: Effect of the Near-field on the Cavitation Limit of Transducers, *J. Acoust. Soc. Am.*, **35**:1409 (1963).
12. Karnovskii, M. L.: Calculations of the Radiation Resistance of Several Types of Distributed Radiator Systems, *Soviet Phys. Acoust.*, **2**:200 (1956).
13. Rusby, J. S. M.: Investigation of a Mutual Radiation Impedance Anomaly between Sound Projectors Mounted in an Array, *Acustica*, **14**:127 (1964).

14. Eichler, E.: Calculation of Coupling Effects in Pritchard's Array, *J. Acoust. Soc. Am.*, **36**:1393 (1964).
15. Arase, E. M.: Mutual Radiation Impedance of Square and Rectangular Pistons in a Rigid Infinite Baffle, *J. Acoust. Soc. Am.*, **36**:1521 (1964).
16. Stumpf, F. B., and F. J. Lukman: Radiation Resistance of Magnetostrictive-stack Transducer in Presence of Second Transducer at Air-Water Surface, *J. Acoust. Soc. Am.*, **32**:1420 (1960).
17. Waterhouse, R. V.: Radiation Impedance of a Source near Reflectors, *J. Acoust. Soc. Am.*, **35**:1144 (1963).
18. Sherman, C. H.: Theoretical Model for Mutual Radiation Resistance of Small Transducers at an Air-Water Surface, *J. Acoust. Soc. Am.*, **37**:532 (1965).
19. Waterhouse, R. V.: Mutual Impedance of Acoustic Sources, Paper J21, 5th International Congress on Acoustics, September, 1965.
20. Carson, D. L.: Diagnosis and Cure of Erratic Velocity Distributions in Sonar Projector Arrays, *J. Acoust. Soc. Am.*, **34**:1191 (1962).
21. Cole, R. H.: "Underwater Explosions," Princeton University Press, Princeton, N.J., 1948.
22. "Underwater Explosions Research," a compendium of British and American Reports, U.S. Navy Office of Naval Research, Washington, D.C., 1950.
23. Slifko, J. P.: Informal communication, Naval Ordnance Laboratory, 1962.
24. Brockhurst, R. R., J. G. Bruce, and A. B. Arons: Refraction of Underwater Explosion Shock Waves by a Strong Velocity Gradient, *J. Acoust. Soc. Am.*, **33**:452 (1961).
25. Arons, A. B., and D. R. Yennie: Energy Partition in Underwater Explosion Phenomena, *Rev. Mod. Phys.*, **20**:519 (1948).
26. Arons, A. B., and D. R. Yennie: Long Range Shock Propagation in Underwater Explosion Phenomena I, *U.S. Navy Dept. Bur. Ordnance Rept.* NAVORD 424, 1949.
27. Arons, A. B., D. R. Yennie, and T. P. Cotter: Long Range Shock Propagation in Underwater Explosion Phenomena II, *U.S. Navy Dept. Bur. Ordnance NAVORD Rept.* 478, 1949.
28. Arons, A. B.: Underwater Explosion Shock Wave Parameters at Large Distances from the Charge, *J. Acoust. Soc. Am.*, **26**:343 (1954).
29. Marsh, H. W., R. H. Mellen, and W. L. Konrad: Anomalous Absorption of Pressure Waves from Explosions in Water, *J. Acoust. Soc. Am.*, **38**:326 (1965).
30. Urick, R. J.: Implosions as Sources of Underwater Sound, *J. Acoust. Soc. Am.*, **35**:2026 (1963).
31. Arons, A. B.: Secondary Pressure Pulses Due to Gas Globe Oscillation in Underwater Explosions: II: Selection of Adiabatic Parameters in the Theory of Oscillation, *J. Acoust. Soc. Am.*, **20**:277 (1948).
32. Weston, D. E.: Underwater Explosions as Acoustic Sources, *Proc. Phys. Soc. London*, **76**(pt. 2):233 (1960).
33. Stockhausen, J. H.: Energy-per-unit-area Spectrum of the Shock Wave from 1-lb TNT Charges Exploded Underwater, *J. Acoust. Soc. Am.*, **36**:1220 (1964).
34. Blaik, M., and E. A. Christian: Near Surface Measurements of Deep Explosions I: Pressure Pulses from Small Charges, *J. Acoust. Soc. Am.*, **38**:50 (1965).
35. Christian, E. A., and M. Blaik: Near Surface Measurements of Deep Explosions II: Energy Spectra of Small Charges, *J. Acoust. Soc. Am.*, **38**:57 (1965).
36. Snay, H. G.: Hydrodynamics of Underwater Explosions, *Naval Hydrodynamics Publ.* 515, chap. XIII, National Academy of Sciences, 1957.
37. Genau, M. A.: unpublished memorandum, Naval Ordnance Laboratory, 1965.

FIVE

Propagation of Sound in the Sea: Transmission Loss, I

5.1 Introduction

Definition of Transmission Loss The sea, together with its boundaries, forms a remarkably complex medium for the propagation of sound. It possesses an internal structure and a peculiar upper and lower surface that create many diverse effects upon the sound emitted from an underwater projector. In traveling through the sea, an underwater sound signal becomes delayed, distorted, and weakened. Transmission loss expresses the magnitude of one of the many phenomena associated with sound propagation in the sea.

The sonar parameter *transmission loss* quantitatively describes the weakening of sound between a point 1 yd from the source and a point at a distance in the sea. More specifically, if I_0 is the intensity at the reference point located 1 yd from the "acoustic center" of the source (10 log I_0 is the source level of the source) and I_1 is the intensity at a distant point, then the transmission loss between the source and the distant point is

$$\mathrm{TL} = 10 \log \frac{I_0}{I_1} \qquad \mathrm{db}$$

Since I_0 and I_1 are intensities, a time average is implied in the definition. For a CW or quasi-CW source, this time average is taken over a long enough interval of time to include all the effects of fluctuation in transmission; for a short transient pulse, the averaging time must be long enough to include the multipath transmission effects that cause distortion of the received signal. For short pulses, a transmission loss equivalent to that for

CW is given by the ratio of the energy flux density at 1 yd from the source E_0 to the energy flux density at the distant point E_1, or

$$TL = 10 \log \frac{E_0}{E_1} \qquad \text{db}$$

Sources of Loss Transmission loss may be considered to be the sum of a loss due to *spreading* and a loss due to *attenuation*. *Spreading loss* is a geometrical effect representing the regular weakening of a sound signal as it spreads outward from the source. As will be shown in Sec. 5.2, spreading loss varies with range according to the logarithm of the range, and can be expressed as a certain number of *decibels per distance doubled*. *Attenuation loss* includes the effects of absorption, scattering, and leakage out of sound channels. As shown in Sec. 5.3, attenuation loss varies linearly with range, and is expressed as a certain number of *decibels per unit distance*. In addition to these two principal kinds of loss, other losses not readily identified with range may occur; examples are the *convergence gain* in sound channels or its converse, *refraction loss* in shadow zones.

Transmission loss, as a single number, summarizes the effects of a variety of propagation phenomena in the sea. These phenomena are many and varied; extensive studies of them can be found in the literature. In the following pages we will approach the complex subject of sound propagation in the sea by beginning with the simplest form of loss—spreading in an ideal medium—and then proceeding to the complications caused by absorption, refraction, reflections from the boundaries, the various kinds of sound channels existing in the sea, and the multipath effects causing fluctuations in sound transmission.

5.2 Spreading Laws

Spherical Spreading Referring to Fig. 5.1*a*, let a small source of sound be located in a homogeneous, unbounded, lossless medium. For this most simple propagation condition, the power generated by the source is radiated equally in all directions so as to be equally distributed over the surface of a sphere surrounding the source. Since there is no loss in the medium, the power P crossing all such spheres must be the same. Since power equals intensity times area, it follows that

$$P = 4\pi r_1^2 I_1 = 4\pi r_2^2 I_2 = \cdots$$

If r_1 be taken as 1 yd, the transmission loss to range r_2 is then

$$TL = 10 \log \frac{I_1}{I_2} = 10 \log r_2^2 = 20 \log r_2$$

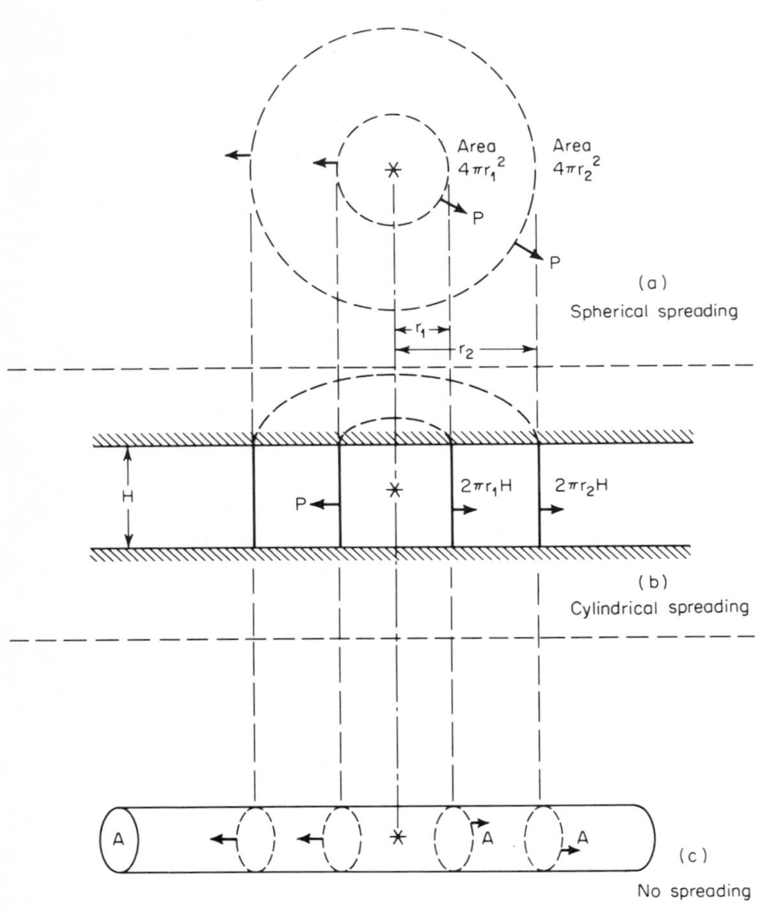

***FIG. 5.1.** Spreading in (a) an unbounded medium, (b) a medium between two parallel planes, and (c) a tube.*

This kind of spreading is called *inverse-square*, or *spherical*, *spreading*. The intensity *decreases* as the square of the range, and the transmission loss *increases* as the square of the range. Another way of describing spherical spreading is to say that the product of the rms sound pressure and the range is a constant.

Cylindrical Spreading When the medium has plane-parallel upper and lower bounds (Fig. 5.1*b*), the spreading is no longer spherical because sound cannot cross the bounding planes. Beyond a certain range, the power radiated by the source is distributed over the surface of a cylinder having a radius equal to the range and a height H equal to the distance between the parallel planes. The power crossing cylindrical surfaces at range r_1 and r_2 is

$$P = 2\pi r_1 H I_1 = 2\pi r_2 H I_2 = \cdots$$

If r_1 be taken as 1 yd, the transmission loss to r_2 then becomes

$$\mathrm{TL} = 10 \log \frac{I_1}{I_2} = 10 \log r_2$$

and the spreading is said to be *inverse first power*, or cylindrical. In cylindrical spreading the product of the rms pressure and the square root of the range is a constant. This type of spreading exists at moderate and long ranges whenever sound is trapped by a sound channel in the sea.

No Spreading As an academic case (Fig. 5.1*c*), we may consider propagation down a lossless tube or pipe of constant cross section. Beyond a certain range, the area over which the power is distributed is a constant, and the pressure, intensity, and transmission loss of a wave propagating down the tube are independent of range.

Time Stretching A special kind of spreading occurs when the signal from a pulsed source is spread out in *time* as the pulse propagates. The pulse becomes elongated by multipath propagation effects and becomes smeared out in time as it travels in range. Such an effect is particularly important in long-range propagation in the deep-ocean sound channel. If the medium were unbounded, and time stretching proportional to range occurred, the intensity would fall off as the inverse cube of the range. In a sound channel, time stretching proportional to range causes the intensity to decrease as the *inverse square* of the range instead of as the *inverse first power*.

Table 5.1 is a summary of the various kinds of spreading and the spreading laws to which they give rise.

TABLE 5.1 Spreading Laws

Type	Intensity varies as	Transmission loss is, db	Propagation in
No spreading*	r^0	0	Tube
Cylindrical	r^{-1}	$10 \log r$	Between parallel planes
Spherical	r^{-2}	$20 \log r$	Free field
Hyperspherical*	r^{-3}	$30 \log r$	Free field with time stretching

* Hypothetical for sonar.

5.3 Absorption of Sound in the Sea

Absorption Loss as a Function of Range Absorption is a form of loss which obeys a different law of variation with range than does the loss due to spreading. It involves a process of conversion of acoustic energy into heat and thereby represents a true loss of acoustic energy to the medium in which propagation is taking place.

When a plane wave travels through an absorbing medium, a certain fraction of its intensity is lost in each unit small distance traveled. If the intensity at some distance is I, the loss of intensity dI in traveling a small additional distance dx is given by

$$\frac{dI}{I} = -n\,dx$$

where n is a proportionality constant, and the minus sign indicates that dI is a *negative* change of intensity. On integration between ranges r_2 and r_1, we find that the intensity I_2 at range r_2 is related to the intensity I_1 at range r_1 by

$$I_2 = I_1 e^{-n(r_2 - r_1)}$$

Taking 10 times the logarithm to the base 10, we obtain

$$10 \log I_2 - 10 \log I_1 = -10n(r_2 - r_1) \log_{10} e$$

Writing $\alpha = 10n \log_{10} e$, the change of level between range r_2 and r_1 becomes

$$10 \log I_2 - 10 \log I_1 = -\alpha(r_2 - r_1)$$

or

$$\alpha = \frac{10 \log I_1 - 10 \log I_2}{r_2 - r_1}$$

The quantity α is the *logarithmic absorption coefficient* (to the base 10), and is expressed in decibels per kiloyard (db/kyd). It means that for each kiloyard traveled, the intensity is diminished by absorption by the amount α db.

Measurements of Absorption The first measurements of the absorption coefficient of seawater were made during the years from 1931 to 1934 by Stephenson (1) using sinusoidal pulses transmitted through the sea between a surface ship and a submerged submarine. By the beginning of World War II, measurements had been made in different ocean areas in the frequency range then of interest to sonar (2). During the wartime years, extensive additional data were obtained, principally by the University of California Division of War Research (3). Later, laboratory measurements were made by Wilson and Leonard (4) using the decay of sound in a glass sphere of seawater excited into vibration in one of its natural modes; still later, a rectangular cavity was used at lower frequencies, both in the laboratory (5) and at sea (6).

Causes of Absorption It was soon clear that the absorption of sound in the sea was unexpectedly high compared with that in pure water, and could not be attributed to scattering, refraction, or other anomalies attributable to propagation in the natural environment. For example, the absorption in seawater at frequencies between 5 and 50 kHz was found to be some 30 times that in distilled water. This excess absorption was attributed by Liebermann (7) to a kind of chemical reaction that occurs under the influence of a sound wave and involves one of the minor dissolved salts in the sea.

In seawater, the absorption of sound is caused by three effects. One is the effect of ordinary or *shear viscosity*, a classical effect investigated theoretically long ago by Rayleigh (8, p. 316). Rayleigh derived an expression for the absorption coefficient equivalent to

$$\alpha = \frac{16\pi^2\mu_s}{3\rho c^3} f^2$$

where α = intensity absorption coefficient, cm^{-1}
μ_s = shear viscosity, poises (about 0.01 for water)
ρ = density, g/cm^3 (about 1 for water)
c = sound velocity (about 1.5×10^5 cm/sec)
f = frequency, Hz

For pure water, the value of α computed by this formula is $6.7 \times 10^{-11} f^2$ db/kyd and amounts to only about one-third of the absorption actually measured in pure (distilled) water. Figure 5.2 shows curves of measured absorption coefficients in seawater and distilled water and the theoretical absorption due to shear viscosity. The additional absorption in pure water over that due to shear viscosity is attributed to a second kind of viscosity, called *volume viscosity*, which produces absorption as a result of a time lag required for the water molecules to "flow" under pressure into lattice "holes" in the crystal structure. The effect adds an additional viscosity term to the above expression. The absorption coefficient due to both kinds of viscosity becomes (9)

$$\alpha = \frac{16\pi^2}{3\rho c^3}\left(\mu_s + \frac{3}{4}\mu_v\right)$$

where μ_v is the volume-viscosity coefficient. The ratio μ_v/μ_s has the value 2.81 for water. The dominant cause of absorption in seawater below 100 kHz is due to *ionic relaxation* of the magnesium sulfate ($MgSO_4$) molecules in seawater. This is a dissociation-reassociation process,

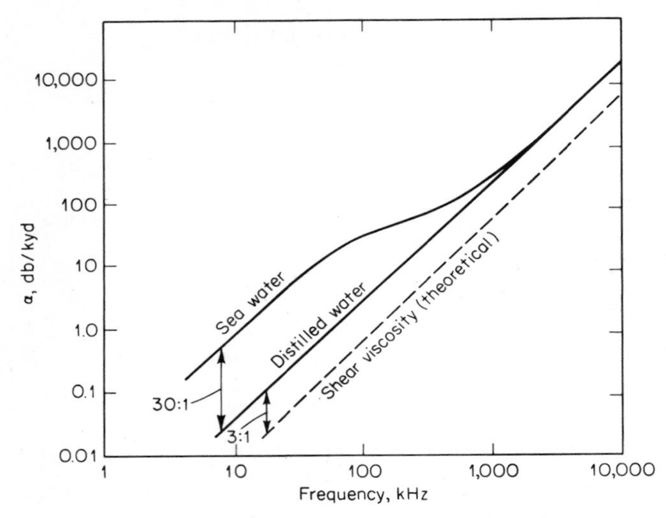

Fig. 5.2. Absorption coefficients in seawater and in distilled water and theoretical absorption due to shear viscosity alone.

involving a finite time interval called the *relaxation time*, in which the $MgSO_4$ ions in solution dissociate under the pressure of the sound wave. Although $MgSO_4$ amounts to only 4.7 percent by weight of the total dissolved salts in seawater, this particular salt (rather than the principal constituent of seawater, NaCl) was found by Leonard, Combs, and Skidmore (10) to be the dominant absorptive constituent of seawater.

Variation with Frequency Liebermann showed theoretically (11) that the ionic relaxation mechanism, together with viscosity, should yield a frequency dependence of the absorption coefficient of the form

$$\alpha = a \frac{f_T f^2}{f_T^2 + f^2} + b f^2$$

where a and b = constants

f_T = relaxation frequency equal to reciprocal of relaxation time

A modified form of this frequency law was subsequently fitted by Schulkin and Marsh (12) to some thirty thousand measurements made at sea between 2 and 25 kHz out to ranges of 24 kyd. The result was

$$\alpha = A \frac{S f_T f^2}{f_T^2 + f^2} + B \frac{f^2}{f_T} \qquad \text{db/kyd}$$

in which S is the salinity in parts per thousand (ppt), A and B are constants found to be equal to 1.86×10^{-2} and 2.68×10^{-2}, respectively, f is the frequency in kilohertz, and f_T is the temperature-dependent relaxation frequency given by

$$f_T = 21.9 \times 10^{6-1,520/(T+273)} \quad \text{kHz}$$

where T is the temperature in degrees centigrade. At low frequencies ($f \ll f_T$), the absorption is dominated by the first term and becomes

$$\alpha = \frac{AS}{f_T} f^2$$

and is thus proportional to the square of the frequency. At high fre-

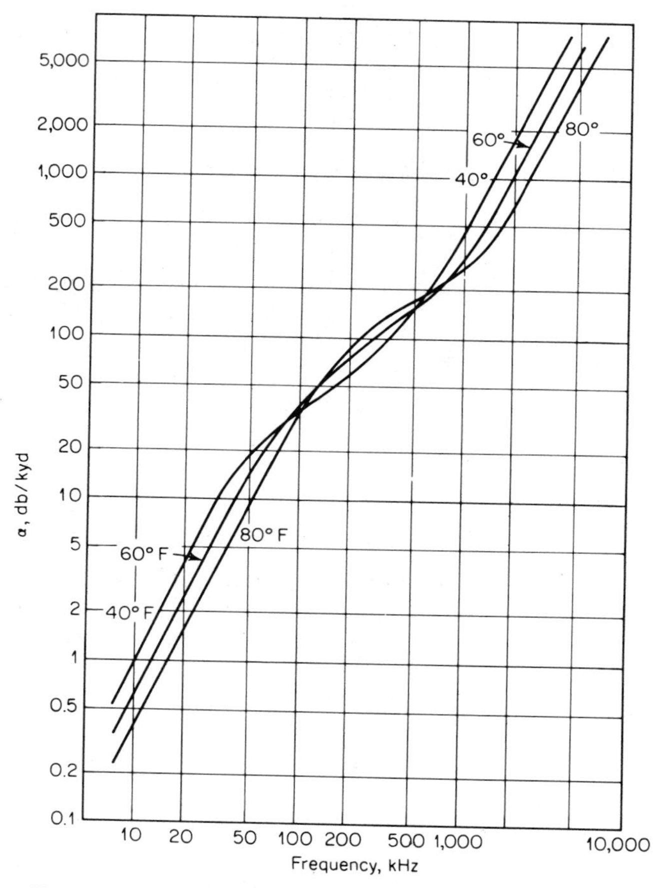

Fig. 5.3. Absorption coefficient in seawater of salinity 35 ppt as a function of frequency at three temperatures.

TABLE 5.2 Temperature Dependence of the Absorption Coefficient

Temperature, °F	Relaxation frequency f_T, kHz	Low-frequency coefficient, $AS/f_T = m$ ($S = 35$ ppt)	High-frequency coefficient, $B/f_T = n$
40	71	9.2×10^{-3}	3.8×10^{-4}
50	91	7.2	3.0
60	121	5.4	2.2
70	145	4.5	1.8
80	182	3.6	1.5
90	223	2.9	1.2

5–40 kHz: α db/kyd = mf^2 kHz
>1000 kHz α db/kyd = nf^2 kHz

quencies ($f \gg f_T$), the second term becomes dominant and

$$\alpha = \frac{B}{f_T} f^2$$

again varies as the square of the frequency. At both extremes the absorption is strongly temperature-dependent, decreasing at 90°F to only about one-third of its value at 40°F. At intermediate frequencies ($f \approx f_T$), the coefficient varies in a complicated way with both frequency and temperature.

Figure 5.3 shows α plotted against frequency for three temperatures at zero depth. For the frequency regions where the coefficient is proportional to the square of the frequency, α may be computed more accurately by the use of Table 5.2.

Variation with Depth The effect of pressure on the absorption has been investigated theoretically and experimentally (14) using the resonator-decay method mentioned above. In the range of hydrostatic pressure found in the sea, the effect of pressure (12) is to reduce the absorption coefficient by the factor $(1 - 6.54 \times 10^{-4}P)$, where P is the pressure in atmospheres. Taking 1 atm as the pressure equivalent of 33.9 ft of water at 39°F, the absorption coefficient at depth d ft in the sea becomes

$$\alpha_d = \alpha_0(1 - 1.93 \times 10^{-5}d)$$

in terms of its value α_0 at zero depth ($d = 0$). The absorption of sound in seawater accordingly decreases by about 2 percent for every increase of 1,000 ft in depth; at a depth of 15,000 ft the absorption coefficient decreases to 71 percent of its value at the surface.

Attenuation at Low Frequencies At frequencies below about 5 kHz the attenuation coefficients measured at sea are much higher than would be expected from the expressions just given. At these frequencies an additional source of loss other than that caused by ionic relaxation of $MgSO_4$ must therefore be predominant. The cause of this low-frequency loss in the sea is still a mystery. Various hypotheses have been advanced, such as the effects of eddy viscosity of oceanic turbulence (13), scattering by index-of-refraction inhomogeneities in the deep sea (15), and finite amplitude.

A compilation of a number of determinations of low-frequency attenuation coefficients made by Thorp (16) yields the solid curve of Fig. 5.4. Included in the Thorp compilation are published data of Sheehy and Halley (17) and Urick (15), as well as various unpublished measurements. Other low-frequency attenuation measurements that may be cited are the original wartime values of Ewing and Worzel (18) and those of Bryan,

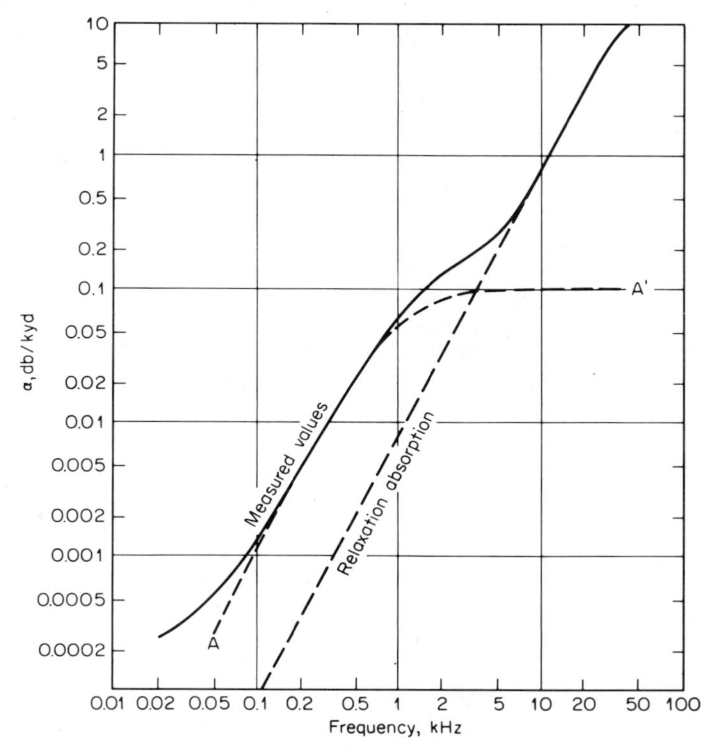

FIG. 5.4. *Low-frequency attenuation coefficients. [Solid curve is the center of measured data compiled by Thorp (16).]*

Truchan, and Ewing (19), Kibblewhite, Denham, and Barker (20), and Urick (21). It is apparent from Fig. 5.4 that the coefficient at low frequencies is some 10 times greater than would be expected from a downward extrapolation of the higher-frequency values. The shape of the curve suggests the existence of another, but yet unknown, low-frequency relaxation process that by itself would result in the curve AA' and would have a relaxation frequency of about 1 kHz. Thorp has fitted the solid curve with the function

$$\alpha = \frac{0.1f^2}{1 + f^2} + \frac{40f^2}{4{,}100 + f^2} \qquad \text{db/kyd}$$

where f is the frequency in kilohertz. Each term represents the contribution of the two relaxation-like processes at a temperature of about 39°F. Alternatively, the excess attenuation may represent merely shock-wave effects inherent in the propagation of the explosive pulses used for most of the measurements.

At the extreme low-frequency end, the measured values tend to be higher still, suggesting the emergence of still another source of loss—perhaps the fact that the deep sound channel of the sea, within which the data were obtained, has become too shallow to trap the long wavelengths associated with these frequencies.

Spherical Spreading and Absorption When propagation measurements are made at sea, it is found that spherical spreading, together with absorption, provides a reasonable fit to the measured data under a wide variety of conditions. That is to say, within the usually wide spread of measured data, the assumption of spherical spreading plus absorption yields reasonable agreement with observations. Indeed, spherical spreading often occurs where it has no right to occur. It often occurs under trapping conditions in sound channels where the gain due to trapping is apparently compensated for by the leakage of sound out of the channel.

When specific propagation conditions are of no interest and only a rough approximation of the transmission loss is adequate, the ubiquitous spherical-spreading law, plus an added loss due to absorption, is a handy working rule. It may be expressed as

$$\text{TL} = 20 \log r + \alpha r \times 10^{-3}$$

where the first term represents spherical spreading and the second term absorption, and where the factor 10^{-3} is required to take care of the fact that r is in units of yards and α is, by custom, stated in units of decibels per kiloyard. Figure 5.5 is a nomogram to facilitate computations using this relationship. The dashed lines in the figure indicate how the nomogram is to be used.

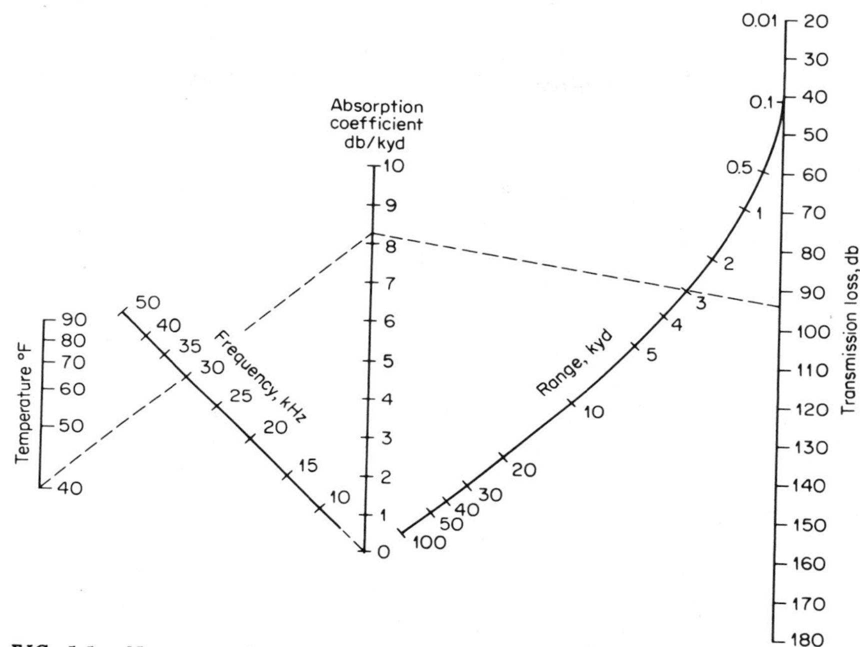

FIG. 5.5. Nomogram for computing the transmission loss due to spherical spreading and absorption for a salinity of 35 ppt and zero depth. The dashed line indicates the method of use of the nomogram: at 40°F and 30 kHz, the absorption coefficient is 8.3 db/kyd and the transmission loss to 3 kyd is 94 db.

Although spherical spreading often sensibly occurs in transmission measurements, free-field conditions seldom actually exist in the sea, except at very short ranges, because of the existence of refraction, scattering, and the pressure of the ocean boundaries. The following pages will center about departures from spherical spreading caused by these and other effects occurring in the natural ocean medium.

5.4 Velocity of Sound in the Sea

Methods of Measurement The velocity of sound* in the sea is an oceanographic variable that determines many of the peculiarities of sound transmission in the medium. It varies with depth, with the seasons, with geographic location, and with time at a fixed location.

* Historically, and improperly, this scalar quantity has been called the "velocity" of sound. Sound "speed," instead of sound "velocity," is coming into common usage.

The history of attempts to measure sound velocity in a natural body of water dates back to 1827, when, as quoted by Wood (22, p. 261), Colladon and Sturm struck a submerged bell in Lake Geneva and simultaneously set off a charge of powder in the air. By timing the interval between the two signals across the lake, they obtained the velocity of sound in water. The result was 1,435 m/sec at 8.1°C, a value surprisingly close to its modern value. Subsequent investigators not only have timed the arrival of sound in the sea to derive the sound velocity (23–25), but have attempted to relate it to the standard and more readily measured oceanographic parameters, such as temperature, salinity, and depth. A relationship of this kind can be established either by means of theory, using certain basic properties of water like its specific volume and ratio of specific heats, or by laboratory measurements of sound velocity over a range of temperatures, salinities, and pressures. By the former, or theoretical, method tables of sound velocity were long ago prepared by Heck and Service (24), Matthews (26), and Kuwahara (27), the latter serving as the standard tables of sound velocity for nearly 20 years. The more modern experimental method yields direct measurements of sound velocity by the use of laboratory techniques of one kind or another under carefully controlled conditions. Examples are the measurements of Weissler and Del Grosso (28), Del Grosso (29), and Wilson (30).

Variation with Temperature, Salinity, and Pressure Both methods yield expressions for the velocity of sound in terms of the three basic quantities: temperature, salinity, and pressure. The following are three formulas for the sound velocity at zero depth or atmospheric pressure:

Kuwahara (27), 1939:

$$c = 1{,}445.5 + 4.664T - 0.0554T^2 + 1.307(S - 35) + \cdots$$

Del Grosso (29), 1952:

$$c = 1{,}448.6 + 4.618T - 0.0523T^2 + 1.25(S - 35) + \cdots$$

Wilson (30), 1960:

$$c = 1{,}449.2 + 4.623T - 0.0546T^2 + 1.391(S - 35) + \cdots$$

In these expressions, c is the sound velocity in meters per second, T is the temperature in degrees centigrade, and S is the salinity in parts per thousand. The symbol $\cdots$ means that higher-order terms follow, such as terms in T^3, $(S - 35)^2$, $(S - 35)T$. The differences between various formulas are discussed by Mackenzie (31), but are less than about 3 m/sec or 0.2 percent at zero depth. Several field experiments have

TABLE 5.3 Approximate Coefficients of Sound Velocity

Variation with	Coefficient	Coefficient
Temperature (near 70°F)	$\dfrac{\Delta c/c}{\Delta T} = +0.001/°\text{F}$	$\dfrac{\Delta c}{\Delta T} = +5$ ft/(sec)(°F)
Salinity	$\dfrac{\Delta c/c}{\Delta S} = +0.0008/\text{ppt}$	$\dfrac{\Delta c}{\Delta S} = +4$ ft/(sec)(ppt)
Depth	$\dfrac{\Delta c/c}{\Delta D} = +3.4 \times 10^{-6}/\text{ft}$	$\dfrac{\Delta c}{\Delta D} = +0.017$ ft/(sec)(ft)

c = velocity, ft/sec
T = temperature, °F
S = salinity, parts per thousand (ppt)
D = depth, ft

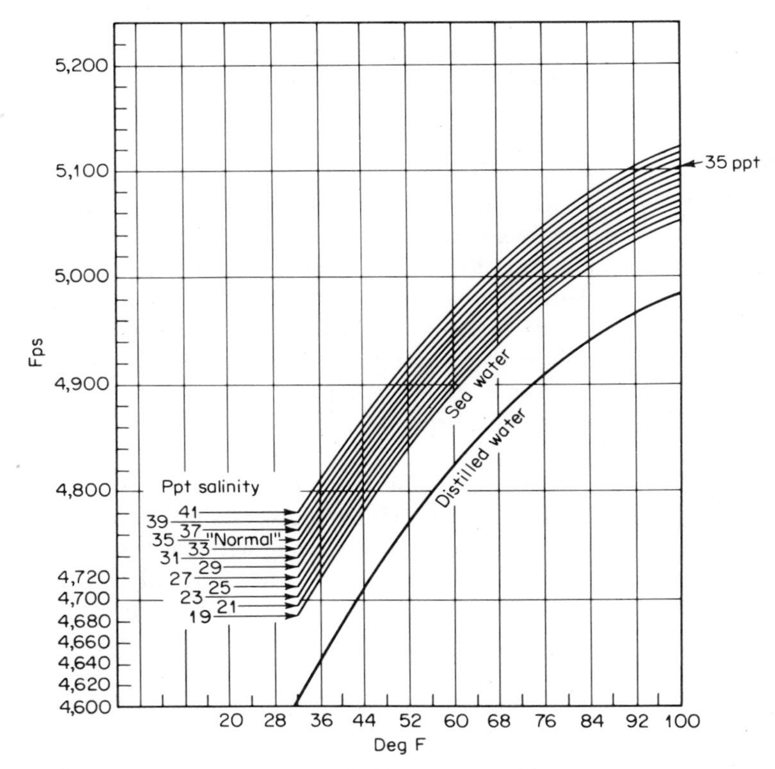

FIG. 5.6. Sound velocity in distilled water and in seawater at zero depth for various salinities and temperatures. [From Del Grosso (29).]

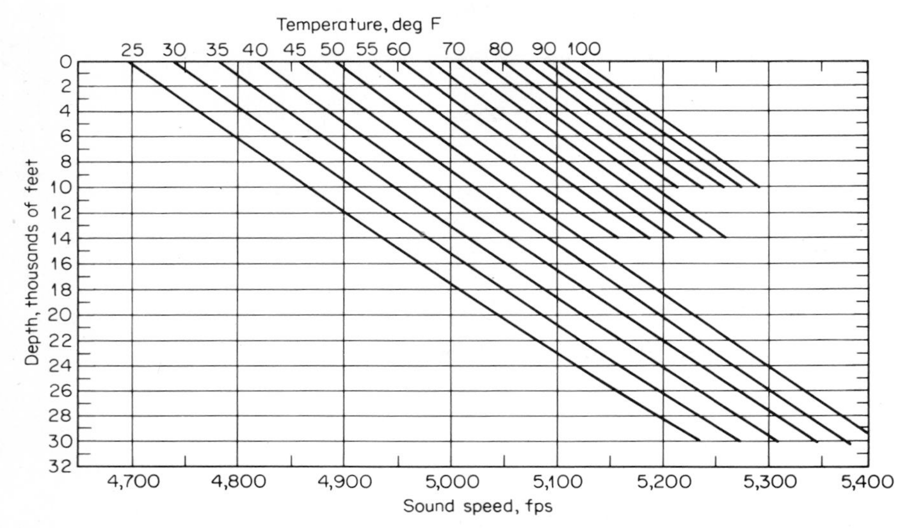

FIG. 5.7. Sound velocity as a function of depth and temperature for a salinity of 35 ppt. [***From Lyons*** (***33***).]

been undertaken (25, 32) to verify these empirical relationships. At depth, the Wilson formula fits the data best.

Kuwahara's tables give a pressure term for the sound velocity which amounts to +0.01815 fps/ft of depth, or +0.01815 $\sec^{-1}$. That is, for a depth increase of 100 ft, the velocity increases by 1.81 fps. The more recent formulas of Wilson give the coefficient +0.017 $\sec^{-1}$. Although apparently small, the difference sometimes results in large changes in ray diagrams, range computations, and travel times in isothermal water where the pressure coefficient controls the variation of sound velocity with depth.

As indicated by the above expressions, the velocity of sound in the sea increases with temperature, salinity, and depth. Approximate coefficients for the rate of change with these quantities are given in Table 5.3. In open deep water, salinity normally has only a small effect on the velocity. Figure 5.6 illustrates the variation of velocity with temperature at zero depth for a number of different salinities. More accurate values may be read from Fig. 5.7, based on the Wilson formulas (33) for the average open-sea salinity value of 35 ppt. When required, a correction for salinity may be made from Fig. 5.8.

Field Observations of Sound Velocity Two devices are commonly used for finding the velocity of sound as a function of depth in the sea. One, called the bathythermograph, measures temperature as a function of

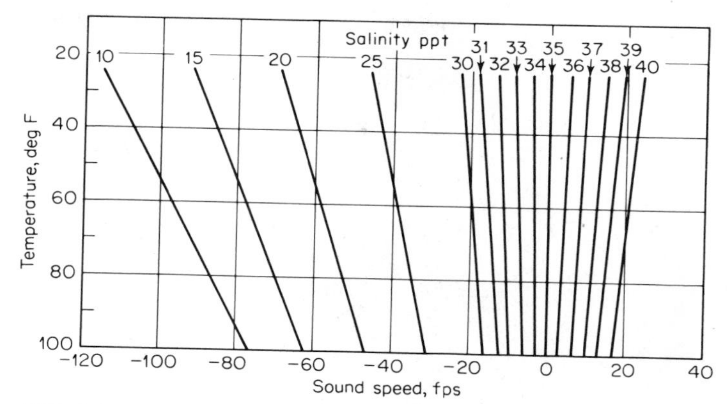

FIG. 5.8. Correction to Fig. 5.7 for various salinities.

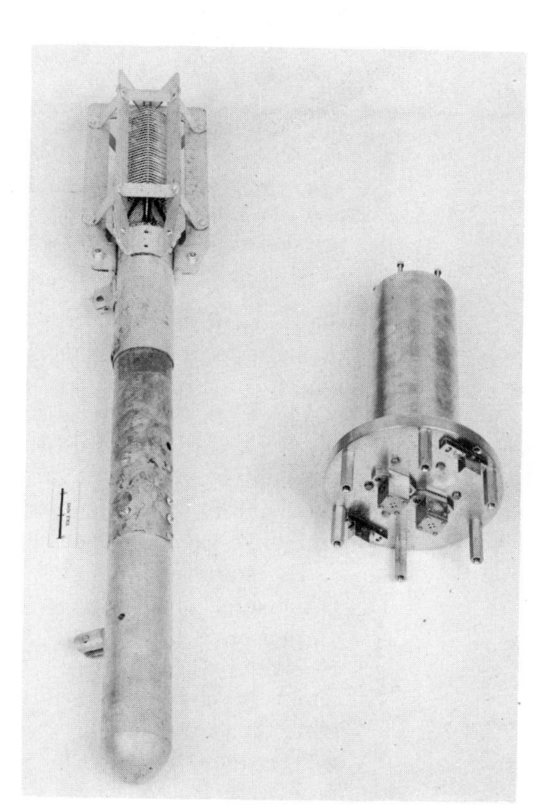

FIG. 5.9. Photographs of a bathythermograph (left) and a velocimeter (right). The velocimeter is shown with its protective end cover removed so that its transducers can be viewed.

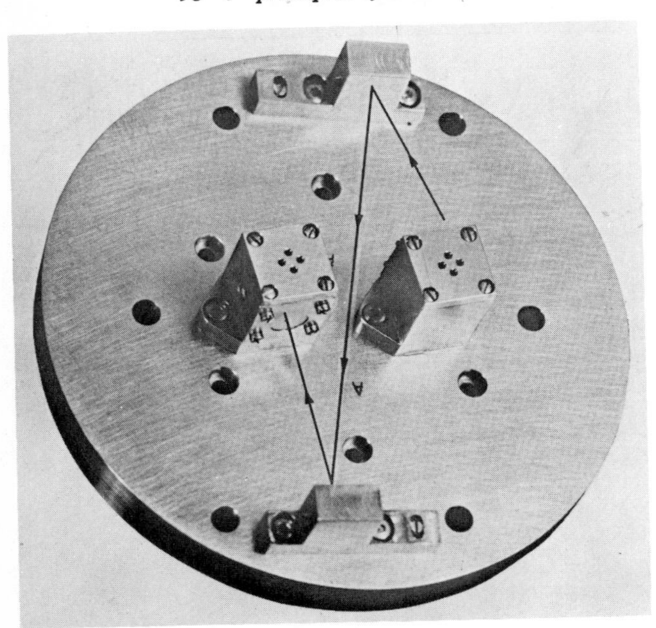

FIG. 5.10. Transducers and acoustic path of a velocimeter. (Photograph by permission, Weston Instruments, Inc.)

depth as it is lowered into the sea; the other, called the *velocimeter*, measures sound velocity directly in terms of the travel time of sound over a constant fixed path. These are shown side by side in Fig. 5.9.

The bathythermograph is a mechanical device which inscribes on a small plated glass slide a trace of temperature against depth. It was frequently used on naval craft during World War II, and has been in extensive use in subsequent years. The device was designed to be used while the vessel is under way.

The sound velocimeter is an acoustic device, originally developed by Greenspan and Tschiegg (34) and now made by several manufacturers. It measures the travel time of short pulses between a projector and a receiver. It operates on the "sing-around," or "howler," principle, in which the arrival of a pulse at the receiver triggers the succeeding pulse from the projector. Since nearly all the time delay between pulses occurs as the acoustic delay in the water between projector and receiver, the repetition frequency of the pulses is determined by the sound velocity of the water between projector and receiver. Figure 5.10 is a close-up of the acoustic path in the velocimeter between the sound projector and receiver.

When the bathythermograph temperature-depth trace is converted to sound velocity, it gives a velocity-depth trace comparable to that of the velocimeter. This was demonstrated during World War II (35), when perhaps the first seagoing velocimeter, a variable-frequency fixed-path interferometer, was lowered into the sea side by side with a bathythermograph. An example of the traces obtained is given in Fig. 5.11. The similarity of the velocity-depth curve obtained with the interferometer to that obtained by converting the bathythermograph trace into sound velocity indicates that temperature and depth are the dominant determinants of sound velocity under normal conditions in the open sea. This same conclusion was reached subsequently by means of comparative measurements (36) with a more modern velocimeter. Under certain conditions, however, as in nearshore waters and under a melting ice cap, salinity changes and suspended material, rather than temperature alone, may play an important part in determining the sound velocity.

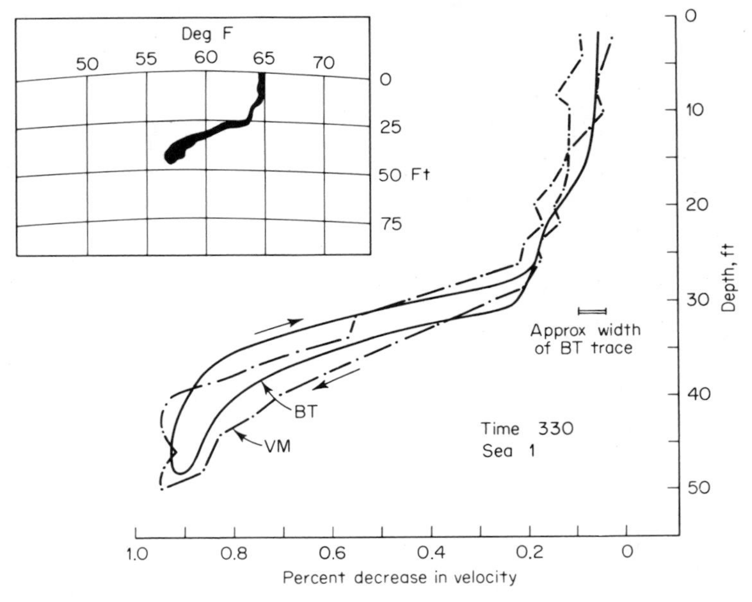

FIG. 5.11. Comparison of a bathythermogram (upper left) and the relative velocity recordings of an early velocimeter (dashed curves, right). The solid curves on the right were obtained by converting the bathythermogram to relative sound velocity and adjusting to agree with the velocimeter data at zero depth. (From Ref. 35.)

5.5 Velocity Structure of the Sea

Layers of the Deep-sea Velocity Profile By "velocity profile" is meant the variation of sound velocity with depth, or the velocity-depth function. In the deep sea, the velocity profile is obtained by the instruments described in the preceding section or, alternatively, by hydrographic observations of temperature, salinity, and depth.

A typical deep-sea profile is shown in Fig. 5.12. The profile may be divided into several layers having different characteristics and occurrence. Just below the sea surface is the *surface layer*, in which the velocity of sound is susceptible to daily and local changes of heating, cooling, and wind action. The surface layer may contain a mixed layer of isothermal water that is formed by the action of wind as it blows across the surface above. Sound tends to be trapped or channeled in this mixed layer. Under prolonged calm, sunny conditions the mixed layer disappears, and is replaced by water in which the temperature decreases with depth. Below the surface layer lies the *seasonal thermocline*—the word "thermocline" denoting a layer in which the temperature changes with depth. The seasonal thermocline is characterized by a *negative* thermal or veloc-

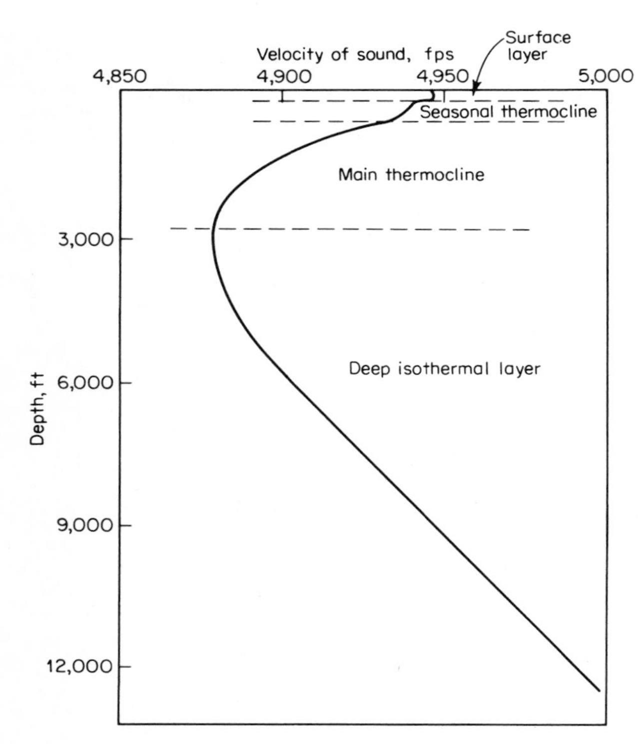

FIG. 5.12. Typical deep-sea velocity profile divided into layers.

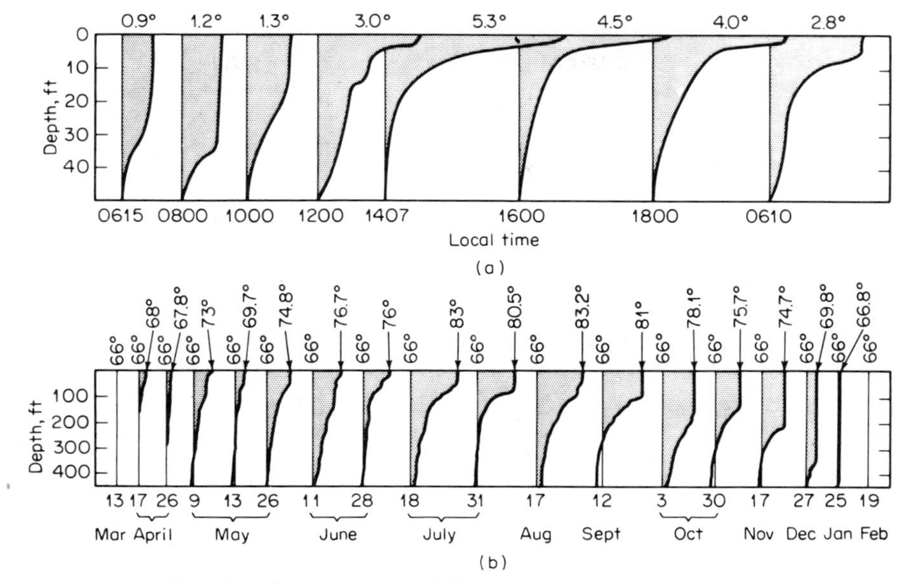

***FIG. 5.13.** Diurnal and seasonal variability of temperature near Bermuda. (a) Temperature profiles at various times of day showing the increase of surface temperature over the temperature at 50 ft. (b) Temperature profiles in different months of the year. (From Ref. 37.)*

ity gradient (temperature or velocity *decreasing* with depth) that varies with the seasons. During the summer and fall, when the near-surface waters of the sea are warm, the seasonal thermocline is strong and well defined; during the winter and spring and in the Arctic, it tends to merge with, and be indistinguishable from, the surface layer. Underlying the seasonal thermocline is the *main thermocline,* which is affected only slightly by seasonal changes. The major increase of temperature over that of the deep cold depths of the sea occurs in the main thermocline. Below the main thermocline and extending to the sea bottom is the *deep isothermal layer* having a nearly constant temperature near 39°F, in which the velocity of sound *increases* with depth because of the effect of pressure on sound velocity. Between the negative velocity gradient of the main thermocline and the positive gradient of the deep layer, there is a velocity minimum toward which sound traveling at great depths tends to be bent or focused by refraction. At high latitudes, the deep isothermal layer extends nearly to the sea surface.

Variation of the Profile with Latitude, Season, and Time of Day The occurrence and thicknesses of these layers vary with latitude, season, time of day, and meteorological conditions. Some examples of this vari-

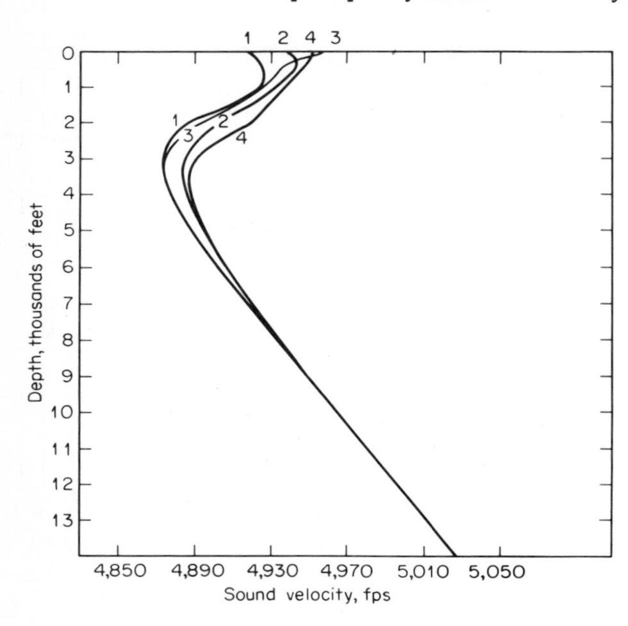

FIG. 5.14. Average velocity profiles in different seasons in an area halfway between Newfoundland and Great Britain. Latitudes 43 to 55°N, longitudes 20 to 40°W. (1) Winter. (2) Spring. (3) Summer. (4) Autumn. (Ref. 38, Region IV-2.)

ability are shown in the following figures. Figure 5.13*a* illustrates the diurnal behavior of the surface layer. It shows a series of bathythermograms taken at different times of day to illustrate how the surface waters of the sea warm up during the daytime hours on a sunny day and cool off at night. These diurnal changes have a profound effect on sound transmission from a surface-ship sonar.* Figure 5.13*b* is a series of bathythermograms taken in the Bermuda area and illustrating the development of the seasonal thermocline during the summer and autumn months. Figure 5.14 shows a number of velocity profiles for the four seasons in a single area. It illustrates how the seasonal effect becomes less at the deeper depths. The effect of latitude on the sound-velocity profile in the deep sea is illustrated by Fig. 5.15, which shows the profile for two locations in the North Atlantic at the same season of the year. At low latitudes, the velocity minimum lies at a depth of about 4,000 ft. In high latitudes, the velocity minimum lies near the sea surface, and the main and seasonal thermoclines tend to disappear from the profile.

Velocity Profile in Shallow Water In the shallow waters of coastal regions and on the continental shelves, the velocity profile tends to be irregular and unpredictable, and is greatly influenced by surface heating and cooling, salinity changes, and water currents. The shallow-water

* Surface-ship echo ranging has long been known to be poorest in the afternoon—a phenomenon called the "afternoon effect."

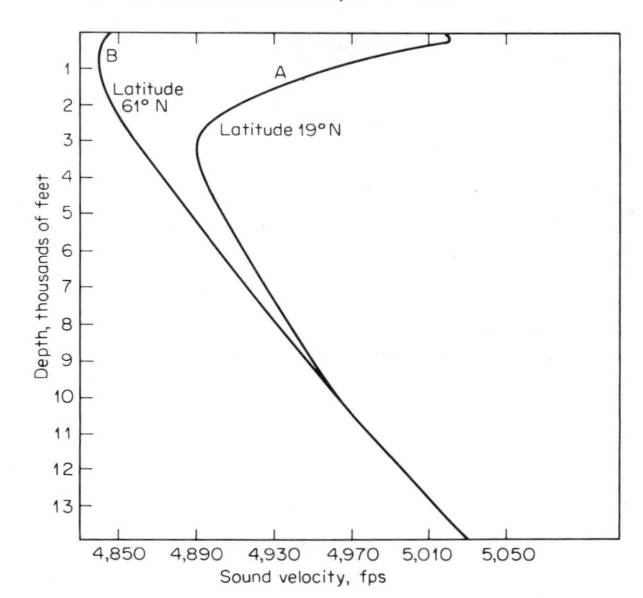

FIG. 5.15. Velocity profiles in different latitudes. A: *18°50′N, 30°01′W, spring.* B: *61°02′N, 34°01′W, spring.* (*Ref. 38, profiles 41 and 143.*)

profile is complicated by the effects of salinity changes caused by nearby sources of fresh water and contains numerous gradient layers of little temporal or spatial stability. Figure 5.16 shows some examples of bathythermograms selected at random for various seasons and locations off the East and Gulf Coasts of the United States.

5.6 Propagation Theory and Ray Tracing

Wave and Ray Theory The propagation of sound in an elastic medium can be described mathematically by solutions of the *wave equation* using the appropriate boundary and medium conditions for a particular problem. The wave equation is a partial differential equation relating the acoustic pressure p to the coordinates x, y, z and the time t, and may be written

$$\frac{\partial^2 p}{\partial t^2} = c^2 \left(\frac{\partial^2 p}{\partial x^2} + \frac{\partial^2 p}{\partial y^2} + \frac{\partial^2 p}{\partial z^2} \right)$$

where c is a quantity which has the general significance of sound velocity and which may vary with the coordinates.

There are two theoretical approaches to a solution of the wave equation. One is called *normal-mode theory,* in which the propagation is described in terms of characteristic functions called *normal modes,* each of which is a solution of the equation. The normal modes are combined additively to satisfy the boundary and source conditions of interest.

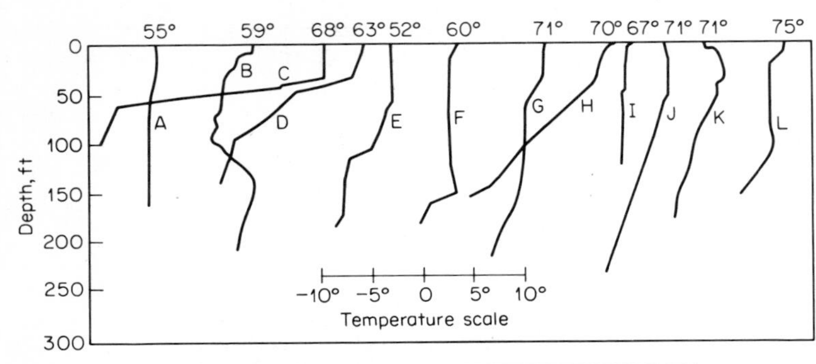

Profile	*Latitude*	*Longitude*	Month
A	41°N	67°W	Aug.
B	40	71	July
C	42	70	Aug.
D	42	69	Aug.
E	42	70	Oct.
F	40	71	Nov.
G	26	80	Feb.
H	26	80	Feb.
I	29	84	Mar.
J	28	90	Apr.
K	40	71	July
L	28	90	Apr.

FIG. 5.16. Temperature profiles in shallow water at various locations of the United States East and Gulf Coasts. Top figures are surface temperatures. (Ref. 39.)

The result is a complicated mathematical function which, though adequate for computations on a digital computer, gives little insight, compared to ray theory, concerning the distribution of the energy of the source in space and time. However, normal-mode theory is particularly suited for a description of sound propagation in shallow water, and will be discussed more fully in that connection.

The other form of solution of the wave equation is *ray theory* and the body of results and conclusions therefrom is called *ray acoustics*. The essence of ray theory is (1) the postulate of *wavefronts*, along which the phase or time function of the solution is constant, and (2) the existence of *rays* that describe where in space the sound emanating from the source is being sent. Like its analog in optics, ray acoustics has considerable intuitive appeal and presents a picture of the propagation in the form of the *ray diagram*. Ray theory has, however, certain shortcomings and does not provide a good solution under conditions where (1) the radius of curvature of the rays or (2) the pressure amplitude changes appreciably over the distance of one wavelength. Practically speaking, ray theory is therefore restricted to high frequencies or short wavelengths; it is useless for predicting the intensity of sound in shadow zones or caustics. The reader interested in the theory of sound propagation in the sea is referred to one of the NDRC Summary Technical Reports (40) and to books by Officer (41) and Brekhovskikh (42).

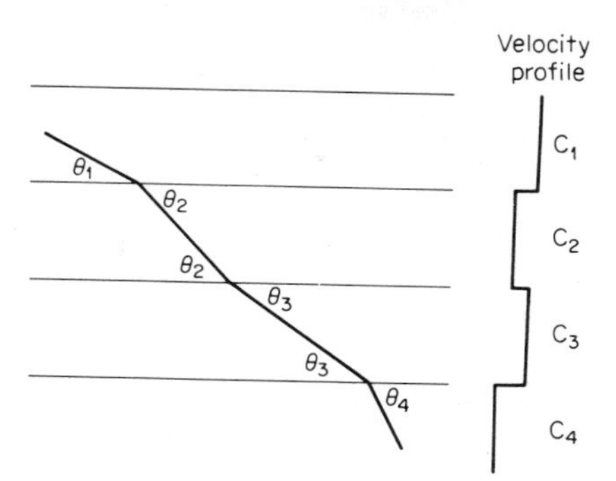

FIG. 5.17. Refraction in a layered medium.

Snell's Law One of the most important practical results of ray theory is *Snell's law*, which describes the refraction of sound rays in a medium of variable velocity. Snell's law states that in a medium consisting of constant velocity (Fig. 5.17), the grazing angles θ_1, θ_2, . . . of a ray at the layer boundaries are related to the sound velocity c_1, c_2, . . . of the layers by

$$\frac{\cos \theta_1}{c_1} = \frac{\cos \theta_2}{c_2} = \frac{\cos \theta_3}{c_3} = \cdots = \text{a constant for any one ray}$$

In this expression the ray constant is the reciprocal of the sound velocity in the layer in which the ray becomes horizontal, that is, where $\cos \theta = 1$. This expression is the basis of ray computation used by most analog and digital computers, since it enables a particular ray to be "traced out" by following it through the successive layers into which the velocity profile may have been divided. In a layered medium having layers of constant velocity, the rays consist of a series of straight-line segments joined together, in effect, by Snell's law.

Linear Variation of Velocity with Depth In a medium in which the velocity of sound changes linearly with depth, the sound rays can be shown to be arcs of circles, that is, to have a constant radius of curvature. A simple and heuristic demonstration of the circularity of rays in a medium with a linear gradient is given by Kinsler and Frey (43). In Fig. 5.18*a*, consider an arc of a circle of radius R connecting two points P_1 and P_2, where the velocity of sound is c_1 and c_2. This circle is horizontal at the depth where the velocity is c_0. Referring to the figure, we observe that

$$d_2 - d_1 = R \cos \theta_2 - R \cos \theta_1 \tag{1}$$

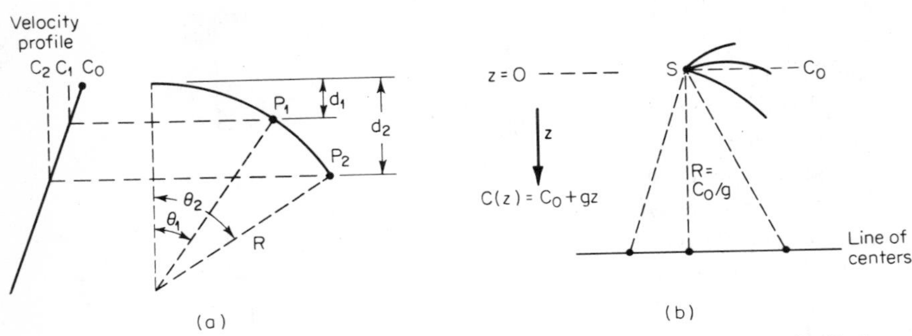

FIG. 5.18. Arcs of circles in a medium in which the velocity is a linear function of depth.

Since the velocity gradient is linear,

$$c_1 = c_0 + gd_1$$
$$c_2 = c_0 + gd_2$$

and

$$d_2 - d_1 = \frac{c_2 - c_1}{g} \tag{2}$$

The circle between P_1 and P_2 will be a *ray* if Snell's law is satisfied. This requires that

$$\cos \theta_1 = \frac{c_1}{c_0} \tag{3}$$

$$\cos \theta_2 = \frac{c_2}{c_0} \tag{4}$$

On eliminating $d_2 - d_1$, $\cos \theta_1$, and $\cos \theta_2$ from Eqs. (1) to (4), one finds that the circle drawn between P_1 and P_2 will be a ray if, and only if,

$$R = \frac{c_0}{g}$$

Since P_2 and P_1 were arbitrarily selected, it follows that *all* rays in the medium will be arcs of circles.*

As illustrated in Fig. 5.18*b*, the centers of curvature of the rays leaving a source S, at which the sound velocity is c_0, lie along a horizontal line at a distance c_0/g above or below the source, depending on the direction in which the velocity decreases with distance. In the sea, the distance c_0/g is normally very large. For isothermal water, such as in the mixed layer or in the deep isothermal layer, in which the velocity increases

* A more formal proof of the circularity relationship may be found in a book by Officer (41, pp. 59–60).

linearly with depth, the line of centers lies at a distance above or below the source equal to

$$R = \frac{c_0}{g} = \frac{5{,}000 \text{ fps}}{0.017 \text{ sec}^{-1}} = 294{,}000 \text{ ft} = 49 \text{ nm}$$

Ray Tracing The principle of circularity of rays in linear gradients is commonly employed in ray-tracing computers. The initial step in programming is to divide the velocity profile into layers of constant linear gradient and to program the computer to follow, by means of Snell's law, the arcs of rays leaving the source at different angles. An analog computer was developed (44) for this purpose as early as 1943, and consisted of a circular slide rule for finding the grazing angles of a ray at the various layers and the corresponding increments of range and depth. Other analog computers may also be used for drawing rays (45).

Ray tracing using Snell's law is easily adapted for high-speed digital computation (46). Some examples of computer-produced ray diagrams will be seen in the figures to follow. The discontinuities of velocity gradient that result from dividing up the profile into linear segments may give rise to spurious caustics in the ray diagram, but may be eliminated, as Pedersen showed (47), by using curved-line segments to approximate the velocity profile.

Transmission Loss from Ray Diagrams The transmission loss between the source and any point in a ray diagram may be readily found in terms of the vertical spacing between rays that are adjacent at the source and pass through the vicinity of the distant point. Figure 5.19 shows a pair of rays separated by a small vertical angle $\Delta\theta$ leaving the source O at angles $\theta_1 + \Delta\theta/2$ and $\theta_1 - \Delta\theta/2$ with the horizontal. After traveling through the sea by any pair of refracted paths, however complex, as

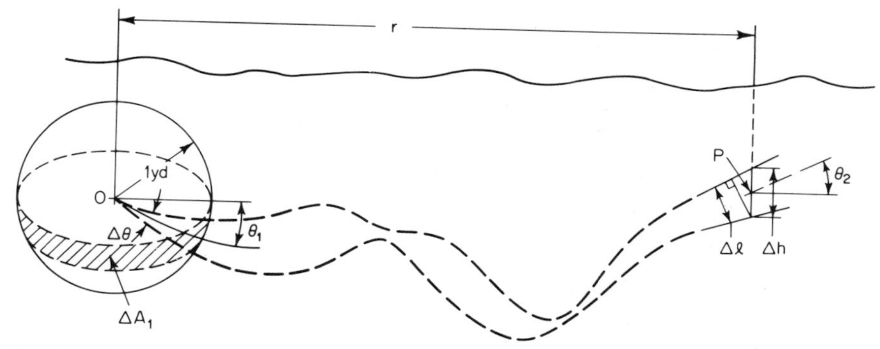

FIG. 5.19. Transmission loss calculated from the spacing of rays in a ray diagram.

shown schematically by the dashed curves, they appear in the neighborhood of the field point P, at an angle θ_2 with the horizontal with a vertical separation Δh.

Within the angle $\Delta\theta$, the source at O radiates a certain amount of power ΔP; by a fundamental property of ray acoustics, this same amount of power must appear within these same rays at P. This is equivalent to saying that energy or power does not "leak out"—as by scattering or diffraction—of the tube formed by the pair of rays. The intensity at 1 yd from the source in the θ_1 direction is

$$I_1 = \frac{\Delta P}{\Delta A_1}$$

where ΔA_1 is the area on a 1-yd sphere subtended by the pair of rays. At P, the intensity is similarly

$$I_2 = \frac{\Delta P}{\Delta A_2}$$

where ΔA_2 is the area included by the ray pair in the vicinity of P, taken normal to the rays. The transmission loss to P is therefore

$$\text{TL} = 10 \log \frac{I_1}{I_2} = 10 \log \frac{\Delta A_2}{\Delta A_1}$$

But

$$\Delta A_1 = 2\pi \cos \theta_1 \, \Delta\theta$$

and

$$\Delta A_2 = 2\pi r \, \Delta l = 2\pi r \, \Delta h \cos \theta_2$$

where r = horizontal distance from O to P
Δl = perpendicular distance between rays at P

The transmission loss is accordingly

$$\text{TL} = 10 \log \frac{r \, \Delta h}{\Delta\theta} \frac{\cos \theta_2}{\cos \theta_1}$$

But by Snell's law in a horizontally stratified medium,

$$\frac{\cos \theta_2}{\cos \theta_1} = \frac{c_2}{c_1}$$

where c_2 = velocity of sound at P
c_1 = velocity at O

Hence we have

$$\text{TL} = 10 \log \frac{r \, \Delta h}{\Delta\theta} \frac{c_2}{c_1}$$

For practical use, r and Δh must be taken in yards, and $\Delta\theta$ must be in radians. In the sea, c_2/c_1 is so nearly unity that it may be neglected. As an example, let the vertical spacing of rays 1° apart at the source be 500 yd at a range of 50,000 yd. The transmission loss then becomes

$$\text{TL} = 10 \log \frac{50{,}000 \times 500}{1/57.4} = 92 \text{ db}$$

For straight-line paths, the expression above can be shown to reduce to TL = 10 log r^2, equivalent to spherical spreading.

The accuracy of this method of finding TL from a ray diagram—aside from the limitations of ray acoustics—depends on the density or number of rays drawn on the ray diagram, that is, on the size of the ray interval $\Delta\theta$. The method, along with ray acoustics generally, fails in the immediate vicinity of caustics, where wave theory (42, pp. 483–492) is required for making an estimate of the intensity. More elaborate treatments of the spreading loss under more general conditions are available (48, 49).

5.7 The Sea Surface

Reflection and Scattering The surface of the sea is both a reflector and a scatterer of sound and has a profound effect on propagation in most applications of underwater sound where source or receiver lie at shallow depth.

If the sea surface were perfectly smooth, it would form an almost perfect reflector of sound. The intensity of sound reflected from the smooth sea surface would be very nearly equal to that incident upon it. The *reflection loss*, equal to 10 log I_r/I_i, where I_r and I_i are the reflected and incident intensities of an incident plane wave, would be closely equal to zero db.

When the sea is rough, as it is to some extent at all times, the loss on reflection is found to be no longer zero. At 25 kHz, Urick and Saxton (50) measured an average sea-surface-reflection loss of about 3 db under conditions of 1-ft waves and grazing angles between 3 and 18°; using another method, Liebermann (51) found a median loss of 3 db at 30 kHz for wave heights of 0.2 to 0.8 ft and a grazing angle of 8°. At lower frequencies, a smaller loss can be expected because of the fact that the sea is becoming smoother relative to a wavelength. This is borne out by the work of Addlington (52), who used explosive pulses reflected from the sea surface. In all octave bands between 400 and 6,400 Hz, over a range of wind speed 5 to 20 knots and grazing angle 10 to 55°, a zero-db median reflection loss was measured. A similar result may be inferred from the transmission measurements of Pedersen (53) at 530 and 1,030 Hz.

Indeed, at a low enough frequency and grazing angle, the sea surface must become effectively smooth and act as a perfect mirror of sound. A smoothness criterion attributed to Rayleigh is that the surface is to be considered smooth if the path difference between rays reflected from the top and bottom of the surface irregularities is less than one-quarter wavelength. In terms of the roughness height h and grazing angle θ, this requires that

$$2h \sin \theta < \frac{\lambda}{4}$$

In terms of the wavelength λ, the surface becomes effectively smooth, and the reflection loss is zero, when

$$\lambda > 8h \sin \theta$$

Fluctuation of Sea Surface Sound Large and rapid fluctuation in amplitude or intensity is produced by reflection at the sea surface. In the above-cited works of Urick and Saxton and of Liebermann, about 10 percent of the measured reflection losses of short pulses were greater by 10 db or more above the average, and 10 percent were smaller by 3 db or more. Similarly large fluctuations in surface-reflected amplitudes were found by Pollak (54) in Chesapeake Bay. Such fluctuations in amplitude may be attributed to reflections from wave facets on the sea surface that present a constantly changing size and inclination to the incident and reflected sound. Surface-reflection fluctuations occur even at low frequencies if the grazing angle is large, as observed, for example, by Brown and Ricard (55) at 168 Hz between a source and receiver deep in the sea.

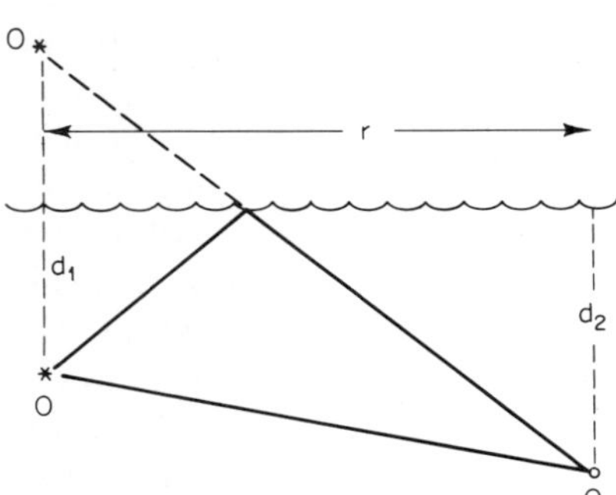

FIG. 5.20. Direct and surface-reflected paths between a source at O and a receiver at Q. O′ is the image of O in the surface.

Image Interference When the sea surface is not too rough, it creates an interference pattern in the underwater sound field. This pattern is caused by constructive and destructive interference between the direct and surface-reflected sound and is called the *Lloyd mirror*, or *image-interference*, effect. In Fig. 5.20, sinusoidal sound travels from O to Q via a direct path and via the surface-reflected path. Let the instantaneous pressure of the direct sound at Q be written

$$p_1 = P \sin \omega t$$

where P = pressure amplitude
$\omega = 2\pi \times$ frequency

The interfering sound pressure arriving by surface reflection will be

$$p_2 = -\mu P \sin \omega(t + \tau)$$

where μ = amplitude reflection coefficient of surface
τ = time delay caused by longer path traveled by surface reflection

The minus sign indicates that the phase of the reflected wave is shifted by 180° on reflection at the water-air interface. The combined pressure will be

$$p = p_1 + p_2$$

and the intensity will be

$$I = \frac{\overline{p^2}}{\rho c}$$

where the bar emphasizes that the intensity is proportional to a time average of the squared pressure. On averaging over a half cycle, we obtain

$$\begin{aligned} I &= \frac{\overline{(p_1 + p_2)^2}}{\rho c} = \frac{1}{\rho c}\frac{1}{\pi}\int_0^\pi (p_1 + p_2)^2\, dt \\ &= \frac{P^2}{2\rho c}(1 + \mu^2 - 2\mu \cos \omega \tau) \\ &= I_0(1 + \mu^2 - 2\mu \cos \omega \tau) \end{aligned}$$

where I_0 is the intensity of the direct sound alone. The term in parentheses accordingly expresses the interference effect of the surface reflection. For straight-line paths and for a horizontal range r appreciably greater than the depths d_1 and d_2, the time delay may be shown to be

$$\tau = \frac{2d_1d_2}{rc}$$

where c is the velocity of sound. Noting that $\omega = 2\pi f = 2\pi c/\lambda$, we obtain finally

$$\frac{I}{I_0} = 1 + \mu^2 - 2\mu \cos \frac{4\pi d_1 d_2}{r\lambda}$$

At a constant receiver depth, a series of peaks and troughs is obtained as r is increased. Figure 5.21 gives a series of such curves for several

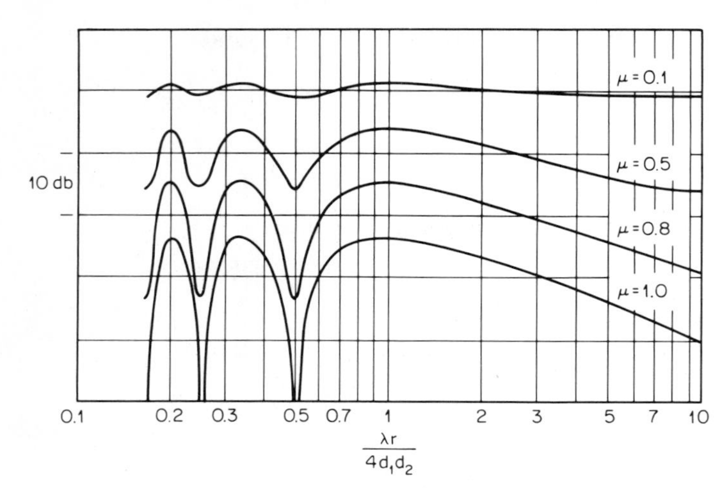

***FIG. 5.21.** Image-interference curves as a function of a normalized range for different values of reflection coefficient of the surface. (**Ref. 40, fig. 10.**)*

values of μ in terms of a normalized range $\lambda r/4d_1d_2$, under the assumption that μ is a constant, independent of angle with the sea surface. This is the "correction" in decibels, relative to the inverse-square spreading loss, that is required in theory to account for surface-reflection interference. In the figure, the curves are displaced by 10 db relative to one another for clarity in display.

Two features of the interference patterns of Fig. 5.21 are noteworthy. One is the decreasing amplitude of the peaks and troughs as the reflection coefficient decreases—an effect that is the basis of the method used by Liebermann (51) for determining the reflection coefficient of the sea surface. The other is the falloff at ranges beyond the last peak. This falloff, for $\mu = 1.0$, is as the inverse square of the range, so that the total falloff, when inverse-square free-field spreading is added, is as the inverse *fourth power* of the range. This is characteristic of dipole sources at ranges great enough for the two sources to begin to cancel one another in phase.

Although of theoretical interest, these Lloyd mirror patterns are not always observed at sea. The rough sea surface commonly obscures the regular pattern at kilohertz frequencies; indeed, it is said that (40, p. 96) "at frequencies greater than several thousand cycles no definite trace of the image effect has been consistently observed in the open sea." At the longer ranges, also, the assumption of straight-line paths is no longer applicable, and even in calm seas a distorted interference pattern occurs under conditions of downward refraction (56). The theoretical pattern is therefore restricted in its actual occurrence at sea to frequencies below a few kilohertz, to ranges less than a few hundred yards for a shallow source and receiver, and to moderately calm seas.

Shadow Zones When a negative gradient exists just beneath the sea surface, a shadow is cast by the surface in the sound field of a shallow source. A *shadow zone* is produced, in which the intensity from the source is very low. An example of a computer-produced ray diagram showing a shadow zone is seen in Fig. 5.22. The velocity profile from which the ray diagram was computed is given on the right. Although no rays enter the shadow, some sound exists inside the shadow zone, and the shadow is not completely acoustically dark. At-sea measurements have shown (40, p. 125) that the intensity of 24-kHz sound inside the shadow is between 40 and 60 db below what it would be in the free field; moreover, severe distortion of sinusoidal and explosive pulses occurs in transmission to a hydrophone located inside the shadow. The shadow boundary is very sharp at high frequencies and causes an attenuation in the vicinity of the boundary amounting to between 20 and 70 db/kyd at 24 kHz.

Several causes of the sound inside a sea-surface shadow zone may be postulated. One is the diffraction of sound into the shadow zone as accounted for by wave theory. This is an effect likely to be of some consequence near the boundary of the shadow zone. Another is the likelihood that the gradient of temperature causing the shadow seldom extends all the way to the sea surface but approaches zero, or isothermal

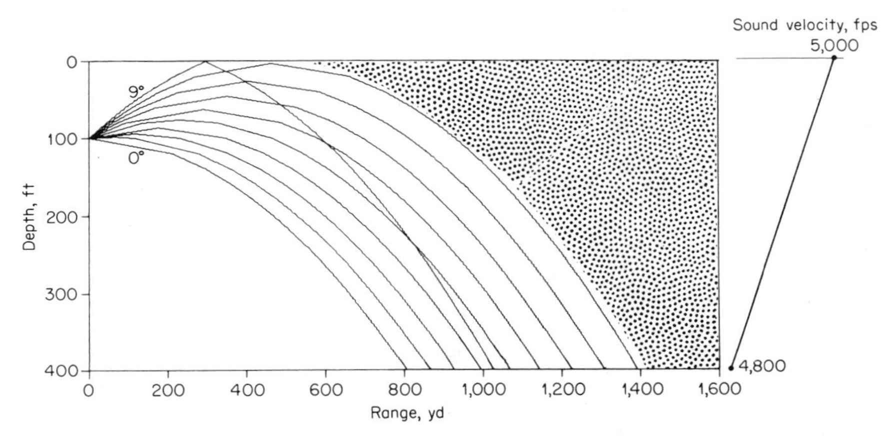

FIG. 5.22. Ray diagram for a source at a depth of 100 ft in the linear gradient shown at the right. Stippled area is the surface shadow zone.

water, just underneath the surface; this effect was considered theoretically by Morse (57) and was found to account for the observed sound levels in shadow zones at 24 kHz. Finally, volume scatterers—especially those in the deep scattering layer—located in the insonified sound field short of the shadow will scatter sound forward into the shadow zone and will tend to "fill it up" with forward-scattered sound analogous to volume reverberation. This is probably the most likely cause of shadow-zone sound. If it is, the propagation is essentially identical to the tropospheric-scatter propagation of radio waves beyond the horizon.

5.8 The Sea Bottom

Comparison with the Sea Surface The sea bottom is a reflecting and scattering boundary of the sea having a number of characteristics similar to the sea surface. However, its effects are more complicated because of its diverse and multilayered composition. An example of this similar behavior is the fact that the sea bottom casts a shadow, or produces a shadow zone, in the upward-refracting water above it in the depths of the deep sea. In Fig. 5.23 the source is taken to be located 100 ft above the bottom; the distance to the beginning of the shadow zone at the bottom is 3,500 yd. Also, a Lloyd mirror effect may be expected in the neighborhood of the seabed, and an interference pattern may be expected to exist in the radiation of a nearby source. Examples of image-inter-

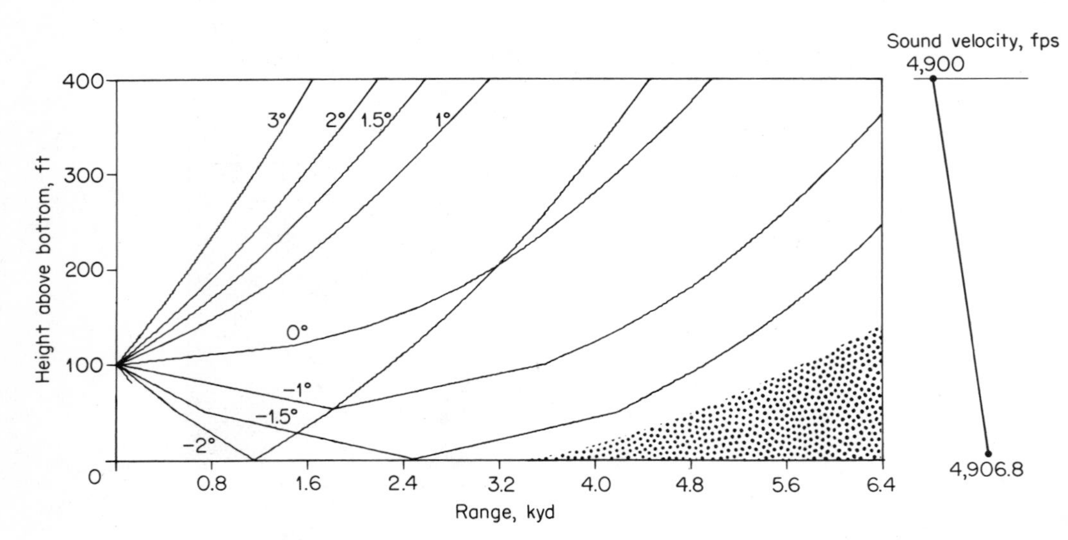

FIG. 5.23. Ray diagram for a source at a height of 100 ft above the bottom in isothermal water. The sound-velocity profile is shown at the right. Stippled area is the bottom shadow zone.

ference directivity patterns caused by bottom reflection have been computed by Mackenzie (61) for a rock, sand, and silt bottom.

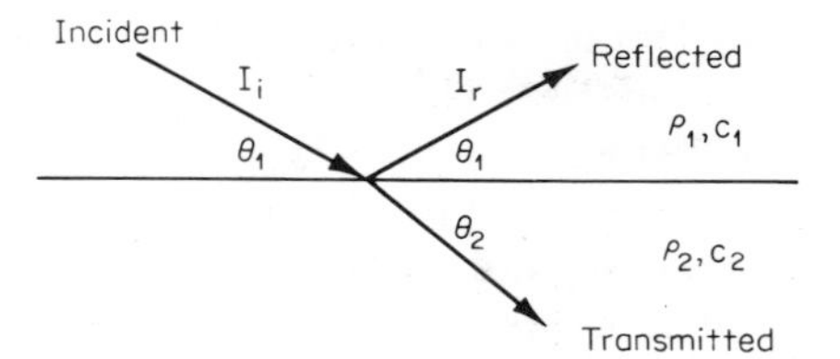

FIG. 5.24. Reflected and transmitted rays at a discontinuity between two mediums.

For two reasons, the reflection of sound from the seabed is vastly more complex than that from the sea surface. First, the bottom is more variable in its acoustic properties because it may vary in composition from hard rock to soft mud. Second, it is often layered, with a density and a sound velocity that change gradually or abruptly with depth. For these reasons, the reflection loss of the seabed is less easily predicted than that of the sea surface.

Bottom-reflection Loss The reflection loss of sound incident at an angle to a plane boundary between two fluids was worked out originally by Rayleigh (8, p. 78). If a plane wave is incident at grazing angle θ_1 upon the boundary between fluids of density ρ_1 and ρ_2 and of sound velocity c_1 and c_2, as indicated in Fig. 5.24, then by the Rayleigh formula the intensity of the reflected wave I_r is related to the intensity of the incident wave I_i by

$$\frac{I_r}{I_i} = \left[\frac{m \sin \theta_1 - n \sin \theta_2}{m \sin \theta_1 + n \cos \theta_2}\right]^2 = \left[\frac{m \sin \theta_1 - (n^2 - \cos^2 \theta_1)^{1/2}}{m \sin \theta_1 + (n^2 - \cos^2 \theta_1)^{1/2}}\right]^2$$

where, following the notation of Brekhovskikh (42, pp. 16–20),

$$m = \frac{\rho_2}{\rho_1}$$

and

$$n = \frac{c_1}{c_2}$$

The reflection loss is the logarithmic expression of the above ratio, or 10 log I_r/I_i db. As a function of grazing angle, the reflection loss is therefore dependent on the ratios m and n. Figure 5.25, adapted from Brekhovskikh, shows the behavior of the loss with grazing angle for four different conditions on m and n. Of these four, the most common condition for natural bottoms is probably that of Fig. 5.25*c*, in which a *critical angle* θ_0 exists such that complete or total reflection occurs (zero loss) at grazing angles less than critical. In many soft mud bottoms, the sound velocity is *less* than that in the water above and an *angle of intromission* θ_B may exist, as in Fig. 5.25*a*.

The density and sound velocity of a great many samples of different naturally occurring sedimentary materials have been compiled by Nafe and Drake (58). Figure 5.26 is based on several hundred measurements on muds, sands, sandstones, limestones, etc.; measured sound velocities plotted against density fall in the shaded area. Although the velocity of sound generally increases with density in bottom sediments, it does not do so uniformly, even when layering is absent. For loosely compacted muds and sands of density only slightly greater than that of water, the sound velocity is apt to be slightly *less* than that of water—an effect explainable by the *mixture theory* of Wood (22, p. 361) and verified in the laboratory (59) for suspensions of finely divided substances. For water-saturated sand sediments, the porosity of the material is the most important factor affecting the velocity (60).

Mackenzie (61) has modified the Rayleigh formula by a theory attributed to Morse to include the effect of loss or attenuation in the bottom material. In a lossy bottom, the critical angle is not as sharp, and a reflection loss occurs at angles less grazing than the critical angle, as indicated by the dashed curve of Fig. 5.25*c*. Reasonable agreement was found by Mackenzie between the theoretical reflection loss and measured losses at 1 kHz for an attenuating bottom material. Other data are best fitted by assuming that the bottom, in addition to being lossy, has

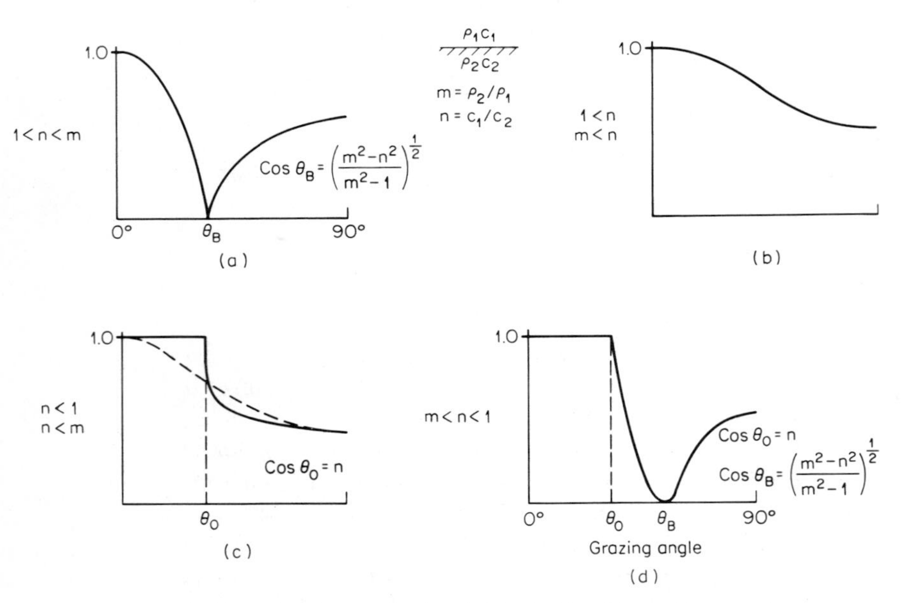

***FIG. 5.25.** Ratio of reflected to incident intensities for four combinations of conditions of sound velocities and densities in lossless mediums separated by a plane interface. The dashed curve in (c) shows the effect of an attenuating lower medium.* [*After Brekhovskikh* (*Ref. 42, Fig. 7*).]

a layered structure, with layers of appropriate thickness, density, and sound velocity, as indicated by core samples or geologic information. In the theory, reflected contributions from each layer are added up, with careful attention paid to phase. Examples of agreement obtained between measured and theoretical losses are those of Cole (62), using field data, and those of Barnard, Bardin, and Hempkins (63), using laboratory measurements.

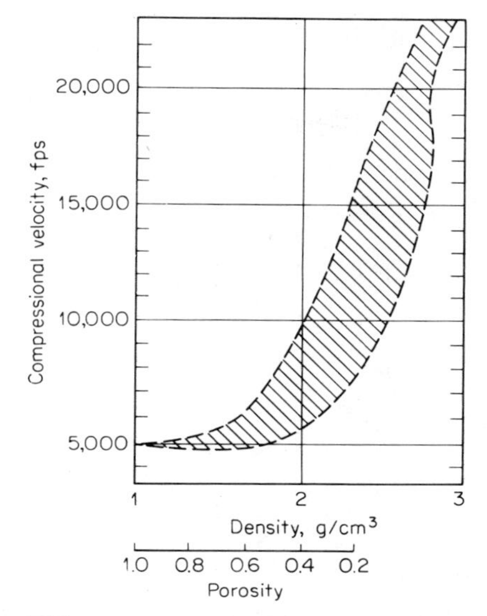

FIG. 5.26. Sound velocity versus density in sedimentary materials. [Based on a compilation by Nafe and Drake (58).] Measured velocities fall in crosshatched area. Porosity is the fraction of the sediment volume occupied by water.

On the other hand, it is clear that the return of sound from the bottom must, for some bottoms, occur by a process of scattering rather than reflection, and must be accounted for in theory by considerations similar to those of surface reverberation. Evidence that some sea bottoms, at least, are scatterers rather than reflectors of sound is the distortion that occurs when an explosive pulse is "reflected" from the seabed, as well as by the investigations of Thompson, Flowers, and Hurdle (64) into the distribution in angle of the return of 19.5-kHz sound from the deep ocean bottom. Undoubtedly all bottoms are to some extent *both* reflectors and scatterers of sound. The true "reflection" loss is likely to be given by correlation measurements between the incident and "reflected" waves, such as those performed by Jones, Leslie, and Barton (65).

Measured Bottom Losses Figure 5.27 is a series of curves of bottom loss as a function of grazing angle for various frequencies, based on many measurements made at deep-sea locations (66). These curves are useful for prediction purposes for "bottom-bounce" sonars using the deep-sea bottom-reflected path. The curves are average losses for abyssal bottoms and are not necessarily representative of any one location; wide variations should be expected for any particular set of measurements. For other types of bottoms, such as those found in coastal locations, Table 5.4 presents the losses measured by Liebermann (67) and by Mackenzie (61). These are examples of the range of losses to be expected for bottoms characterized by words such as "sand" or "mud" that denote the particle size of the sedimentary bottoms to which they refer.

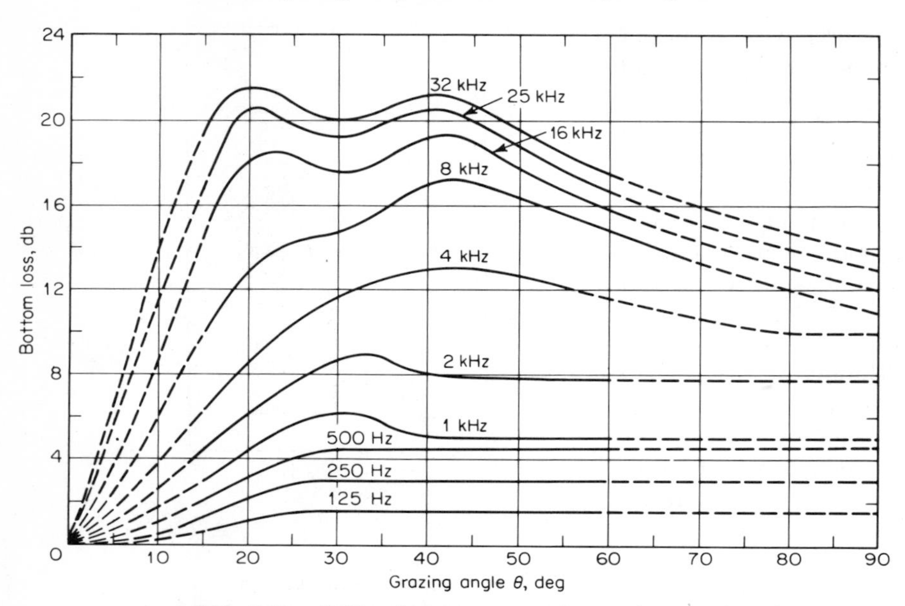

FIG. 5.27. Bottom loss versus grazing angle at various frequencies. [*From Marsh* (*66*).]

TABLE 5.4 Measured Reflection Losses for Different Bottom Types

24 kHz, 10° Grazing Angle (Ref. 67) 17 Stations	
Mud	16 db
Mud-sand	10
Sand-mud	6
Sand	4
Stony	4

Normal Incidence (Ref. 61)
7 Stations

	4 kHz	7.5 kHz	16 kHz
Sandy silt (3 stations)	14	14	13
Fine sand (1 station)	7	3	6
Coarse sand (1 station)	7	8	8
Medium sand with rock (1 station)	8	6	10
Rock with some sand (1 station)	5	4	10

REFERENCES

1. Stephenson, E. B.: Transmission of Sound in Sea Water: Absorption and Reflection Coefficients and Temperature Gradients, *U.S. Naval Res. Lab. Rept.* S-1204, 1935.
2. Stephenson, E. B.: Absorption Coefficients of Supersonic Sound in Open Sea Water, *U.S. Naval Res. Lab. Rept.* S-1549, 1939.
3. Principles of Underwater Sound, *Natl. Defense Res. Comm. Sum. Tech. Rept.* 7, sec. 3.4, National Research Council, 1946. Also, Physics of Sound in the Sea, *Natl. Defense Res. Comm. Sum. Tech. Rept.* 8, sec. 5.2.2, National Research Council, 1946.
4. Wilson, O. B., and R. W. Leonard: Measurements of Sound Absorption in Aqueous Salt Solutions by a Resonator Method, *J. Acoust. Soc. Am.*, **26**:223 (1954).
5. Hansen, P. G.: Measurements of Low Frequency Sound in Small Samples of Sea Water, *U.S. Navy Electron. Lab. Rept.* 1135, 1962.
6. Glotov, V. P.: Reverberation Tank Method for the Study of Sound Absorption in the Sea, *Soviet Phys. Acoust.*, **4**:243 (1958).
7. Liebermann, L. N.: Sound Propagation in Chemically Active Media, *Phys. Rev.*, **76**:1520 (1949).
8. Rayleigh, Lord: "The Theory of Sound," vol. II, Dover Publications, Inc., New York, 1945.
9. Mason, W. P. (ed.): "Physical Acoustics," vol. II, part A, pp. 293–295, Academic Press Inc., New York, 1965.
10. Leonard, R. W., P. C. Combs, and L. R. Skidmore: Attenuation of Sound in Synthetic Sea Water, *J. Acoust. Soc. Am.*, **21**:63 (1949).
11. Liebermann, L. N.: Origin of Sound Absorption in Water and in Sea Water, *J. Acoust. Soc. Am.*, **20**:868 (1948).
12. Schulkin, M., and H. W. Marsh: Absorption of Sound in Sea Water, *J. Brit. IRE*, **25**:493 (1963). Also, Sound Absorption in Sea Water, *J. Acoust. Soc. Am.*, **34**:864 (1962).
13. Schulkin, M.: Eddy Viscosity as a Possible Acoustic Absorption Mechanism in the Ocean, *J. Acoust. Soc. Am.*, **35**:253 (1963).
14. Fisher, R. H.: Effect of High Pressure on Sound Absorption and Chemical Equilibrium, *J. Acoust. Soc. Am.*, **30**:442 (1958).
15. Urick, R. J.: Low Frequency Sound Attenuation in the Deep Ocean, *J. Acoust. Soc. Am.*, **35**:1413 (1963).
16. Thorp, W. H.: Deep Ocean Sound Attenuation in the Sub- and Low-kilocycle-per-second Region, *J. Acoust. Soc. Am.*, **38**:648 (1965).
17. Sheehy, M. J., and R. Halley: Measurement of the Attenuation of Low-frequency Underwater Sound, *J. Acoust. Soc. Am.*, **29**:464 (1957).
18. Ewing, M., and J. L. Worzel: Long Range Sound Transmission, article in "Propagation of Sound in the Ocean," p. 11, Geological Society of America Memoir 27, 1948.
19. Bryan, G. M., M. Truchan, and J. Ewing: Long Range SOFAR Studies in the South Atlantic Ocean, *J. Acoust. Soc. Am.*, **35**:273 (1963).
20. Kibblewhite, A. C., R. N. Denham, and P. H. Barker: Long Range Sound Propagation Study in the Southern Ocean—Project Neptune, *J. Acoust. Soc. Am.*, **38**:629 (1965).
21. Urick, R. J.: A Long-range Deep Sea Attenuation Measurement, *J. Acoust. Soc. Am.*, **39**:904 (1966).
22. Wood, A. B.: "A Textbook of Sound," The Macmillan Company, New York, 1941.
23. Stephenson, E. B.: Velocity of Sound in Sea Water, *Phys. Rev.*, **21**:181 (1923).
24. Heck, N. H., and J. H. Service: Velocity of Sound in Sea Water, *U.S. Coast Geodetic Surv. Spec. Publ.* 108, 1924.

25. Garrison, G. R., P. C. Kirkland, and S. R. Murphy: Long Range Measurement of Sound Speed in Sea Water, *J. Acoust. Soc. Am.*, **33**:360 (1961).
26. Matthews, D. J.: Tables of the Velocity of Sound in Pure Water and Sea Water for Use in Echo Sounding and Sound Ranging, *Admiralty Hydrograph. Dept. Publ.* 282, 1939.
27. Kuwahara, S.: Velocity of Sound in Sea Water and Calculation of the Velocity for Use in Sonic Sounding, *Hydrograph. Rev.*, **16**:123 (1939).
28. Weissler, A., and V. A. Del Grosso: The Velocity of Sound in Sea Water, *J. Acoust. Soc. Am.*, **23**:219 (1951).
29. Del Grosso, V. A.: Velocity of Sound in Sea Water at Zero Depth, *U.S. Naval Res. Lab. Rept.* 4002, 1952.
30. Wilson, W. D.: Speed of Sound in Sea Water as a Function of Temperature, Pressure and Salinity, *J. Acoust. Soc. Am.*, **32**:641 (1960). Also, see extensions and revised formulas in *J. Acoust. Soc. Am.*, **32**:1357 (1960) and **34**:866 (1962).
31. Mackenzie, K. V.: Formulas for the Computation of Sound Speed in Sea Water, *J. Acoust. Soc. Am.*, **32**:100 (1960). Also, Sound Speed Measurements Using the Bathyscaphe "Trieste," *J. Acoust. Soc. Am.*, **33**:1113 (1961).
32. Hays, E. E.: Comparison of Directly Measured Sound Velocities with Values Calculated from Hydrographic Data, *J. Acoust. Soc. Am.*, **33**:85 (1961).
33. Lyons, A. M.: Sea Water Sound Speed Expressed in English Units, *U.S. Naval Ordnance Lab. Tech. Rept.* 63–168, 1963.
34. Greenspan, M., and C. E. Tschiegg: Sing-around Ultrasonic Velocimeter for Liquids, *Rev. Sci. Instr.*, **28**:897 (1957). Also, *J. Acoust. Soc. Am.*, **31**:1038 (1959).
35. Urick, R. J.: An Acoustic Interferometer for the Measurement of Sound Velocity in the Ocean, *U.S. Navy Radio Sound Lab. Rept.* S-18, 1944.
36. Brown, R. K.: Measurement of Sound Velocity in the Ocean, *J. Acoust. Soc. Am.*, **26**:64 (1954).
37. Application of Oceanography to Subsurface Warfare, *Natl. Defense Res. Comm. Div. 6 Sum. Tech. Rept.*, vol. 6A, figs. 17, 32, 1946.
38. An Interim Report on the Sound Velocity Distribution in the North Atlantic Ocean, *U.S. Navy Oceanog. Office Tech. Rept.* 171, 1965.
39. Woollard, G. P.: "Sound Transmission Measurements at 12 kc and 24 kc in Shallow Water," Woods Hole Oceanographic Institution unpublished memorandum, Contract NObs-2083, May 1, 1946.
40. Physics of Sound in the Sea, *Natl. Defense Res. Comm. Natl. Res. Council Div. 6 Sum. Tech. Rept.* 8, chaps. 2–4, 1946.
41. Officer, C. B.: "Introduction to the Theory of Sound Transmission," McGraw-Hill Book Company, New York, 1958.
42. Brekhovskikh, L. M.: "Waves in Layered Media," Academic Press Inc., New York, 1960.
43. Kinsler, L. E., and A. R. Frey: "Fundamentals of Acoustics," 2d ed., p. 466, John Wiley & Sons, Inc., New York, 1962.
44. "Calculation of Ray Paths Using the Refraction Slide Rule," U.S. Navy Department, Bureau of Ships, May, 1943.
45. Galkin, O. P., and V. S. Grigorev: Instruments for Plotting Refracted Rays, *Soviet Phys. Acoust.*, **6**:20 (1960).
46. Dosso, H. W., et al.: Ray Tracing with an LGP-30: Multiple Velocity Profiles, Multiple Segmented Bottom Profiles, Bottom and Surface Reflections, Travel Time, *Pacific Naval Lab. (Can.) Rept.* 60-3, 1960.
47. Pedersen, M. A.: Acoustic Intensity Anomalies Introduced by Constant Velocity Gradients, *J. Acoust. Soc. Am.*, **33**:465 (1961).

48. Anderson, G., R. Gocht, and D. Sirota: Spreading Loss in an Inhomogeneous Medium, *J. Acoust. Soc. Am.*, **36**:140 (1964).
49. Eby, E. S., and L. T. Einstein: General Spreading Loss Expression, *J. Acoust. Soc. Am.*, **37**:933 (1965).
50. Urick, R. J., and H. L. Saxton: Surface Reflection of Short Supersonic Pulses in the Ocean, *J. Acoust. Soc. Am.*, **19**:8 (1947).
51. Liebermann, L. N.: Reflection of Underwater Sound from the Sea Surface, *J. Acoust. Soc. Am.*, **20**:498 (1948).
52. Addlington, R. H.: Acoustic Reflection Losses at the Sea Surface Measured with Explosive Sources, *J. Acoust. Soc. Am.*, **35**:1834 (1963).
53. Pedersen, M. A.: Comparison of Experimental and Theoretical Image Interference in Deep Water Acoustics, *J. Acoust. Soc. Am.*, **34**:1197 (1962).
54. Pollak, M. J.: Surface Reflections of Sound at 100 kc, *J. Acoust. Soc. Am.*, **30**:343 (1958).
55. Brown, M. V., and J. Ricard: Fluctuations in Surface Reflected Pulsed CW Arrivals, *J. Acoust. Soc. Am.*, **32**:1551 (1960).
56. Young, R. W.: Image Interference in the Presence of Refraction, *J. Acoust. Soc. Am.*, **19**:286(A) (1947).
57. Morse, R. W.: Dependence of Shadow Zone Sound on the Surface Sound Velocity Gradient, *J. Acoust. Soc. Am.*, **22**:857 (1950).
58. Nafe, J. E., and C. L. Drake: Physical Properties of Marine Sediments, *Lamont Geol. Obs. Columbia Univ. Tech. Rept.* 2, 1961.
59. Urick, R. J.: A Sound Velocity Method for Determining the Compressibility of Finely Divided Substances, *J. Appl. Phys.*, **18**:983 (1947).
60. Brandt, H.: Factors Affecting Compressional Wave Velocity in Unconsolidated Marine Sand Sediments, *J. Acoust. Soc. Am.*, **32**:171 (1960).
61. Mackenzie, K. V.: Reflection of Sound from Coastal Bottoms, *J. Acoust. Soc. Am.*, **32**:221 (1960).
62. Cole, B. F.: Marine Sediment Attenuation and Ocean-bottom Reflected Sound, *J. Acoust. Soc. Am.*, **36**:1993(A) (1964).
63. Barnard, G. R., J. L. Bardin, and W. B. Hempkins: Underwater Sound Reflection from Layered Media, *J. Acoust. Soc. Am.*, **36**:2119 (1964).
64. Thompson, K. P., K. D. Flowers, and B. G. Hurdle: Scattered Fields from the Ocean Bottom, *J. Acoust. Soc. Am.*, **38**:932(A), 1965.
65. Jones, J. L., C. B. Leslie, and L. E. Barton: Acoustic Characteristics of Underwater Bottoms, *J. Acoust. Soc. Am.*, **36**:154 (1964).
66. Marsh, H. W.: Reflection and Scattering of Sound by the Sea Bottom, *J. Acoust. Soc. Am.*, **36**:2003(A) (1964). Complete Paper Avco Corporation, Marine Electronics Office, New London, Conn.
67. Liebermann, L. N.: Reflection of Sound from Coastal Bottoms, *J. Acoust. Soc. Am.*, **20**:305 (1948).

SIX

Propagation of Sound in the Sea: Transmission Loss, II

Sound always travels to long distances in the sea by some form of ducted propagation. When sound travels in a *duct*, or *sound channel*, it is prevented from spreading in all directions, and remains confined between the boundaries of the sound channel. A number of kinds of ducts of common occurrence in the sea are the *mixed-layer sound channel*, the *deep sound channel*, and, in shallow water, the *shallow-water sound channel*.

6.1 The Mixed-layer Sound Channel

Sound Trapping in the Mixed Layer The fact that the upper portions of the sea often form a trap, or duct, in which relatively good propagation takes place was recognized long ago. In 1937 Steinberger (1) made propagation tests off Guantánamo, Cuba, when a mixed layer was present and correctly accounted for the good transmission as the result of ducting caused by the upward refraction of sound in the layer. The effect was all but forgotten during the World War II years and was essentially rediscovered shortly thereafter during propagation measurements at the lower frequencies then becoming of interest. The mixed-layer sound channel, like its radio analog, the ground-based duct, is now recognized as a regular characteristic of the medium in which transmission takes place.

In the cloudy, windy ocean areas of the world, the temperature profile regularly shows the presence of an isothermal layer just beneath the sea surface. This layer of isothermal water is created and maintained by turbulent wind mixing of the near-surface water of the sea. Within the layer—called simply the *mixed layer* from its

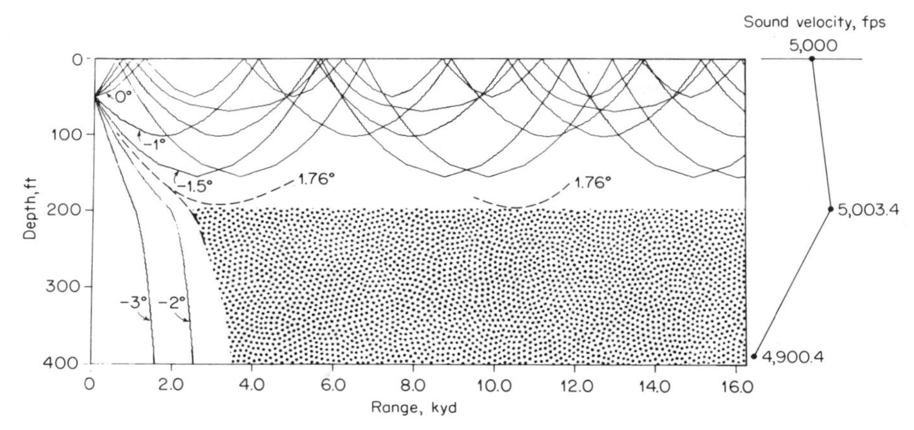

***FIG. 6.1.** Ray diagram for sound transmission from a 50-ft source in a 200-ft mixed layer. Rays are drawn at 1° intervals, with 1.5 and 1.76° added, for the profile at the right.*

manner of origin—the velocity of sound increases with depth because of the pressure effect on sound velocity. The upward refraction that results keeps a portion of the acoustic energy emitted by a shallow source close to the sea surface. The sound trapped in the layer propagates to long ranges by successive reflections from the sea surface along ray paths that are long arcs of circles between their encounters with the sea surface.

A computer-produced ray diagram for a source in a typical mixed layer is shown in Fig. 6.1.* Under the conditions for which the diagram was drawn, the ray leaving the source at an angle of 1.76° becomes horizontal at the base of the layer; rays leaving the source at smaller angles remain in the layer; rays leaving the source at greater angles are sent downward into the abyssal depths of the sea. A shadow zone is produced beneath the layer at ranges beyond the direct or close-in sound field. This shadow is not complete, but is insonified by scattered sound from the sea surface and by diffusion of sound out of the channel, caused by the nature of the lower boundary. By these two processes, sound "leaks out" of the channel at a rate given by a "leakage coefficient" that expresses the attenuation, in decibels per kiloyard, of sound trapped in the channel. The leakage coefficient varies with surface roughness, duct thickness, gradient below the layer, and frequency.

Variation of Transmission Loss with Depth With a source in the mixed layer, the intensity of sound tends to fall off with depth within the layer

* The discontinuous form of the rays in this and other ray diagrams is caused by the way the velocity profile was subdivided in the computer program.

and to fall off faster with depth as the lower boundary of the mixed layer is crossed and as the shadow is entered.

Figure 6.2 shows measured transmission data for a source depth of 20 ft and a mixed layer 100 ft thick (2). In tests at sea, sound levels at 8 and 16 kHz were measured at five depths and at various ranges from a CW source. Measured levels have been plotted relative to those that would occur with spherical spreading and absorption, and are indicated by the shaded areas to the left (lower levels) or to the right (higher levels) of a vertical line at the measurement range. Of particular note are the higher levels obtained when the receiver is placed within the layer—indicating the existence of mixed-layer sound channeling. At the same

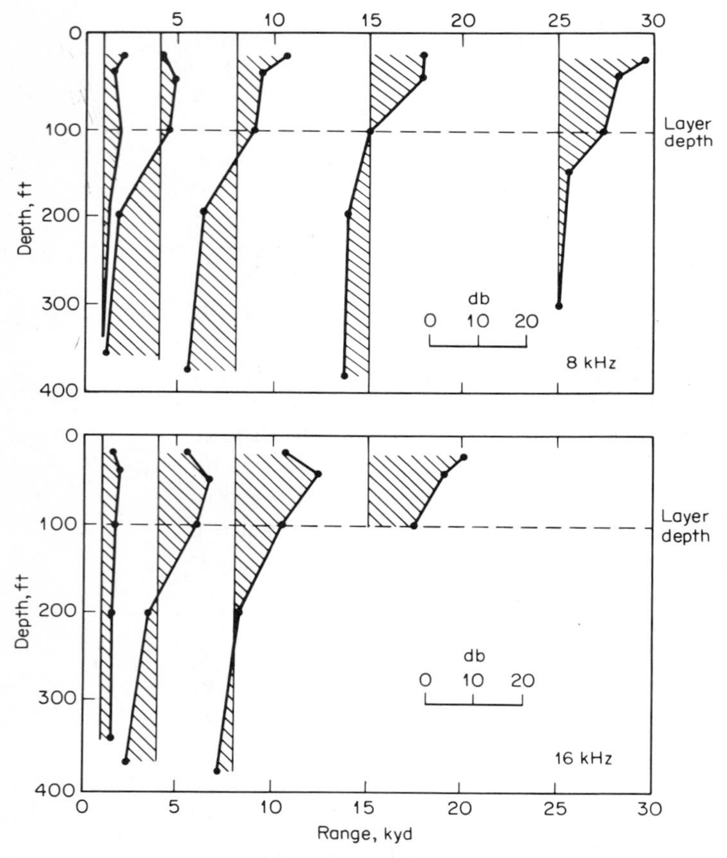

FIG. 6.2. Observed variation of 8 and 16 kHz sound level with depth in and below a mixed layer for a source at 20 ft. (Ref. 2.)

time, levels lower than those that would occur with spherical spreading and absorption exist when the receiver is in the shadow zone beneath the layer. At long ranges, the shadow tends to fill up with bottom-reflected sound.

The goodness of transmission in the layer and the acoustic darkness of the shadow beneath it vary greatly with the thickness of the layer, the conditions at its boundaries, and the frequency. Qualitatively, Fig. 6.3 shows two intensity-depth "profiles" to indicate the variation of transmitted sound level with depth for two contrasting sets of conditions. Profile AA is drawn for the conditions of high frequency, calm sea, strong negative velocity gradient beneath the layer, and moderate ranges. Profile BB is for low frequencies, rough seas, weak negative gradient, and long ranges.

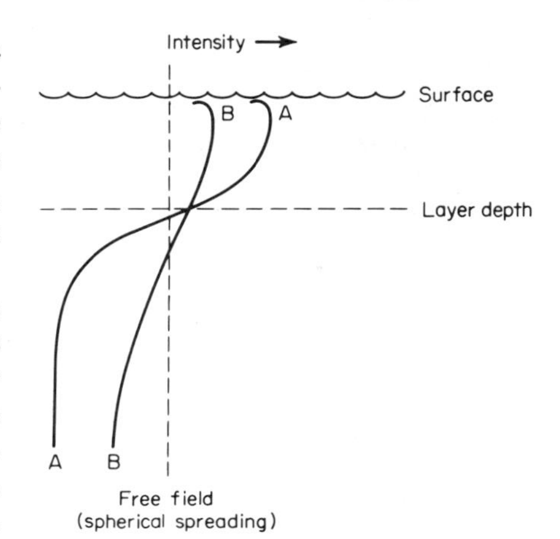

FIG. 6.3. ***Intensity versus depth in and below the mixed layer for two contrasting sets of conditions.***

Low-frequency Cutoff At very low frequencies, sound ceases to be trapped in the mixed layer or, indeed, in any sound channel. This occurs when the frequency approaches the cutoff frequency for the first mode of normal-mode theory, when, in a sense, the wavelength has become too large to "fit" in the duct. This maximum wavelength for duct transmission may be found from the theory of radio propagation in ground-based radio ducts to be (3)

$$\lambda_{\max} = \tfrac{8}{3}\sqrt{2}\int_0^H \sqrt{N(z) - N(H)}\, dz$$

where $N(z)$ = index of refraction at any depth z in duct
$N(H)$ = index of refraction at base of duct

Using values of velocity and velocity gradient appropriate for sound transmission in the mixed layer, one obtains

$$\lambda_{\max} = 4.7 \times 10^{-3} H^{3/2}$$

for the maximum wavelength, in feet, trapped in a mixed-layer duct H ft thick. For example, for a mixed layer 100 ft thick, the maximum trapped wavelength $\lambda_{\max}$ is 4.7 ft, corresponding to a frequency of 1,100 Hz. Although this does not represent a sharp cutoff, wavelengths much longer than this are strongly attenuated; wavelengths much shorter are attenuated by absorption and leakage. For a mixed layer of a given

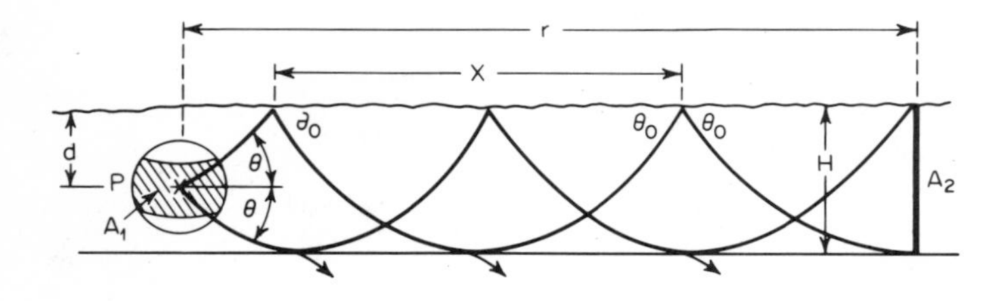

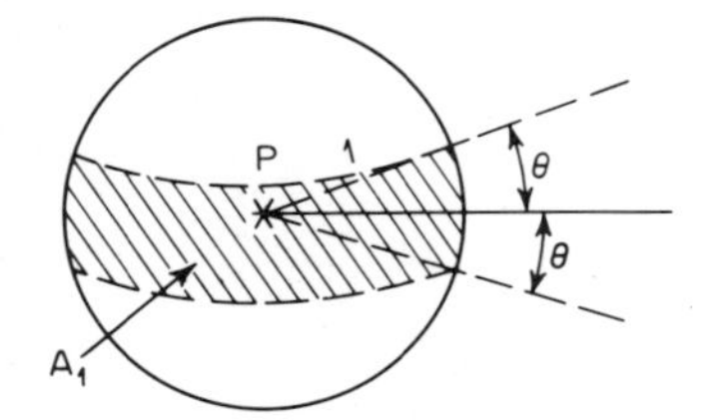

FIG. 6.4. Transmission loss in a mixed-layer channel.

thickness, it follows that there is an optimum frequency of best transmission, at which the leakage of sound out of the layer is a minimum.

Transmission-loss Model An expression for the transmission loss in a mixed layer may be obtained through simple considerations. In Fig. 6.4, let a nondirectional source of sound be located at P in a mixed layer. Of all the rays leaving the source, only those within a certain limiting angle 2θ remain in the channel. At a distance of 1 yd, the power contained in this bundle of rays is distributed over a portion of a spherical surface A_1. At a long distance r, this same amount of power—in the absence of absorption and leakage—is distributed over a cylindrical surface A_2. Because the power crossing areas A_1 and A_2 is the same, the transmission loss to range r, averaged over the duct thickness H, is

$$\mathrm{TL} = 10 \log \frac{A_2}{A_1}$$

By geometry

$$A_2 = 2\pi r H$$

and

$$A_1 = 2\pi \int_{-\theta}^{\theta} \cos \theta \, d\theta = 4\pi \sin \theta$$

Therefore, for the mixed-layer duct,

$$\mathrm{TL} = 10 \log \frac{rH}{2 \sin \theta} = 10 \log r\, r_0 = 10 \log r_0^2 \frac{r}{r_0}$$

where $r_0 = H/2 \sin \theta$

θ = inclination at source of maximum trapped ray

The quantity $r_0^2 r/r_0$ indicates that the transmission to range r may be viewed as the result of spherical spreading out to a *transition range* r_0, followed by cylindrical spreading from r_0 to r. When the attenuation due to absorption and leakage is added, the duct-transmission-loss expression becomes

$$\mathrm{TL} = 10 \log r_0 + 10 \log r + (\alpha + \alpha_L) r \times 10^{-3}$$

where α = absorption coefficient, db/kyd
α_L = leakage coefficient, db/kyd, expressing rate at which acoustic energy leaks out of duct
r = range, yd

For a layer of constant velocity gradient, in which the rays are arcs of circles of radius of curvature $R = c_0/g$, the following relationships may be found by geometry for the conditions of large R ($R \gg H$) and small θ_0 ($\sin \theta_0 \ll 1$):

$$x = \sqrt{8RH} = \text{skip distance of limiting ray}$$

$$\theta_0 = \sqrt{\frac{2H}{R}} = \text{maximum angle of limiting ray, rad}$$

$$\theta = \sqrt{\frac{2(H-d)}{R}} = \text{angle of limiting ray at source depth, rad}$$

$$r_0 = \sqrt{\frac{RH}{2}} \sqrt{\frac{H}{H-d}} = \text{transition range}$$

$$= \frac{x}{4} \sqrt{\frac{H}{H-d}}$$

As an example, for a typical mixed layer of thickness H = 300 ft and for a source at the surface ($d = 0$), we would have

$$R = \frac{5{,}000}{0.017} = 98{,}000 \text{ yd}$$

x = 8,800 yd, $\theta_0 = \theta = 2°$, and r_0 = 2,200 yd.

Leakage of Sound from the Layer The leakage coefficient for mixed-layer transmission may be conceived as being due to two causes. One is scattering out of the duct by the rough sea surface. The attenuation caused by rough-surface scattering has been considered theoretically by Marsh, Schulkin, and Kneale (4), who found that the leakage coefficient could be expressed as a function of the product of wave height and frequency. Figure 6.5 shows the surface loss as derived by them from theoretical considerations in terms of decibels of loss per "bounce" of the limiting ray, or in decibels per distance x of Fig. 6.4. The other component of the leakage has been called "transverse diffusion" out of the mixed layer (5) and depends upon the sharpness of the discontinuity between the mixed layer and the thermocline below, as well as upon the velocity gradient in the thermocline.

Typical Transmission Curves On the more practical side, one must turn to an analysis (6) of a vast quantity of data obtained at sea in 1953 to 1954. The analysis led to a series of formulas from which the curves of Fig. 6.6 and Fig. 6.7 were computed. These give the transmission loss at 2 and 8 kHz for various combinations of source and receiver depths, with layer depth as a parameter. They apply for low sea states and for an average velocity gradient below the mixed layer. They show average transmission-loss values to be expected in the presence of a wind-mixed isothermal layer at the two frequencies 2 and 8 kHz.

Occurrence of the Layer Statistical data on the existence and thickness of the mixed layer in the North Atlantic is given in Fig. 6.8. These distribution curves, based on numerous bathythermograph observations, show the percentage of the time that the layer thickness exceeds the value shown as ordinate for two bands of latitude and for the four seasons of the year.

Internal Waves Over short distances, the thickness of the mixed layer may vary because of internal waves in the thermocline below. Internal waves are oscillations of the thermocline that are manifested by a changing temperature at a fixed point in the sea (7). In one series of observations (8) at a location just off the California shore, these oscillations were found to have periods of 4 to 8 min and to have heights of about 2 to 8 ft. Internal waves propagate along the density discontinuity between the mixed layer and the water below, with a velocity, as measured in the study just referred to, of about 0.3 knot and a wavelength of

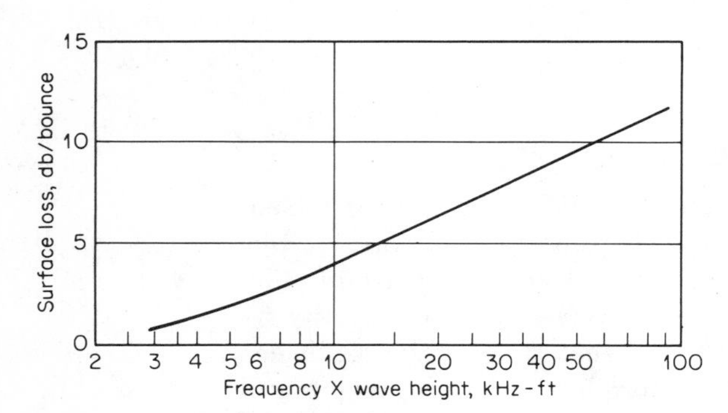

FIG. 6.5. ***Reflection loss from the sea surface in terms of the product of frequency and wave height, crest to trough.*** *(Ref. 4.)*

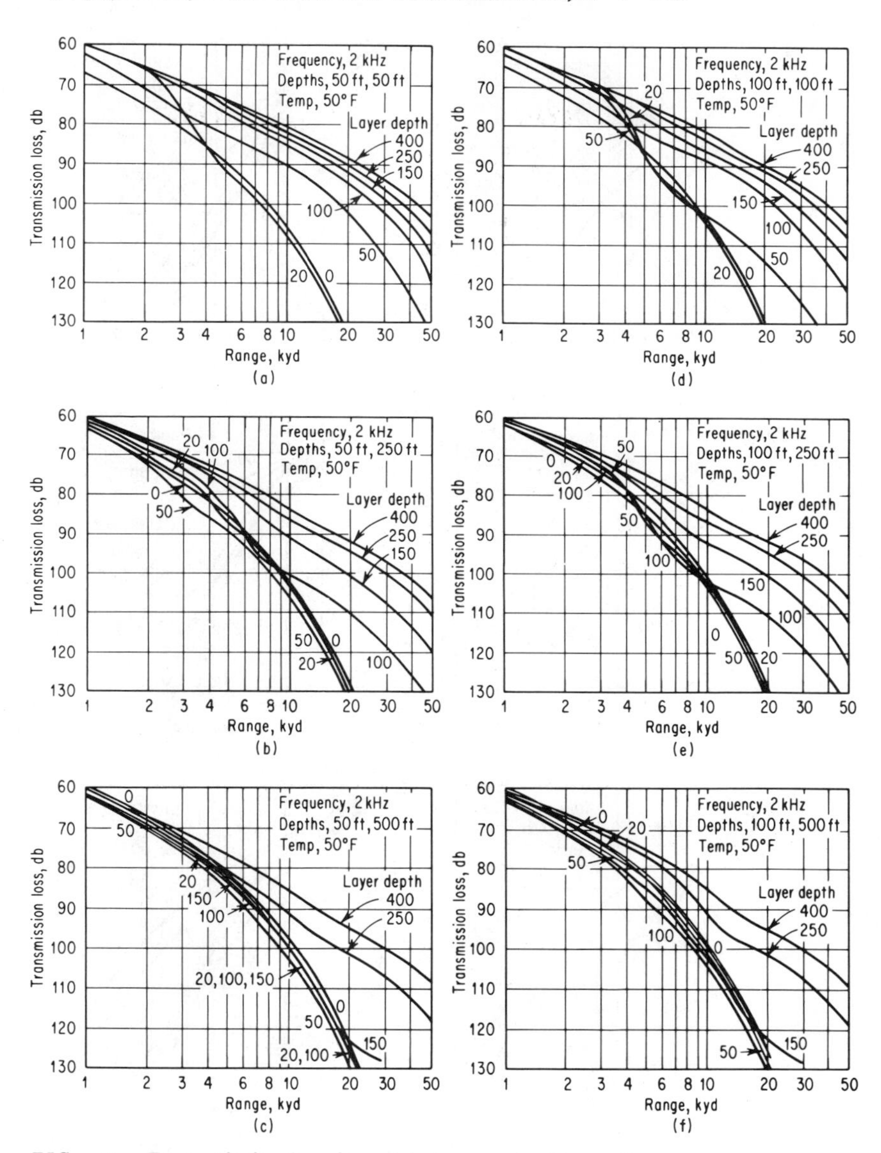

FIG. 6.6. Transmission loss in and below a mixed layer for different depth combinations, 2 kHz. Layer depths given in feet. (*Ref. 6.*)

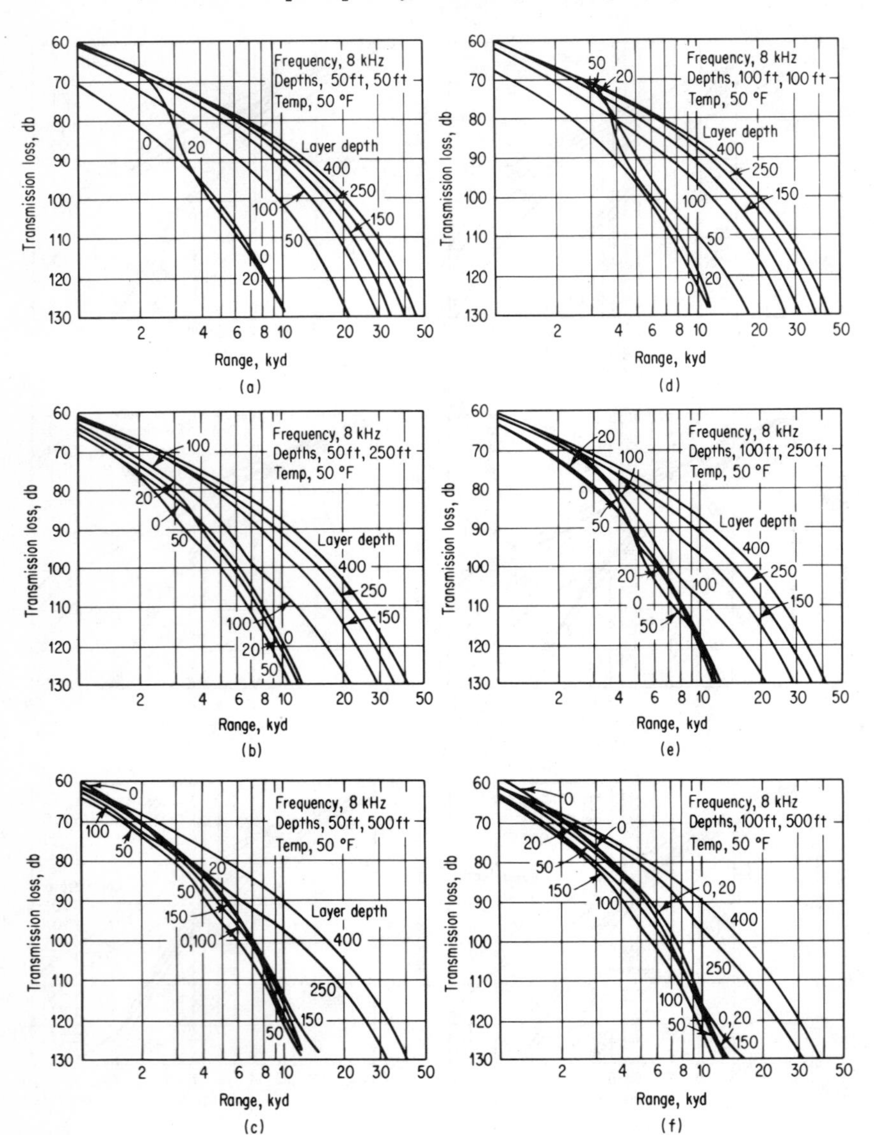

FIG. 6.7. Transmission loss in and below a mixed layer for different depth combinations, 8 kHz. Layer depths given in feet. (Ref. 6.)

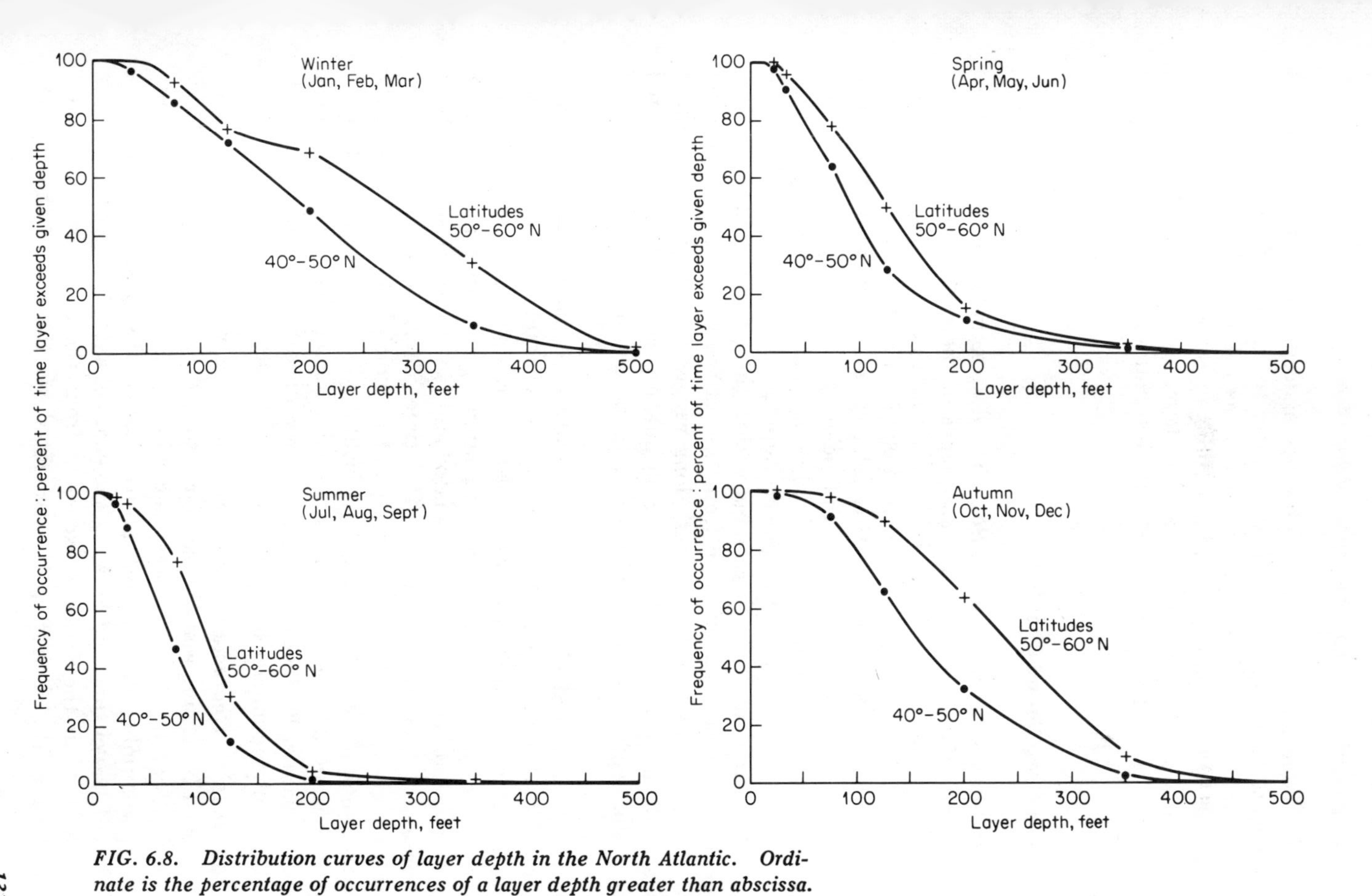

FIG. 6.8. Distribution curves of layer depth in the North Atlantic. Ordinate is the percentage of occurrences of a layer depth greater than abscissa. Based on original data provided by the Navy Oceanographic Office.

50 yd. Longer wavelengths are observed in deep water. Internal waves complicate the propagation of underwater sound by causing variations in transmission loss over short distances, as has been demonstrated by means of ray diagrams (9). Fluctuations of echoes between a surface-ship source and a target below the mixed layer are doubtless attributable, in part at least, to the existence of such internal waves in the thermocline (10).

6.2 The Deep Sound Channel

Sound Channeling in the Deep Sea The deep sound channel, sometimes called the sofar channel, is a consequence of the characteristic velocity profile of the deep sea. It will be recalled that this profile has a minimum at a depth which varies from about 4,000 ft in midlatitudes to near the surface in the polar regions. This velocity minimum causes the sea to act like a kind of lens; above and below the minimum, the velocity gradient continually bends the sound rays toward the depth of minimum velocity. A portion of the power radiated by a source in the deep sound channel accordingly remains within the channel and encounters no acoustic losses by reflection from the surface and bottom. Because of the low transmission loss, very long ranges can be obtained from a source of moderate acoustic power output, especially when it is located near the depth of minimum velocity. This depth is called the *axis* of the sound channel.

The deep sound channel was originally investigated by Ewing and Worzel (11). Its remarkable transmission characteristics were utilized in the sofar system for the rescue of aviators downed at sea. In sofar (the letters denote *so*und *f*ixing *a*nd *r*anging) a small explosive charge is dropped at sea by a downed aviator and is received at shore stations thousands of miles away (12). The time of arrival at two or more stations gives a "fix" locating the point at which the detonation of the charge took place. More recently, the ability to measure accurately the arrival time of explosive signals traveling along the axis of the deep sound channel has been used for geodetic distance determinations (13) and for missile-impact location (14).

Transmission Characteristics Figure 6.9 is a ray diagram, originally published by Ewing and Worzel, for a sound source on the channel axis at a depth of 4,000 ft. Between the source and a point a great distance away, a number of refracted propagation paths exist, each having a different travel time and crossing the channel axis at different intervals. The path with the greatest excursion from the axis has the shortest travel time; the shortest path, straight down the axis of the channel,

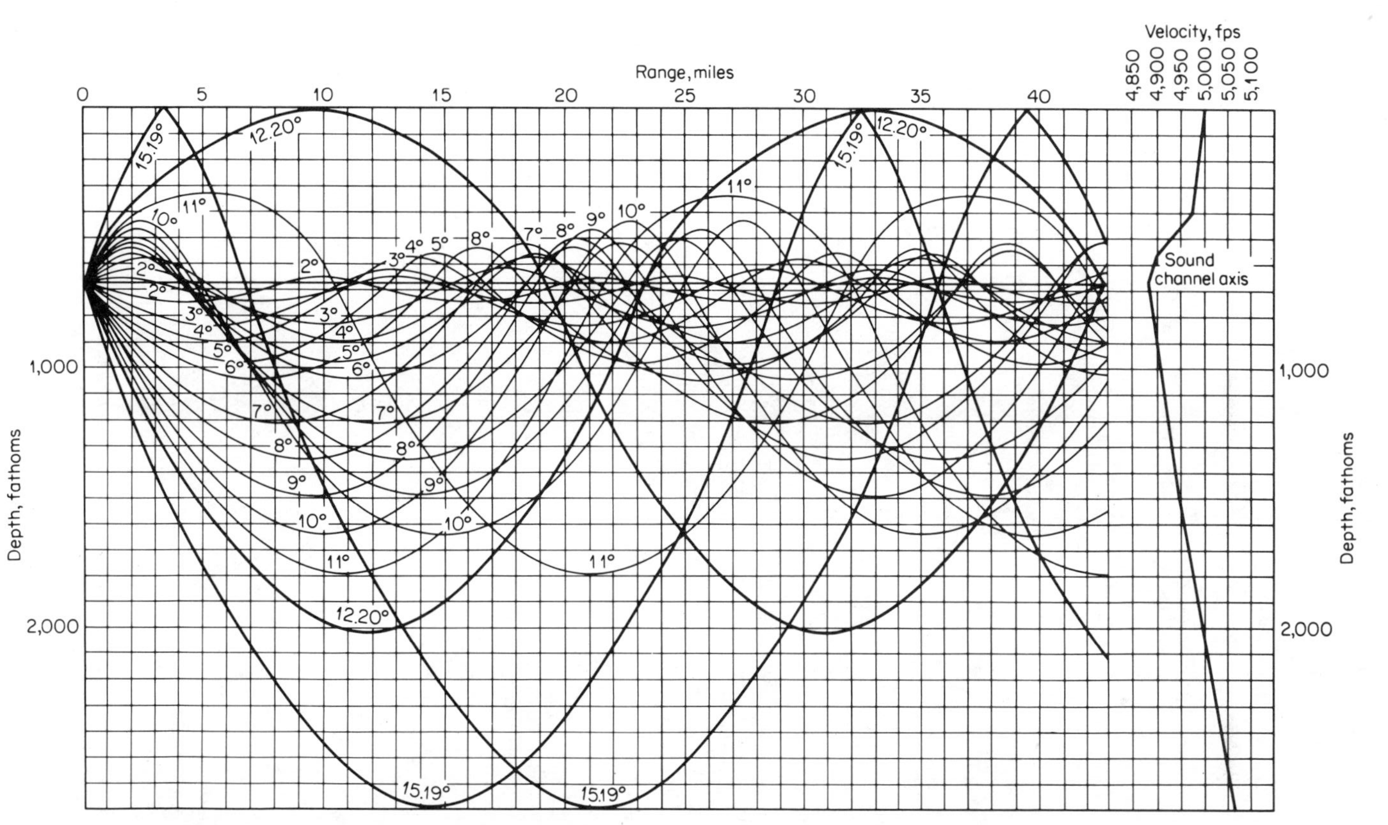

FIG. 6.9. ***Ray diagram of transmission in the deep sound channel for a source on the axis.*** ***(Ref. 11.)***

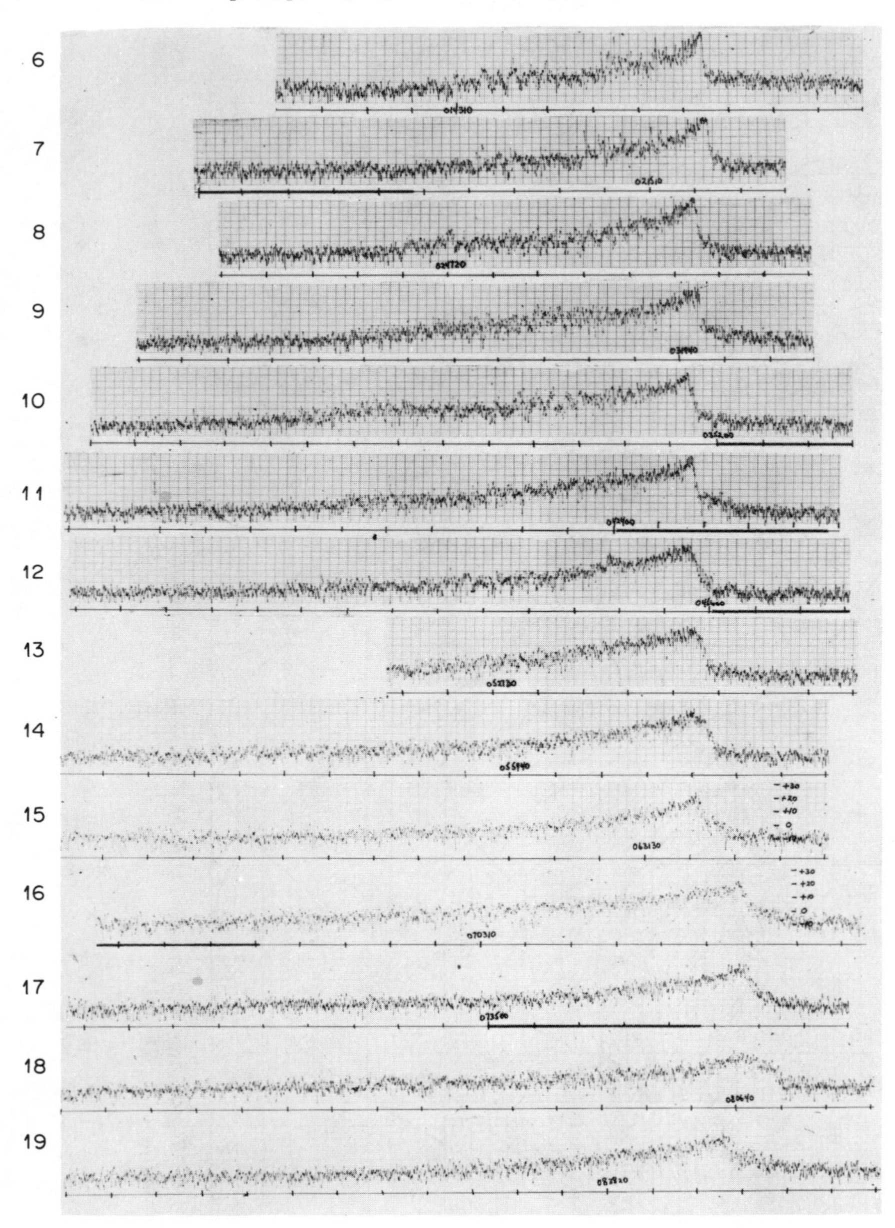

FIG. 6.10. Sofar signal envelopes at ranges about 100 miles apart. Horizontal scale has 1-sec time ticks; vertical scale is in decibels; total width of each recording, 50 db.

has the longest travel time, since it lies entirely at the depth of minimum velocity. Of all these paths, the axial path carries the greatest amount of acoustic energy and is the path of minimum transmission loss.

Because of these travel-time and intensity characteristics, an explosion on the channel axis is heard at a distance as a long-drawn-out signal. Its amplitude rises slowly to a climax and has an abrupt cutoff at the instant of arrival of sound traveling along the channel axis. Examples of these characteristic sofar-type signatures observed when both source and receiver are on the channel axis are shown in Fig. 6.10. This is a sequence of recordings, made with a hydrophone on the axis at Bermuda, of the sound of 4-lb explosive sofar signals dropped from an aircraft and detonating on the axis at different distances in a northeasterly direction from Bermuda. The distances vary from 43 miles (Drop 6) to 1,632 miles (Drop 19). Of particular note is the severe distortion accompanying deep-sound-channel transmission. The time stretching of the initially short explosive pulse in this series of signals amounts to 9.4 sec/1,000 miles of travel. However, when both source and receiver are *not* located on the axis of the channel, the signature no longer shows the smooth buildup to its climax, but consists of a series of discrete short pulses that can be accounted for by accurate travel time and ray computation on a digital computer (15).

Transmission-loss Model In a way similar to that already described for the mixed layer, the transmission loss in the deep sound channel can be conceived to be due to spherical spreading out to a transition range r_0, with cylindrical spreading beyond, plus a loss proportional to range. Because of the signal distortion in long-range propagation, the transmission loss is more properly described in terms of energy. This model gives

$$\mathrm{TL} = E_0 - E_r = 10 \log r_0 + 10 \log r + \alpha r \times 10^{-3}$$

where E_r = energy-flux-density level at r yd
E_0 = energy-flux-density level of source at 1 yd
α = attenuation coefficient, db/kyd

For explosives, values for E_0 are given in Figs. 4.15 and 4.16. Values of the attenuation coefficient α appropriate for deep-sound-channel transmission are those given previously in Fig. 5.4; indeed, most of such data were obtained with explosive sources received over long distances in the deep sound channel. The magnitude of r_0 appears to vary greatly from one determination to another. Values between 1,600 and 4,000 yd were obtained in one series of measurements (Chap. 5, Ref. 15); in another (Chap. 5, Ref. 16), r_0 was found to range from 40,000 to 150,000 yd. Determination of r_0 in field trials is particularly sensitive to calibration and measurement errors.

Ray Paths in the Deep Sound Channel In terms of the velocity profile, the upper and lower limits of the channel are defined by the two depths of equal maximum velocity in the profile between which a velocity minimum exists. In Fig. 6.11 these limits of the deep sound channel are the depths AA' in the velocity profile. Different ray paths from a source in the channel exist, depending on whether or not the channel extends to the sea surface or bottom. In Fig. 6.11*a*, the velocity at the surface and bottom is the same. All depths in the sea lie within the channel, and sound is propagated via paths that are either *refracted* (path 1) or *reflected* (with consequent losses) at the sea surface and bottom (path 2). In Fig. 6.11*b*, the upper bound of the deep sound channel lies at the sea surface. Here, in addition to the two types of paths 1 and 2, *refracted surface-reflected* (RSR) paths occur (path 3) involving losses intermediate between those of paths 1 and 2. In Fig. 6.11*c*, the channel is cut off by the sea bottom and *refracted bottom-reflected* (RBR) paths exist (path 4). The entirely refracted paths and the low transmission losses associated with these paths do not exist when the source or the receiver is outside the depth limits AA' of the channel.

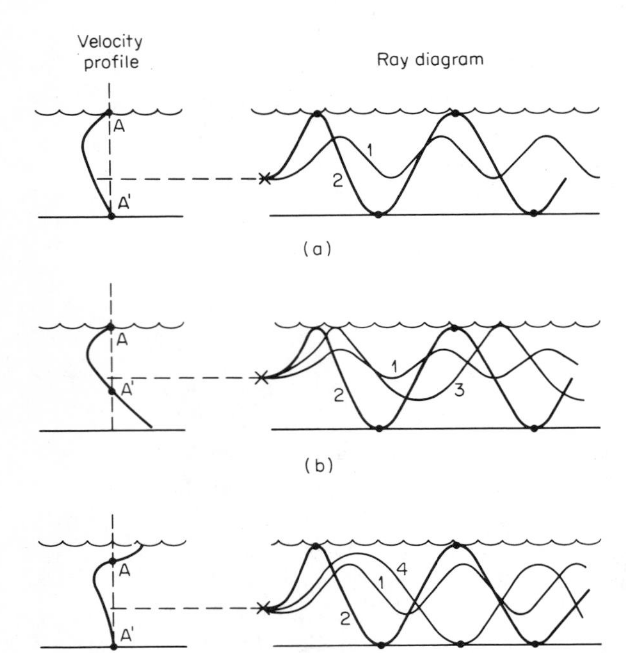

FIG. 6.11. Ray paths for a source in the deep sound channel. In (a) the channel extends between the sea surface and bottom; in (b) and (c) it is cut off by the sea surface and by the sea bottom, respectively.

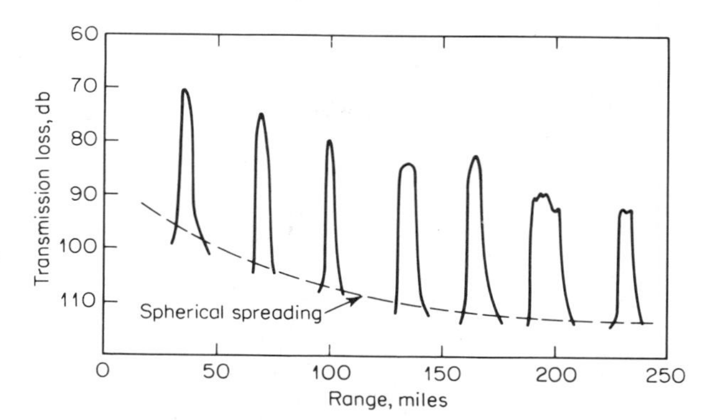

FIG. 6.12. Transmission loss versus range for a shallow source and receiver, showing convergence zones at 35-mile intervals. (*Ref. 16.*)

Caustics and Convergence Zones Within the channel, *caustics* and *convergence* zones are formed in which relatively high sound intensities occur. In a ray diagram, a caustic is the envelope formed by the intersection of adjacent rays. It is a region of partial focusing, in which wave theory is required to find the intensity—for example, the wave theory of Brekhovskikh (Chap. 5, Ref. 42, pp. 483–496). When a caustic intersects the sea surface, the region at or near the surface, within which high sound levels occur, is known as a *convergence zone.*

The characteristics of convergence zones for a shallow source were described in a paper by Hale (16). At intervals of 30 to 35 miles in temperate and tropical latitudes and under conditions when the deep sound channel extends up to the surface (Fig. 6.11*a* and *b*), successive narrow zones of high intensity are encountered as the range from a shallow source to a shallow receiver is increased. An extreme example of a series of convergence zones, as reported by Hale from observed data, is given in Fig. 6.12. The peaks here extend about 25 db above the level that would be found with spherical spreading and absorption; this increase has been called the *convergence gain.* Convergence gains of 10 to 15 db are more common. The width of the zones is of the order of 5 to 10 percent of the range, with the first zone about 3 miles in width. In between the zones, only the relatively low levels typical of bottom-reflected signals occur.

For a deep source or receiver, the single convergence zone found when both source and receiver are near the surface splits into two *half zones.* These convergence half zones are regions near the surface where deep-going rays are brought by refraction to shallow depths. The ranges to the half zones vary with the depth of the source or receiver. With increasing depth, the *inner* half zone moves *inward* in range and the *outer* half zone moves *outward.* This migration with depth may be seen by examination of the ray diagrams of Fig. 6.13 for source depths of

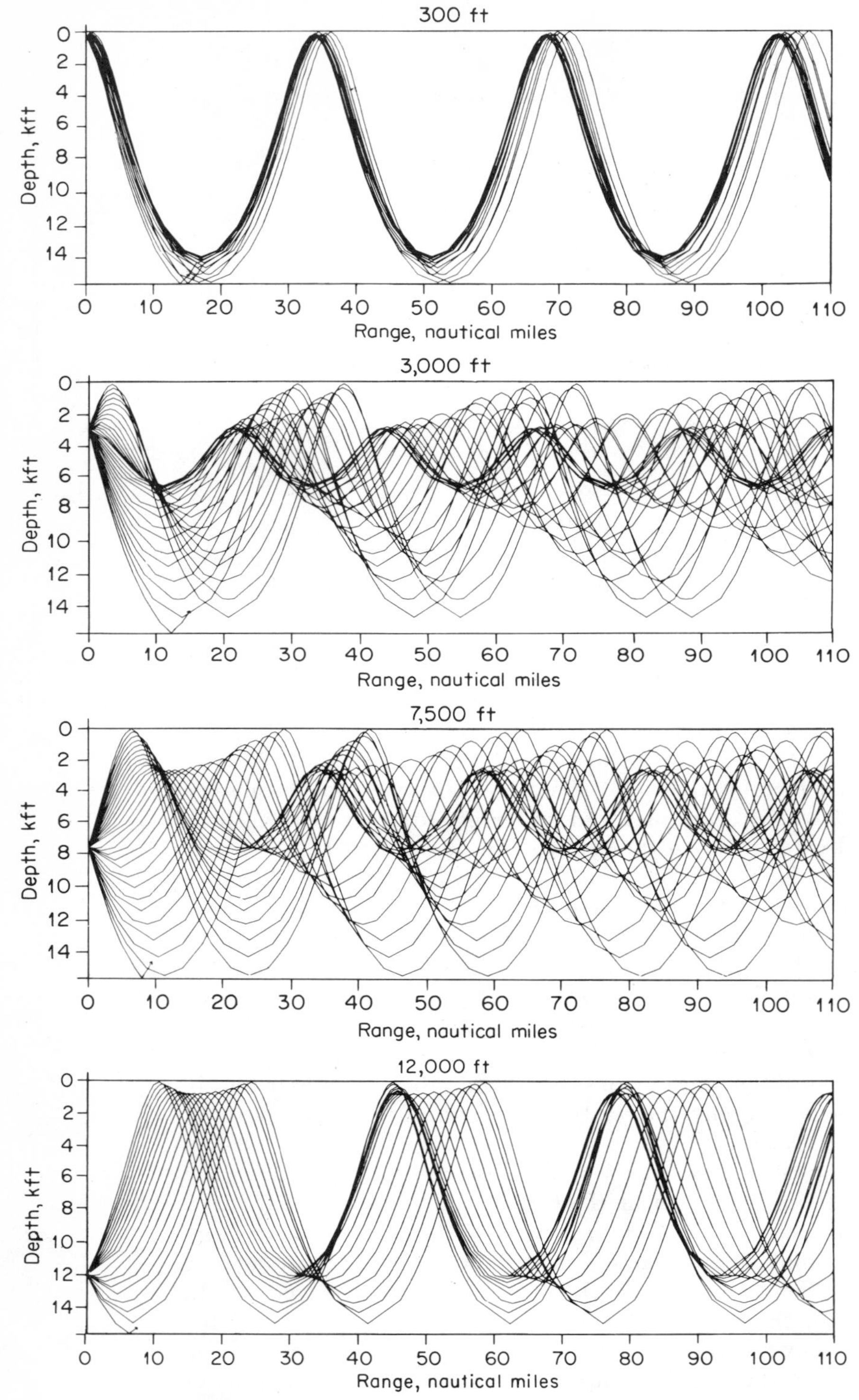

FIG. 6.13. Ray diagrams for a source at four source depths in the deep sea.

300, 3,000, 7,500, and 12,000 ft for a velocity profile having nearly equal sound velocity at the surface and at 16,000 ft. This zone splitting and migration with depth was verified in field trials using explosive sound signals and a deep hydrophone (17); half-zone widths of 2 to 5 miles were found, with convergence gains to 5 to 10 db in the half zones.

This behavior is a result of a changing pattern of caustics in the deep sound channel with changing source depth. Figure 6.14 shows diagrammatically the pattern of caustics for a sound source above, on, and below the axis. The arrows indicate the movement of the various caustics with increasing depth of the source.

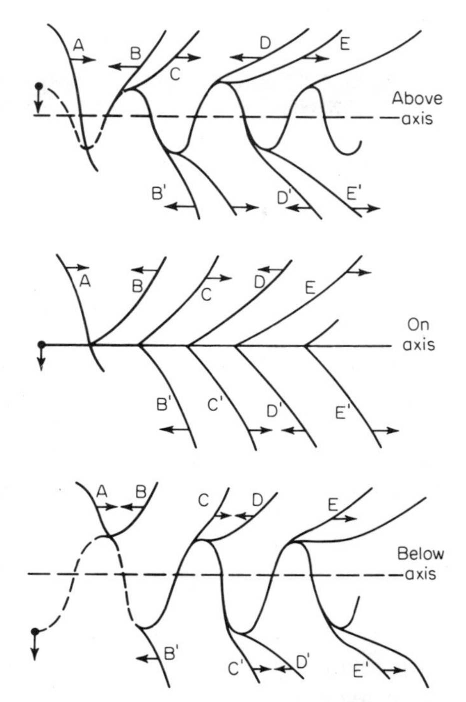

FIG. 6.14. Caustics in the deep sound channel. These should be compared with the ray diagrams of the preceding figure.

Arctic Propagation In the Arctic regions of the world, the axis of the deep sound channel lies at, or near, the ice-covered surface. Sound reaches long ranges by repeated reflections from the undersurface of the ice, to which it is repeatedly returned by upward refraction. This is shown in the ray diagram of Fig. 6.15 for a velocity profile typical of the Arctic.

Under an ice cover, the combination of upward refraction and downward reflection from the rough underice surface creates a number of unique propagation effects. One peculiarity of Arctic transmission is its similarity to a bandpass filter. Both high and low frequencies are rapidly attenuated, the former by reflection losses from the ice cover, the latter by the fact that very low frequencies are not effectively trapped in the channel. In the Arctic, the best propagation has been found to occur in the octave 15 to 30 Hz, approximately. Measured data show rapid attenuation with increasing frequency above 30 Hz, as well as a higher loss at 10 Hz (the lower limit of measurement thus far) than at 20 Hz. Within the passband, dispersion occurs. Wave packets of different frequency from an explosive shot are observed to have different group velocities, with low frequencies traveling faster than high. In

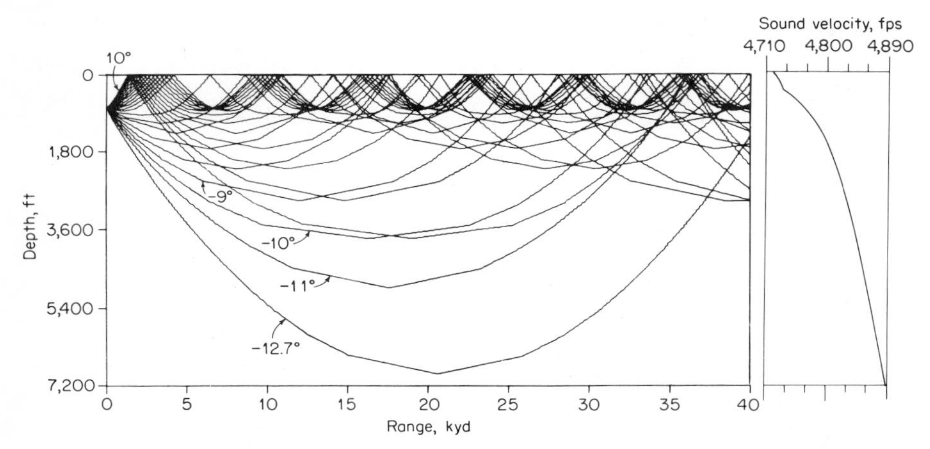

***FIG. 6.15.** Ray diagram for transmission in the Arctic. Ray interval 1°, with 12.7° added. The velocity profile is shown at the right.*

addition, the time stretching that is a dominant characteristic of sofar signals occurs as well, causing, in one experiment (18), an increase of duration of 15 sec/1,000 miles of travel. Because of these two effects—dispersion in both frequency and time—distant explosion in Arctic waters is received as a long-drawn-out pulse of several seconds' duration, with frequencies near the lower end of the Arctic passband appearing at the beginning of the signal and higher frequencies appearing at the end.

The transmission loss in the ice-covered Arctic has been measured by filtering, squaring, and integrating the signal from an explosive source recorded at different ranges. The somewhat discordant results of meas-

TABLE 6.1 Arctic Transmission Loss Measurements

Quantity	Marsh & Mellen	Buck & Greene
Water depth, ft	6,000–9,000 ±	11,100
Measurement distance, nautical miles	430	115–517
r_o, yd	5,000	14,000
α,20 kHz, db/kyd	0.7×10^{-2}	1.8×10^{-2}
40	2.8	2.6
80	6.3	3.6
200		5.0
500		8.5
1,000		11.5
Reference	18	19

urements reported by Marsh and Mellen (18) and Buck and Greene (19) are summarized in Table 6.1 in terms of the transition range r_0 and the attenuation coefficient α. The rapid increase of α with frequency is noteworthy. Other measurements, extending to higher frequencies at a range of 90 km (49 miles), have been reported by Milne (20). In a summary by Mellen and Marsh (21) of transmission-loss measurements in the Arctic, it is concluded that the available data can be related to the skip distance r_0' between successive encounters with the sea surface of the limiting ray that becomes horizontal at the bottom, the frequency f in hertz, and the rms roughness height h of the underice surface by the expression

$$\text{TL} = 10 \log r_0' + 10 \log r + 1.10 \times 10^{-3} f^{3/2} h^{8/5} N \sin\theta$$

where θ = angle with horizontal of limiting ray at source
N = number of reflections taken by limiting ray to reach range r

Hence, $r = Nr_0'$. In this expression the distances r, r_0', and h are in meters.

6.3 The Shallow-water Channel

Sound Channeling by Shallow Water In the present context, "shallow" means a water depth in which sound is propagated to a distance by repeated reflections from both surface and bottom. In water that is thus acoustically shallow, the acoustic characteristics of both surface and bottom are important determinants of the sound field. Although no precise definition can be given, shallow water, acoustically speaking, may be said to mean propagation to distances at least several times the water depth, under conditions where both boundaries have an effect on transmission. In a geographic sense, shallow water refers to the inland waters of bays and harbors and to coastal waters less than 100 fathoms deep, which often extend outward to the edge of the continental shelf. Water that is "shallow" forms a sound channel between the surface and bottom, in which, as in other kinds of channels, sound is trapped between the upper and lower channel boundaries.

Of the many aspects of sound propagation in the sea, the propagation of sound in shallow water has received perhaps the greatest amount of attention. The subject is characterized by both theoretical complexity and difficulty in formulating a mathematical description of the acoustic conditions existing at the boundaries. Unlike the propagation of sound in deep water, the propagation of sound in shallow water is the subject of an abundant literature.

The initial studies of the subject—both theoretical and with field observations—were made by Ide, Post, and Fry (22), Roe (23), and

Pekeris (24). Good theoretical summaries may be found in books by Brekhovskikh (Chap. 5, Ref. 42) and Officer (Chap. 5, Ref. 41).

Two theoretical approaches exist for describing the shallow-water sound field. Both involve functions that are solutions of the wave equation and are added up with the correct coefficients to satisfy the acoustic conditions at the boundaries and at the source.

Ray Theory Ray theory represents the sound field as a sum of *ray* contributions, each ray emanating from the source or its *image* in the surface and the bottom. In Fig. 6.16, a source is located at a point O in a shallow-water channel, and a receiver is at P. To satisfy the boundary condition at the surface (zero pressure at the pressure-release surface), an image O_{10} of the source in the surface is added, with a source strength and phase shift (180°) to account for the reflection coefficient of the sea surface. To match the reflectivity of the sea bottom, a pair of images O_{01} and O_{11} are added in the bottom. But now the boundary condition at the sea surface is no longer satisfied, and to correct for the presence of O_{01} and O_{11}, their images O_{11} and O_{21} in the sea surface must be added. But now the bottom condition is upset, and it is necessary to add still another pair of images in the bottom. By this back-and-forth process an infinite string of images is built up, with higher-order images tending to become insignificant because of weakening by repeated reflec-

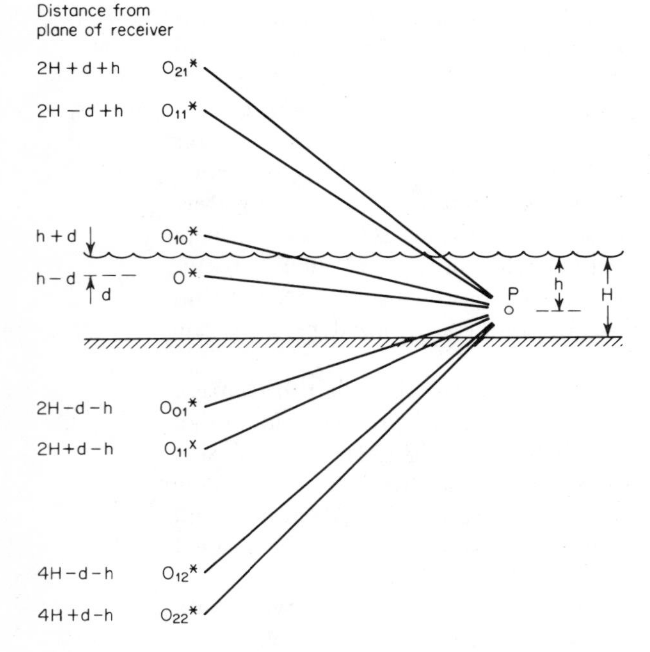

FIG. 6.16. Images in shallow-water transmission.

tions and by greater distance. At a distant point P the sound pressure is the vector sum of the pressure contributions of the images, and may be written

$$P = p_0 \sum_{m=0}^{\infty} \frac{R_m e^{ikr_m}}{r_m}$$

where p_0 = source pressure (time factor omitted)
r_m = distance of mth image from P
R_m = amplitude reflection coefficient appropriate for mth image
k = wave number equal to $2\pi/\lambda$

The term e^{ikr_m} is the phase factor for waves of wavelength λ propagating over the distance r_m. The reflection coefficient R_m varies with angle and will, in general, be complex, with a real and imaginary part, because of phase shifts on reflection.

Normal-mode Theory An alternate expression of the shallow-water sound field can be given in terms of *normal modes.* These are complicated functions, each representing a wave traveling outward from the source with an amplitude that is a function of the source and receiver depths. Various expressions for the normal-mode solution of the wave equation are given by different authors. Brekhovskikh (Chap. 5, Ref. 42, eq. 26.21) gives the solution for the case of a pressure-release surface and a perfectly rigid bottom as:

$$\Psi = \frac{2\pi i}{h} \sum_{l=0}^{\infty} \cosh b_l z_0 \cosh b_l z H_0^{(1)}(x_l r)$$

$$b_l = \frac{i(l + \frac{1}{2})\pi}{h}$$

$$hx_l = h(b_l^2 + k^2)^{\frac{1}{2}} = [(kh)^2 - (l + \tfrac{1}{2})^2\pi^2]^{\frac{1}{2}}$$

$$l = 0, 1, 2, \ldots$$

$$k = \frac{2\pi}{\lambda}$$

where h = water depth
z_0 = source depth
z = receiver depth
$H_0^{(1)}$ = Hankel function of the first kind

Modes propagating without attenuation are those for which the argument $(x_l r)$ of the Hankel function is real; these are the modes for which $kh > \pi/2$ or $h > \lambda/4$, that is, those for which the water depth is greater than one-quarter wavelength. The frequency corresponding to $h = \lambda/4$ is termed the cutoff frequency; frequencies lower than the cutoff frequency are propagated in the channel only with attenuation and are not effec-

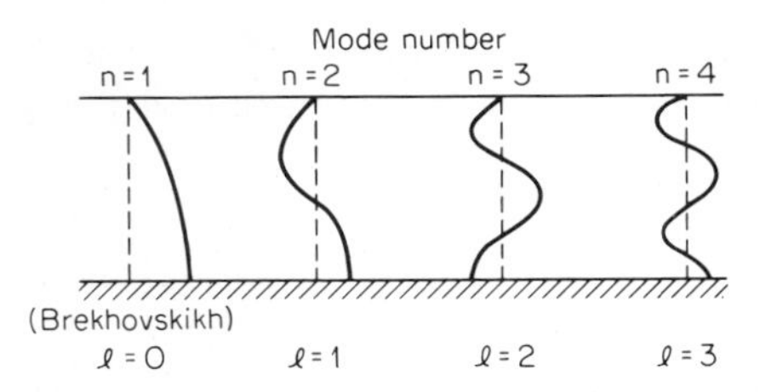

FIG. 6.17. Sound pressure versus depth for the first four modes for a pressure-release (soft) surface and a rigid (hard) bottom.

tively trapped in the duct. When the lower boundary is not rigid but is the boundary of a fluid medium of sound velocity c_2, the cutoff frequency becomes

$$f_0 = \frac{c_1}{4h} \sqrt{\frac{1}{1 - (c_1/c_2)^2}}$$

where c_1 is the velocity in the upper water layer of thickness h. This reduces to the former condition for $c_2 \gg c_1$. The variation of pressure with depth of the first four modes, for the conditions of a free (pressure-release) surface and a rigid bottom, is shown in Fig. 6.17.

Each mode can be conceived to correspond to a pair of plane waves incident upon the boundaries at an angle β and propagating in a zigzag fashion by successive reflections. For the first two modes these equivalent plane waves are shown in Fig. 6.18, where the lines denote the pressure nodes (zero pressure) and the + and − signs give the polarity of the pressure between the pressure nodes. The relation between the grazing angle β and the mode number n for a pressure-release surface and a rigid bottom is

$$\sin \beta = \left(n - \frac{1}{2}\right) \frac{\lambda}{d}$$

The higher modes thus correspond to steeper angles. Modes for which the grazing angle of the equivalent plane wave is less than the critical angle at the bottom will be propagated with relatively low attenuation.

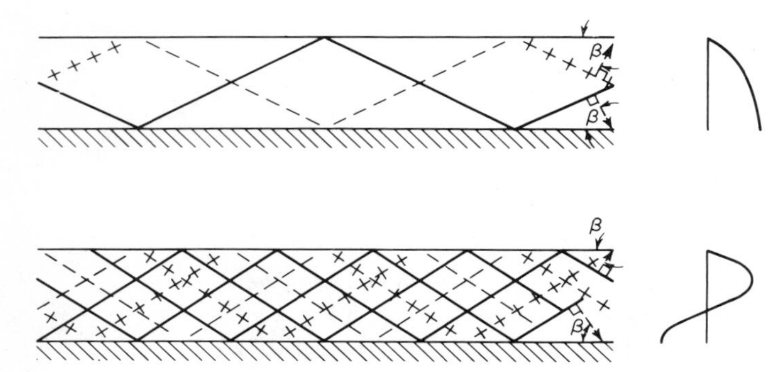

FIG. 6.18. Plane-wave equivalents of the first and second modes in a duct with a pressure-release surface and a rigid bottom.

Because the angle varies with the mode number, different normal modes will propagate with different phase velocities.

Comparison of Ray and Mode Theories Ray theory is more convenient to use at *short* ranges, where the higher-order images rapidly die out because of reflection losses and an increasingly great distance from the field point. Accordingly, only a few images need ordinarily to be summed at short ranges. Normal-mode theory is more appropriate at *long* ranges because of the greater attenuation with distance of the higher-order modes; only a few modes need to be taken to describe the transmission. A "crossover" range between the regions of convenient usefulness of the two theories is given by the expression (25)

$$r = \frac{H^2}{\lambda}$$

where H = water depth
λ = wavelength

With modern digital-computer techniques, however, the number of summations required is no longer a serious restriction, and both theories can be successfully used to describe the sound field at long ranges. Examples are the work of Mackenzie (26) on ray summation (with phase neglected) to provide a basis of comparison between theory and field data at frequencies of 200, 600, and 1,000 Hz and to ranges of about 30 miles, and of Bucker and Morris (27) on normal-mode theory, where some 152 modes were summed to account for the transmission of $1\frac{1}{2}$-kHz sound out to 25 miles.

Modifications to the Simple Theories Complications arise when the simple two-fluid theories are modified to allow for more realistic conditions of the surface and bottom and of the water medium. Normal-mode theory for a multilayered lossless fluid bottom has been given in a series of papers by Tolstoy (28–30) and by Clay (31). Ray theory for a multilayered lossless fluid bottom has been given by McLeroy (32). The effect of attenuation in the bottom is considered by Kornhauser and Raney (33) and by Eby, Williams, Ryan, and Tamarkin (34), and the effect of refraction in the fluid layer on the normal modes is given by Williams (35). Propagation over an elastic bottom supporting both shear and compressional waves is considered by Spitznogle and McLeroy (36) and Victor, Spitznogle, and McLeroy (37), and the effects of a variable water depth over the length of the propagation path—such as a sloping bottom—has been dealt with by Weston (38). Measurements in shallow water beneath an ice-covered surface have been reported by Milne (39), and the effect of surface roughness on shallow-water propagation has been dealt with by Clay (40). In most of the papers just cited, agreement is claimed between theory and data measured in the field or in laboratory tanks. All in all, the transmission of sound in a liquid layer overlying an elastic, absorbing layered bottom is a remarkably complex subject, as was dramatically demonstrated by the sound-field patterns obtained by Wood (41) in tanks of water.

A further complication concerns the dispersion of sound in shallow water. Because different modes propagate with different phase velocities, peculiar interference patterns are observed from a broadband source. The signal from an explosive source in shallow water is an extended pulse containing different frequency components at different instants during its duration interval. The unique dispersive effects peculiar to shallow water were pointed out long ago in the classic paper by Pekeris (24) and have been more recently observed under a shallow-water ice cover in the Arctic (42).

Transmission Loss in Shallow Water The transmission loss for sound propagation in shallow water depends upon many natural variables of the sea surface, water medium, and bottom. Because of its sensitivity to these variables, the transmission loss in shallow water is only approximately predictable in the absence of specific knowledge of variables—especially, the sound velocity and density structure of the bottom. For rough prediction purposes, the semiempirical expressions published by Marsh and Schulkin (43) are particularly useful. These are based on some 100,000 measurements in shallow water in the frequency range 0.1 to 10 kHz. Three equations are used in different ranges of a parameter H defined by $H = [\frac{1}{8}(D + L)]^{1/2}$, where D is the water depth in feet, L is the layer depth in feet, and H is in kiloyards. At short ranges, such that the range r is less than H, the transmission loss is in decibels,

$$\mathrm{TL} = 20 \log r + \alpha r + 60 - k_L$$

where r is in kiloyards, α is the absorption coefficient of seawater in decibels per kiloyard (Sec. 5.3) and k_L is a "near-field anomaly" dependent on sea state and bottom type, as given in Table 6.2*a*. At intermediate ranges, such that $H \leq r \leq 8H$, the transmission loss is

$$\mathrm{TL} = 15 \log r + \alpha r + a_T \left(\frac{r}{H} - 1\right) + 5 \log H + 60 - k_L$$

where r and H are in kiloyards and a_T is a shallow-water attenuation coefficient listed in Table 6.2*b*. For long ranges, such that $r > 8H$,

$$\mathrm{TL} = 10 \log r + \alpha r + a_T \left(\frac{r}{H} - 1\right) + 10 \log H + 64.5 - k_L$$

These equations verify expected transition between spherical spreading near the source and cylindrical spreading at great distances. Table 6.2*c* gives the probable errors of the transmission loss computed by these expressions relative to the individual observations on which they are based.

At higher frequencies, an extensive series of transmission measurements were made during World War II (Chap. 5, Ref. 39). Some 617 runs were made at many locations off the East and Gulf Coasts of the United

TABLE 6.2 Factors and Error Estimate for Shallow-water Transmission Loss*

(*a*) Near-field anomaly k_L, db

Sea state	0		1		2		3		4		5	
f, kHz	Sand	Mud	Sand	Mud	Sand	Mud	Sand	Mud	Sand	Mud	Sand	Mud
0.1	7.0	6.2	7.0	6.2	7.0	6.2	7.0	6.2	7.0	6.2	7.0	6.2
0.2	6.2	6.1	6.2	6.1	6.2	6.1	6.2	6.1	6.2	6.0	6.2	6.0
0.4	6.1	5.8	6.1	5.8	6.1	5.8	6.1	5.8	6.1	5.8	4.7	4.5
0.8	6.0	5.7	6.0	5.6	5.9	5.6	5.3	5.0	4.3	3.9	3.9	3.6
1.0	6.0	5.6	5.9	5.5	5.7	5.3	4.6	4.2	4.1	3.7	3.8	3.4
2.0	5.8	5.4	5.3	4.9	4.2	3.8	3.8	3.4	3.5	3.1	3.1	2.8
4.0	5.7	5.1	3.9	3.5	3.6	3.1	3.2	2.8	2.9	2.4	2.6	2.2
8.0	4.3	3.8	3.3	2.8	2.9	2.5	2.6	2.2	2.3	1.9	2.1	1.7
10.0	3.9	3.4	3.1	2.6	2.7	2.2	2.4	2.0	2.2	1.7	2.0	1.6

(*b*) Attenuation factor a_T, db

Sea state	0		1		2		3		4		5	
f, kHz	Sand	Mud	Sand	Mud	Sand	Mud	Sand	Mud	Sand	Mud	Sand	Mud
0.1	1.0	1.3	1.0	1.3	1.0	1.3	1.0	1.3	1.0	1.3	1.0	1.3
0.2	1.3	1.7	1.3	1.7	1.3	1.7	1.3	1.7	1.3	1.7	1.4	1.7
0.4	1.6	2.2	1.6	2.2	1.6	2.2	1.6	2.2	1.7	2.4	2.2	3.0
0.8	1.8	2.5	1.8	2.5	1.9	2.6	2.2	3.0	2.4	3.8	2.9	4.0
1.0	1.8	2.7	1.9	2.7	2.1	2.9	2.6	3.7	2.9	4.1	3.1	4.3
2.0	2.0	3.0	2.4	3.5	3.1	4.4	3.3	4.7	3.5	5.0	3.7	5.2
4.0	2.3	3.6	3.5	5.2	3.7	5.5	3.9	5.8	4.1	6.2	4.3	6.4
8.0	3.6	5.3	4.3	6.3	4.5	6.7	4.7	6.9	5.0	7.3	5.1	7.5
10.0	4.0	5.9	4.5	6.8	4.8	7.2	5.0	7.5	5.2	7.8	5.3	8.0

(*c*) Probable error (semi-inter quartile range) of computed loss, db

Range, kyd	Frequency, Hz 112	446	1,120	2,820
3	2	4	4	4
9	2	4	5	6
30	4	9	11	11
60	5	9	11	12
90	6	9	11	12

[* From Marsh and Schulkin (43).]

States. During these runs, the sound level from a CW source at 12 and 24 kHz was measured out to distances between 2 and 10 miles, depending on the quality of the propagation. Examples of good and poor transmission are shown in Fig. 6.19, giving the measured excess of loss over spherical spreading for some typical conditions. The "good" condition corresponds to well-mixed, nearly isothermal water overlying a sand bottom; the "poor" condition, to water of negative temperature gradient (downward refraction) over a mud bottom. At these frequencies it was found that the transmission improved with (1) decreasing frequency, (2) smaller negative temperature gradient, (3) increasing water depth (between 60 and 600 ft), and (4) increasing depth of the receiving

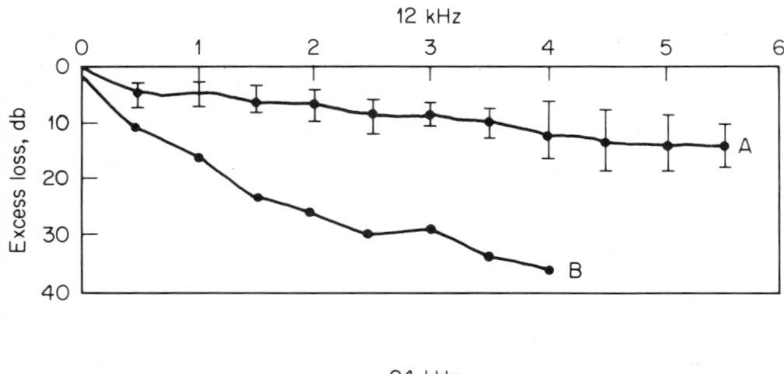

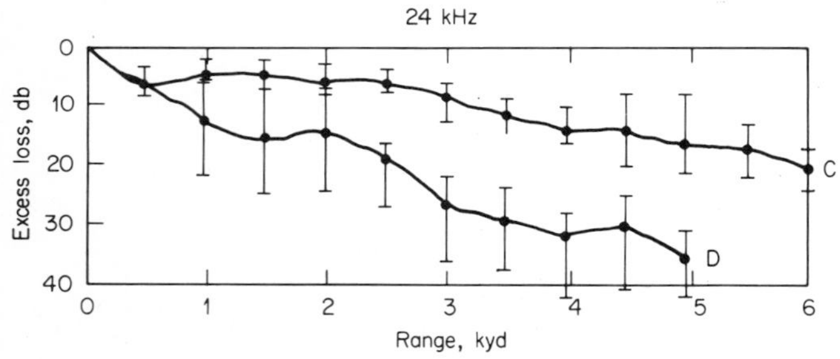

Curve	Bottom	Temperature gradient	Hydrophone depth	Windforce (Beaufort scale)	No. runs	No. areas
A	Sand	None or small	Shallow	1–3	28	14
B	Mud	Negative	Middepth	0–2	3	2
C	Sand	None or small	Shallow	1	18	7
D	Mud	Negative	Shallow	0–2	15	4

FIG. 6.19. Shallow-water transmission at 12 and 24 kHz. Vertical lines show ±1 standard deviation of individual measurements from the mean. (Ref. 39, Chap. 5.)

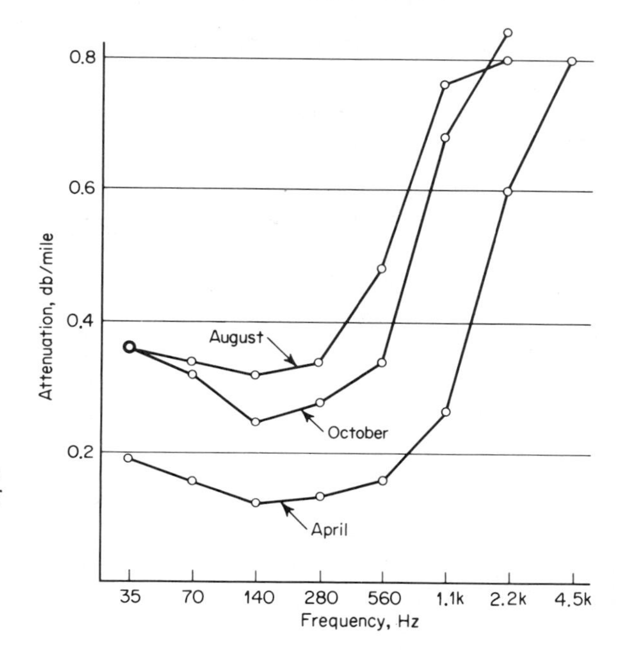

FIG. 6.20. Attenuation coefficient in the North Sea for different months of the year. (*Ref. 25.*)

hydrophone in a negative gradient; the transmission was best over sand bottoms and poorest over mud bottoms. More recently, the sensitivity of shallow-water transmission to temperature gradients has been noted by Macpherson and Fothergill (44) in measurements off the Scotian shelf at different seasons; good transmission was found in winter, when the temperature—and therefore the sound velocity—increased with depth; poor transmission occurred in summer when downward refraction existed. Similar effects have been found by Weston (25) in the North Sea. For the central North Sea, Fig. 6.20 shows the attenuation coefficient in decibels per mile, as measured at different frequencies and at different times of year in an area of exceptionally constant water depth (40 fathoms).

6.4 Fluctuation of Transmitted Sound

Causes of Fluctuation Observations show that a signal received by a distant hydrophone from a steady source of sound constantly fluctuates with time. When the source or receiver happens to be mounted on a ship, much of the fluctuation is due to the irregular motion of the platform, especially when the geometry is such that interference by the nearby sea surface can occur. But amplitude fluctuation occurs even when both ends of the transmission path are held fixed, as between two

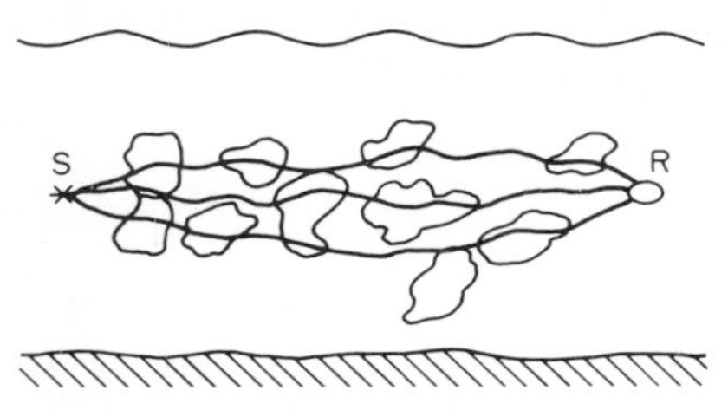

FIG. 6.21. Rays between a source and a receiver in a medium containing blobs of inhomogeneity.

fixed points on the seabed. This fluctuation is due to the existence of random inhomogeneities in sound velocity in the body of the sea and to the fact that these inhomogeneities are in motion relative to the source and receiver. Figure 6.21 shows a segment of an inhomogeneous ocean between a source S and a receiver R. Here the inhomogeneities are crudely visualized as irregular blobs of water of different sound velocity or of different index of refraction. The receiver at R will receive sound from S via a number of refracted and scattered paths, such as the three paths shown in the figure. If the transmitting medium were stationary so that everywhere the blobs of inhomogeneity were "frozen" in place, no temporal fluctuation would occur. But because of turbulence and water currents, a time-varying signal due to interference effects among the various propagation paths is observed at R. This fluctuation causes sonar pings and echoes to have a ping-to-ping and an echo-to-echo variability, even in the absence of interfering propagation paths reflected from the sea surface or bottom.

Thermal Microstructure of the Sea When a sensitive thermometer is towed through the sea at a constant depth, small temperature fluctuations amounting to a few thousandths to a few tenths of a degree are observed. This temperature microstructure has a thermal and spatial size that can be quantitatively described in terms of a mean-square temperature deviation and a "patch size." The latter quantity is the autocorrelation distance of the microstructure, within which the temperature variations maintain coherence. Two correlation functions are useful for theoretical investigations: one exponential, in which the autocorrelation function of temperature $\rho(d)$ is expressed in terms of distance d by

$$\rho(d) = e^{-d/a}$$

the other Gaussian,

$$\rho(d) = e^{-(d/a)^2}$$

in which a is the patch size. Of the two, the latter is the more physically reasonable since the former involves discontinuous changes in the temperature fluctuations. Figure 6.22 shows sample records of the temperature fluctuations in the mixed layer and in the thermocline below recorded by Urick and Searfoss (45) by means of a sensitive submarine-mounted

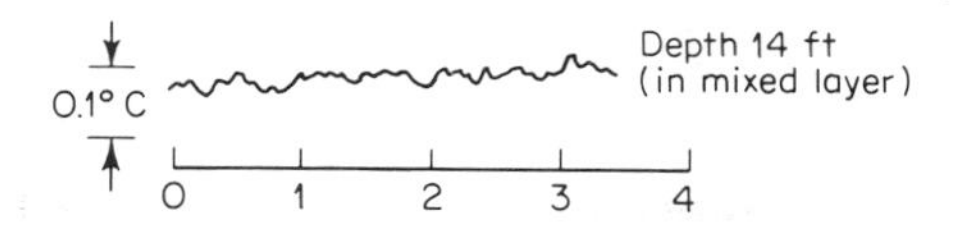

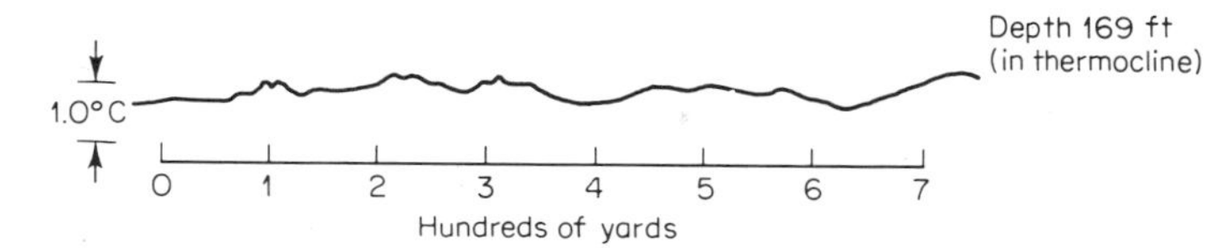

FIG. 6.22. Temperature fluctuations observed with a sensitive thermometer aboard a moving submarine at two different depths. (Ref. 45.)

thermopile carried through the sea by the submarine's motion. The larger and longer fluctuations in the thermocline are probably associated with internal waves. Other observations of microstructure were made by Liebermann (46) using the same method, but with a recording thermometer of more rapid response.

Such data yield values for the patch size a and for a quantity μ describing the relative change in sound velocity, as measured by temperature, along the path on which measurements were made. The rms deviation μ of the relative sound velocity is defined by

$$\mu = \left[\overline{\left(\frac{\Delta c}{c}\right)^2}\right]^{1/2}$$

where $\overline{(\Delta c)^2}$ = mean-square deviation of velocity of sound
c = mean value of velocity of sound

Urick and Searfoss found $\mu^2 = 8 \times 10^{-10}$ and $a = 500$ cm at a depth of about 20 ft in the mixed layer off Key West, Florida; Liebermann observed $\mu^2 = 5 \times 10^{-9}$ and $a = 60$ cm at a depth of 50 m off the California coast.

Fluctuation Characteristics The fluctuation in transmitted sound produced by the microstructure is commonly expressed as the coefficient of variation of the amplitudes of a series of short pulses. If P is the absolute magnitude of the acoustic pressure of a transmitted pulse, then the coefficient of variation V is the fractional standard deviation of the pressure amplitude, defined as

$$V = \left[\frac{\overline{(P - \bar{P})^2}}{(\bar{P})^2}\right]^{1/2} = \left[\frac{\overline{P^2} - (\bar{P})^2}{(\bar{P})^2}\right]^{1/2}$$

where the bars represent averages of a large number of short pulses, or of a large number of readings of the envelope of a continuous signal.

Theoretical studies by Mintzer (47) and others (46, 48–50) have related the fluctuation described by the quantity V to the microstructure as described by the quantities a and μ. At short ranges and high frequencies, such that $r \ll ka^2$, where $k = 2\pi/\lambda = 2\pi f/c$ and r is the range, *ray theory* yields

$$V = \left(\frac{4}{15}\pi^{1/2}\frac{\mu^2 r^3}{a^3}\right)^{1/2}$$

At such short ranges, the dominant physical process producing the fluctuations may be considered to be the focusing and defocusing of sound by the patches of inhomogeneity. On the other hand, at long ranges and low frequencies such that $r \gg ka^2$, *wave theory* gives the result

$$V = \left(\frac{\sqrt{\pi}}{2}\mu^2 k^2 ar\right)^{1/2}$$

At long ranges, forward scattering by the inhomogeneities is the principal cause of the fluctuation. Of particular note is the dependence of V on $r^{3/2}$ at short ranges and on $r^{1/2}$ at long ranges, a dependence that has been repeatedly verified experimentally, both for underwater sound (51–53) and for acoustic and electromagnetic propagation in the atmosphere.

Examples of the observed fluctuation as a function of range of short pulses transmitted through the sea are shown in Fig. 6.23. These are based on averages of the amplitude of a series of short pulses at different ranges. Figure 6.23*a* shows the average and the spread of the fluctuations found by Sheehy (54) at 24 kHz, Fig. 6.23*b* shows similar data reported by Whitmarsh, Skudryzk, and Urick (55) from a different series

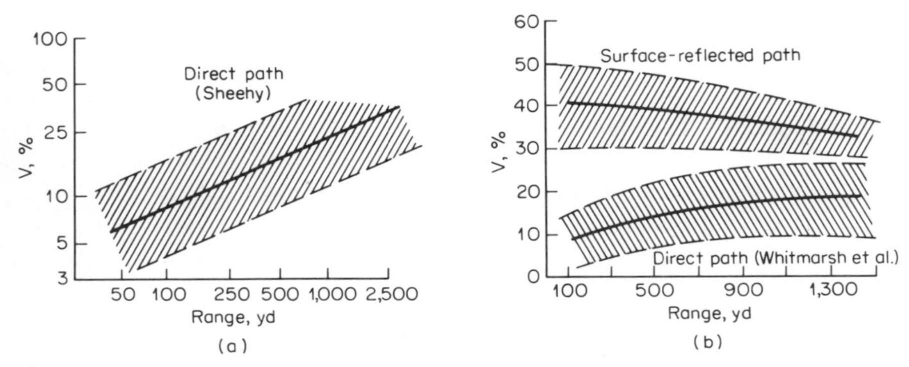

FIG. 6.23. Standard deviation of the pressure amplitude V as a function of range, observed in two different field experiments. [(a) Ref. 54. (b) Ref. 55.] Shaded area shows limits of determinations from different groups of pulses.

of observations at 25 kHz. In both cases, the coefficient of variation V for transmission through the body of the sea was found to follow the $r^{1/2}$ relationship predicted by theory for long ranges.

Fluctuation of Surface-reflected Sound The fluctuation of pulses reflected from the sea surface obeys a different law with range, as can be seen from Fig. 6.23*b*. At short ranges, the fluctuation of the reflection from the rough sea surface is greater, but tends to decrease with increasing range as the grazing angle on the surface becomes less and as the sea surface begins to act more and more like a perfect mirror. At long ranges, the fluctuation of sound traveling along the two paths tends to become the same and tends to be determined by the microstructure in the body of the sea. The fluctuation of surface-reflected sound in the sea has been the subject of a number of more recent investigations (56–59) and has been briefly described above (Sec. 5.7) in connection with sound reflection from the sea surface.

Although surface-reflection interference, when it occurs, is likely to be the dominant source of amplitude and phase fluctuation, recent investigations have shown remarkable stability in transmission through the sea, even over long propagation paths. Steinberg and Birdsall (60), for example, have reported transmission measurements of 420-Hz sound over 40 miles across the Straits of Florida. A comparison of the received signal with the transmitted signal showed phase variations of less than 100° over intervals of ½ to 1 hr; intervals of 1 to 4 hr occurred in which the phase varied less than 360°. The reasons for this unexpected phase stability in the sea are not entirely clear. The finding lends promise to the use of large arrays and long processing times for detection and communication over long distances.

Other Effects Other important aspects of the fluctuation of transmitted sound are the variation in apparent bearing of a distant target and the correlation of received signals between separated receivers (61, 62). The latter effect is important in determining the array gain (Sec. 3.8) of an array of hydrophones in the sea. These and other manifestations of transmission fluctuation are lucidly described in English translation books by Chernov (63) and Tatarski (64) and in a nonmathematical survey paper by Brekhovskikh (65).

Fluctuation in Shallow Water In shallow water, extreme intensity fluctuations are observed because of propagation by repeated reflections and scattering from the surface and bottom. For example, Mackenzie (66) observed variations of as much as 50 db in the intensity of a CW tone transmitted over ranges of a few kiloyards in shallow water. In addition, frequency spreading of an initially pure tone—amounting to a few tenths of a hertz at 1 kHz—was observed to occur, presumably because of the repeated encounters with the constantly changing sea surface.

6.5 Deep Sea Paths and Losses: A Summary

Propagation Paths in the Deep Sea Figure 6.24 illustrates the different types of propagation paths that exist between a source and a receiver in the deep sea. At short ranges *A*, there is a nearly straight-line path between the two, and the transmission loss is determined by spherical spreading (Sec. 5.2), plus a loss due to absorption if the frequency is high enough (Sec. 5.3), and modified by interference effects of the surface reflection for a near-surface source and receiver (Sec. 5.7).

At longer ranges, propagation occurs in the mixed-layer channel *B*, involving repeated surface reflections and leakage out of the channel to a receiver below (Sec. 6.1). Propagation may occur also via "bottom-bounce" reflected paths *C*, which suffer a reflection loss at the sea bottom and along which the transmission loss is determined by spherical spreading plus absorption and by the bottom-reflection loss (Sec. 5.8). At still greater ranges, convergence-zone transmission *D* takes place where, within a zone a few miles wide, convergence gains between 5 and 20 db occur, depending on depth (Sec. 6.2). Beyond the first pair of convergence half zones, the transmission between a shallow source and a shallow receiver involves multiple reflections from the bottom *E*. In the deep

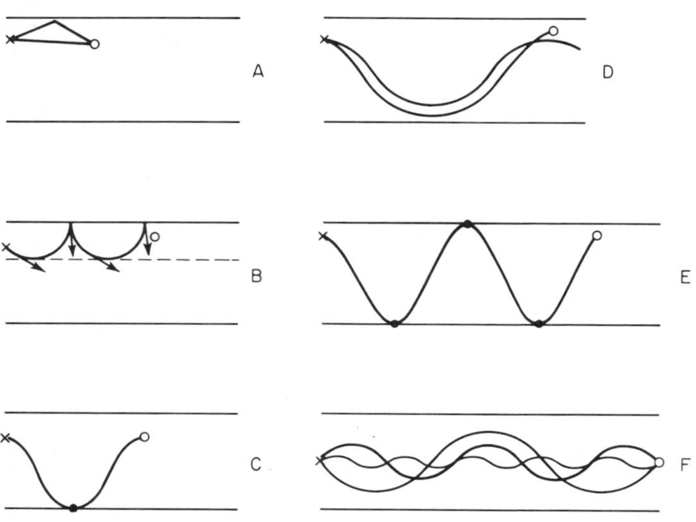

FIG. 6.24. Propagation paths between a source and a receiver in deep water.

sound channel, long-range propagation occurs over internally refracted paths F and is especially good when both source and receiver lie on the channel axis (Sec. 6.2).

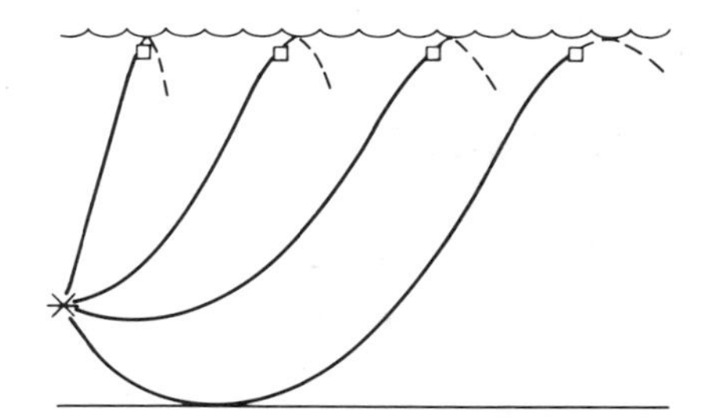

FIG. 6.25. "Reliable" acoustic paths from a deep source to a shallow receiver in the deep sea.

The vagaries of transmission between a shallow source and a shallow receiver may be avoided by placing the source or the receiver deep in the sea. Propagation to moderate ranges then can take place via the so-called *reliable acoustic paths*, a number of which are shown in Fig. 6.25. Such paths are "reliable" because they are sensitive neither to near-surface effects nor to the varying losses on reflection that characterize "bottom-bounce" propagation (C, Fig. 6.24).

Transmission Loss as a Function of Range As an illustration of the transmission losses associated with some of these paths, Fig. 6.26 shows computed curves for conditions of (1) frequency 2 kHz, (2) source depth

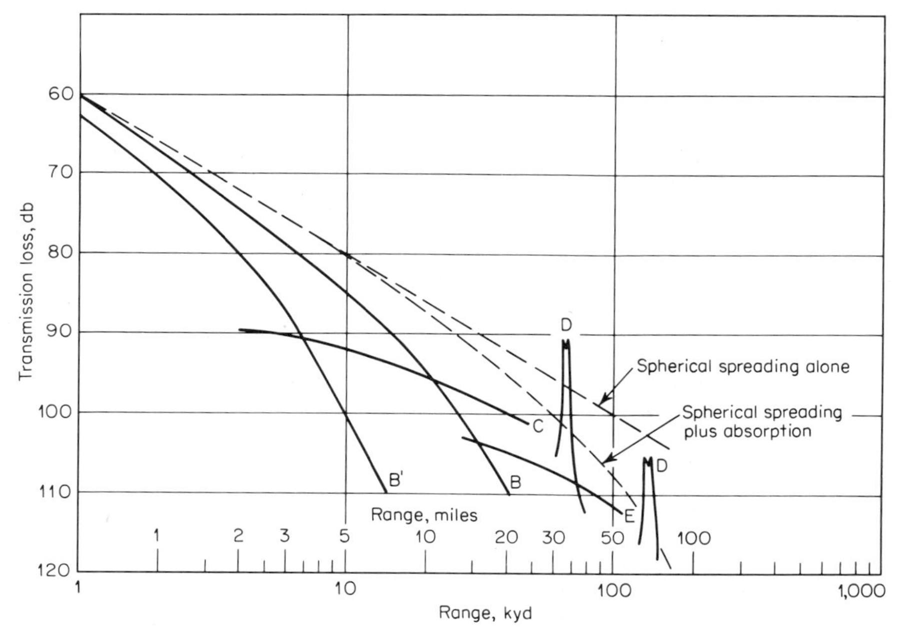

FIG. 6.26. Transmission loss versus range for different propagation paths. The letters on each curve correspond to the paths of Fig. 6.24.

50 ft, (3) receiver depth 500 ft, (4) mixed-layer thickness 100 ft, and (5) water depth 15,000 ft. Curves B and B' are taken from Fig. 6.6 for receiver depths of 50 and 500 ft, respectively. Curve C is the loss by bottom reflection obtained by adding (1) the loss due to spherical spreading over the slant range to the bottom, (2) a loss due to absorption and equal to 0.08 db/kyd, and (3) the reflection loss for the appropriate grazing angle from Fig. 5.27. The peaks marked D illustrate the expected losses to the first and second convergence zones at 65 and 130 kyd, respectively, and E is the transmission loss via the second bottom reflection. The dashed curves illustrate the degree to which spherical spreading and spherical spreading plus absorption approximate the computed losses over the various paths.

Concluding Remarks The example just given illustrates the fact that, almost always, there is more than one propagation path between a source and receiver in the sea. Often one path will be dominant, and the transmission loss corresponding to it will be a minimum compared with other possible paths. But under some conditions, multipath propagation occurs and more than one acoustic path of comparable loss exists between the two points, depending on the source and receiver depths and on the acoustic conditions in the body of the sea and in its boundaries.

Because of the complex nature of sound propagation in the sea, the selection of a reasonably satisfactory value of transmission loss for use in a practical problem is often a vexatious problem in itself. Indeed, of all the sonar parameters, transmission loss presents the greatest difficulty from a quantitative standpoint, since the choice of a numerical value must rest on a clear understanding of the transmission characteristics of the medium. Furthermore, other characteristics, such as the distortion, fluctuation, and coherence of signals transmitted through the sea, are often of comparable importance to the transmission loss itself. In short, the propagation vagaries of sound in the sea are such that, in spite of the extent to which they have been studied, they still present formidable problems, both to the engineer interested in a sonar application, and to the research scientist.

REFERENCES

1. Steinberger, R. L.: "Underwater Sound Investigation of Water Conditions, Guantanamo Bay Area, Feb. 1937," Navy Yard Sound Laboratory, Washington, D.C., May 3, 1937.
2. Urick, R. J.: Sound Transmission Measurements in the Long Island-Bermuda Region, Summer 1949, *NRL Rept.* 3630, 1950.
3. Kerr, D. E. (ed.): "Propagation of Short Radio Waves," M.I.T. Radiation Laboratory Series, vol. 13, McGraw-Hill Book Company, New York, 1951.

4. Marsh, H. W., M. Schulkin, and S. G. Kneale: Scattering of Underwater Sound by the Sea Surface, *J. Acoust. Soc. Am.*, **33**:334 (1961). Also, Marsh, H. W.: Sound Reflection and Scattering from the Sea Surface, *J. Acoust. Soc. Am.*, **35**:240 (1963).
5. Brekhovskikh, L. M., and I. D. Ivanov: Concerning One Type of Attenuation of Waves Propagating in Inhomogeneously Stratified Media, *Soviet Phys. Acoust.*, **1**:23 (1955).
6. Condron, T. P., P. M. Onyx, and K. R. Dickson: Contours of Propagation Loss and Plots of Propagation Loss vs Range for Standard Conditions at 2, 5, and 8 kc, *U.S. Navy Underwater Sound Lab. Tech. Mem.* 1110-14-55, 1955.
7. Hill, M. N. (ed.): Internal Waves, chap. 22, in E. C. Lafond, "The Sea," Interscience Publishers (Division of John Wiley & Sons, Inc.), New York, 1962.
8. Lafond, E. C., and O. S. Lee: Internal Waves in the Ocean, *Navigation*, **9**:231 (1962).
9. Lee, O. S.: Effect of an Internal Wave on Sound in the Ocean, *J. Acoust. Soc. Am.*, **33**:677 (1961).
10. Barakos, P. A.: On the Theory of Acoustic Wave Scattering and Refraction in Internal Waves, *U.S. Navy Underwater Sound Lab. Rept.* 649, 1965.
11. Ewing, M., and J. L. Worzel: Long-range Sound Transmission, *Geol. Soc. Am. Mem.* 27, 1948.
12. Stifler, W. W., and W. F. Saars: Sofar, *Electron.*, **21**:98 (1948).
13. Bryan, G. M., M. Truchan, and J. I. Ewing: Long Range SOFAR Studies in the South Atlantic Ocean, *J. Acoust. Soc. Am.*, **35**:273 (1963).
14. Baker, H. H.: Missile Impact Locating System, *Bell Telephone Lab. Record,* **39**:195 (1961).
15. Urick, R. J.: Ray Identification in Long Range Sound Transmission, *U.S. Naval Ordnance Lab. Tech. Rept.* 65-104, 1965.
16. Hale, F. E.: Long Range Sound Propagation in the Deep Ocean, *J. Acoust. Soc. Am.*, **33**:456 (1961).
17. Urick, R. J.: Caustics and Convergence Zones in Deep-water Sound Transmission, *J. Acoust. Soc. Am.*, **38**:348 (1965).
18. Marsh, H. W., and R. H. Mellen: Underwater Sound Propagation in the Arctic Ocean, *J. Acoust. Soc. Am.*, **35**:552 (1963).
19. Buck, B. M., and C. R. Greene: Arctic Deep-water Propagation Measurements, *J. Acoust. Soc. Am.*, **36**:1526 (1964).
20. Milne, A. R.: A 90-km Sound Transmission Test in the Arctic, *J. Acoust. Soc. Am.*, **35**:1459 (1963).
21. Mellen, R. H., and H. W. Marsh: "Underwater Sound in the Arctic Ocean," Avco Corporation, Marine Electronics Office, August, 1965.
22. Ide, J. M., R. F. Post, and W. J. Fry: Propagation of Underwater Sound at Low Frequencies as a Function of the Acoustic Properties of the Bottom, *U.S. Naval Res. Lab. Rept.* S-2113, 1943.
23. Roe, G. M.: Propagation of Sound in Shallow Water, *U.S. Navy Bur. Ships Minesweeping Branch Rept.* 65, 1943. Also, Addendum, *Rept.* 88, 1946.
24. Pekeris, C. L.: Theory of Propagation of Explosive Sound in Shallow Water, *Geol. Soc. Am. Mem.* 27, 1948.
25. Weston, D. E.: Propagation of Sound in Shallow Water, 1962 Sonar Systems Symposium, University of Birmingham, England, *J. Brit. IRE*, **26**:329 (1963).
26. Mackenzie, K. V.: Long-range Shallow-water Transmission, *J. Acoust. Soc. Am.*, **33**:1505 (1961).
27. Bucker, H. P., and H. E. Morris: Normal Mode Calculations for a Constant-depth Shallow Water Channel, *J. Acoust. Soc. Am.*, **38**:1010 (1965).
28. Tolstoy, I.: Resonant Frequencies and High Modes in Layered Wave Guides, *J. Acoust. Soc. Am.*, **28**:1182 (1956).

29. Tolstoy, I.: Dispersion and Simple Harmonic Point Sources in Wave Ducts, *J. Acoust. Soc. Am.*, **27**:897 (1955).
30. Tolstoy, I.: Guided Waves in a Fluid with Continuously Variable Velocity Overlying an Elastic Solid: Theory and Experiment, *J. Acoust. Soc. Am.*, **32**:81 (1960).
31. Clay, C. S.: Propagation of Band-limited Noise in a Layered Wave Guide, *J. Acoust. Soc. Am.*, **31**:1473 (1959).
32. McLeroy, E. G.: Complex Image Theory of Low Frequency Sound Propagation in Shallow Water, *J. Acoust. Soc. Am.*, **33**:1120 (1961).
33. Kornhauser, E. T., and W. P. Raney: Attenuation in Shallow Water Propagation Due to an Absorbing Bottom, *J. Acoust. Soc. Am.*, **27**:689 (1955).
34. Eby, R. K., A. O. Williams, R. P. Ryan, and P. Tamarkin: Study of Acoustic Propagation in a Two-layered Model, *J. Acoust. Soc. Am.*, **32**:88 (1960).
35. Williams, A. O.: Some Effects of Velocity Structure on Low Frequency Propagation in Shallow Water, *J. Acoust. Soc. Am.*, **32**:363 (1960).
36. Spitznogle, F. R., and E. G. McLeroy: Propagation at Short Ranges of Elastic Waves from an Impulsive Source near a Fluid Two-layer Solid Interface, *J. Acoust. Soc. Am.*, **35**:1808 (1963).
37. Victor, A. S., F. R. Spitznogle, and E. G. McLeroy: Propagation at Short Ranges of Elastic Waves from an Impulsive Source in a Shallow Fluid Overlying a Layered Elastic Solid, *J. Acoust. Soc. Am.*, **37**:894 (1965).
38. Weston, D. E.: Guided Propagation in a Slowly-varying Medium, *Proc. Phys. Soc. London*, **73**:365 (1958).
39. Milne, A. R.: Shallow Water Under-ice Acoustics in Barrow Strait, *J. Acoust. Soc. Am.*, **32**:1007 (1960).
40. Clay, C. S.: Effect of a Slightly Irregular Boundary on the Coherence of Waveguide Propagation, *J. Acoust. Soc. Am.*, **36**:833 (1964).
41. Wood, A. B.: Model Experiments on Sound Propagation in Shallow Seas, *J. Acoust. Soc. Am.*, **31**:1213 (1959).
42. Hunkins, K., and H. Kutschale: Shallow-water Propagation in the Arctic Ocean, *J. Acoust. Soc. Am.*, **35**:542 (1963).
43. Marsh, H. W., and M. Schulkin: Shallow Water Transmission, *J. Acoust. Soc. Am.*, **34**:863 (1962).
44. Macpherson, J. D., and N. O. Fothergill: Study of Low Frequency Sound Propagation in the Hartlen Point Region of the Scotian Shelf, *J. Acoust. Soc. Am.*, **34**:967 (1962).
45. Urick, R. J., and C. W. Searfoss: The Microthermal Structure of the Ocean near Key West, Florida: Part I—Description; Part II—Analysis, *U.S. Naval Res. Lab. Repts.* S-3392, 1948, and S-3444, 1949.
46. Liebermann, L.: Effect of Temperature Inhomogeneities in the Ocean on the Propagation of Sound, *J. Acoust. Soc. Am.*, **23**:563 (1951).
47. Mintzer, D.: Wave Propagation in a Randomly Inhomogeneous Medium, I: *J. Acoust. Soc. Am.*, **25**:922 (1953), II: **25**:1107 (1953); III: **26**:186 (1954).
48. Skudryzk, E.: Scattering in an Inhomogeneous Medium, *J. Acoust. Soc. Am.*, **29**:50 (1957).
49. Potter, D. S., and S. R. Murphy: On Wave Propagation in a Random Inhomogeneous Medium, *J. Acoust. Soc. Am.*, **29**:197 (1957).
50. Knollman, G. C.: Wave Propagation in a Medium with Random Spheroidal Inhomogeneities, *J. Acoust. Soc. Am.*, **36**:681 (1964).
51. Stone, R. G., and D. Mintzer: Range Dependence of Acoustic Fluctuations in a Randomly Inhomogeneous Medium, *J. Acoust. Soc. Am.*, **34**:647 (1962).
52. Whitmarsh, D. C.: Underwater Acoustic Transmission Measurements, *J. Acoust. Soc. Am.*, **35**:2014 (1963).

53. Sagar, F. H.: Acoustic Intensity Fluctuations and Temperature Microstructure in the Sea, *J. Acoust. Soc. Am.*, **32**:112 (1960).
54. Sheehy, M. J.: Transmission of 24 kc Underwater Sound from a Deep Source, *J. Acoust. Soc. Am.*, **22**:24 (1950).
55. Whitmarsh, D. C., E. Skudryzk, and R. J. Urick: Forward Scattering of Sound in the Sea and Its Correlation with the Temperature Microstructure, *J. Acoust. Soc. Am.*, **29**:1124 (1957).
56. Brown, M. V., and J. Ricard: Fluctuations in Surface Reflected Pulsed CW Arrivals, *J. Acoust. Soc. Am.*, **32**:1551 (1960).
57. Clay, C. S.: Fluctuations of Sound Reflected from the Sea Surface, *J. Acoust. Soc. Am.*, **32**:1547 (1960).
58. Scrimger, J. A.: Signal Amplitude and Phase Fluctuations Induced by Surface Waves in Ducted Sound Propagation, *J. Acoust. Soc. Am.*, **33**:239 (1961).
59. Beckerly, J. C.: Effects of Ocean Waves on Acoustic Signals to Very Deep Hydrophones, *J. Acoust. Soc. Am.*, **35**:267 (1963).
60. Steinberg, J. C., and T. G. Birdsall: Underwater Sound Propagation in the Straits of Florida, *J. Acoust. Soc. Am.*, **39**:301 (1966).
61. Gershman, S. G., and Yu. I. Tuzhilkin: Measurements of the Transverse Correlation Coefficient of a Continuous Sound Signal in the Sea, *Soviet Phys. Acoust.*, **6**: 291 (1961).
62. Gulin, E. P., and K. I. Malyshev: Experiments in the Spatial Correlation of the Amplitude and Phase Fluctuations of Acoustic Signals Reflected from a Rough Ocean Surface, *Soviet Phys. Acoust.*, **10**:365 (1965).
63. Chernov, L. A.: "Wave Propagation in a Random Medium," translation, McGraw-Hill Book Company, New York, 1960.
64. Tatarski, V. I.: "Wave Propagation in a Turbulent Medium," translation, McGraw-Hill Book Company, New York, 1961.
65. Brekhovskikh, L. M.: Propagation of Sound in Inhomogeneous Media: A Survey, *Soviet Phys. Acoust.*, **2**:247 (1956).
66. Mackenzie, K. V.: Long Range Shallow Water Signal Level Fluctuations and Frequency Spreading, *J. Acoust. Soc. Am.*, **34**:67 (1962).

SEVEN

The Noise Background of the Sea: Ambient-noise Level

In a loose manner of speaking, ambient noise may be said to be the noise of the sea itself. It is that part of the total noise background observed with a nondirectional hydrophone which is not due to the hydrophone and its manner of mounting called "self-noise," nor due to some identifiable localized source of noise. It is what is "left over" after all identifiable noise sources are accounted for.

According to Webster's Collegiate Dictionary, the word *ambient* means "encompassing, surrounding on all sides." In the present context, ambient noise truly surrounds the hydrophone on all sides, though unequally and in a nonisotropic manner. In deep water, the ambient noise of the sea has a directionality of its own, which will be described later on.

The ambient-noise level, as a sonar parameter, is the intensity, in decibels, of the ambient background measured with a nondirectional hydrophone and referred to the intensity of a plane wave having an rms pressure of 1 dyne/cm^2. Although measured in different frequency bands, ambient levels are always reduced to a 1-Hz frequency band and are then called *ambient-noise spectrum levels*.

The ambient background of the sea often presents difficult measurement problems. For a valid measurement of ambient noise, all possible sources of self-noise must be eliminated, or at least reduced to an insignificant contribution to the total noise level. Self-noise sources, such as cable strumming, splashes of waves against the hydrophone cable, 60-Hz-hum pickup, and sometimes even crabs crawling on the hydrophone, must be absent. Also, identifiable distant noise sources, such as individual ships, must not contribute to the noise background. Thus, ambient noise is the residual noise background in the absence of individual identifiable sources that may be considered the natural noise environment of the hydrophone. To the ear ambient noise often sounds like a low rumble at low frequencies and a nontonal frying, crackling hiss at high frequencies. Sometimes peculiar sounds are heard as part of the ambient background, like the noise of marine animals or the noise made by cracks forming in the ice sheet over a hydrophone.

7.1 Sources of Ambient Noise in Deep Water

The studies of ambient noise made in recent years, largely with deep-bottomed hydrophones, have vastly improved and extended our knowledge about the characteristics and sources of deep-sea ambient noise. Measurements have been made over a frequency range extending from below 1 Hz up to about 100 kHz. Over this broad frequency range, the data show that ambient noise has different characteristics at different frequencies, with a different spectral slope and a different behavior with varying conditions, such as wind speed, in different parts of the spectrum. Because of this, it follows that the noise must be due to a variety of different sources. In any one region of the spectrum, one or more of these are apt to be dominant over the others. In the following sections, the principal sources of noise, as presently understood, will be briefly described. We will begin with those sources likely to be important at the low end of the spectrum and end with the source of noise that completely dominates all the others at high kilohertz frequencies. Some of the noise sources that we will consider are illustrated diagrammatically in Fig. 7.1.

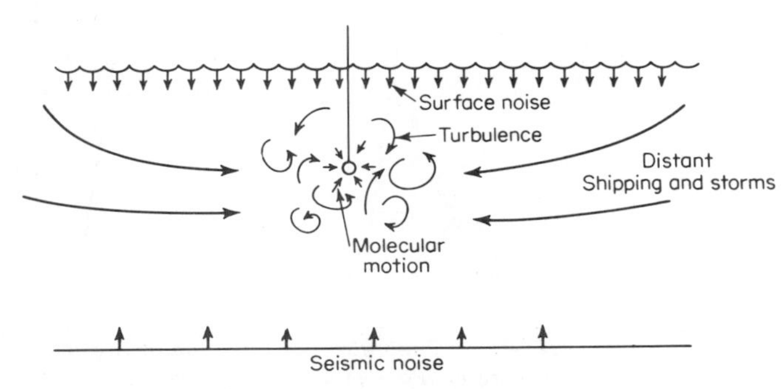

FIG. 7.1. Conceptual diagram showing some of the sources of deep-water ambient noise.

Tides and the Hydrostatic Effects of Waves Most underwater sound hydrophones are pressure-sensitive and will respond to changes in ambient pressure, whether of acoustic origin or not. Tides and waves cause hydrostatic pressure changes of relatively large amplitude at the low-frequency end of the spectrum. The tides, for example, may be imagined to create line components, or tones, in the pressure spectrum at frequencies corresponding to 1 or 2 cycles/day. The magnitude of these tidally produced pressure changes is, of course, enormous; the pressure equivalent of a 1-ft head of water, for example, amounts to 3×10^4 dynes/cm^2. Fortunately such pressure changes lie far below the frequencies of interest to underwater sound and are restricted to narrow regions of the spectrum. At high frequencies, however, tidal currents contribute to the background of oceanic turbulence.

Surface waves are also a source of hydrostatic pressure changes at a depth in the sea. However, they have a pressure amplitude which falls off rapidly with increasing depth and with decreasing wavelength of the surface waves. Figure 7.2 shows how the hydrostatic pressure produced

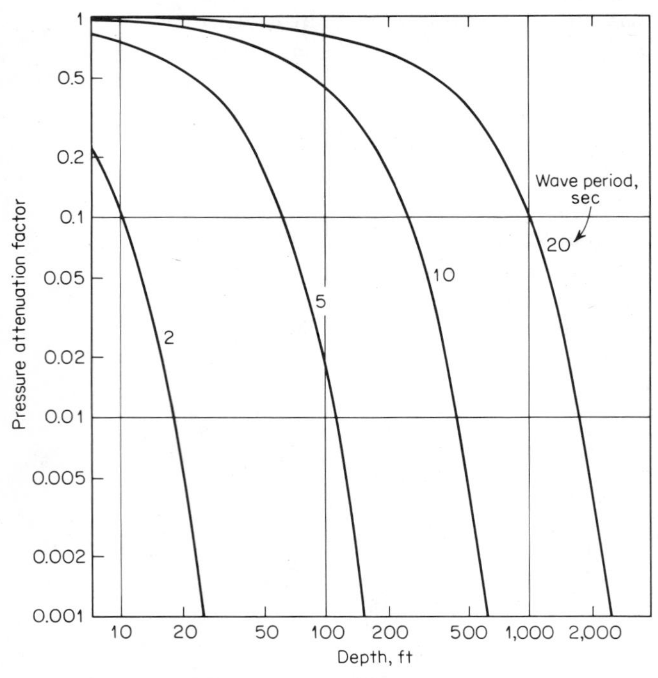

FIG. 7.2. Theoretical attenuation of the pressure changes produced by surface waves of a given period at a depth below the surface.

by waves of a given wave period becomes attenuated with depth below the surface. This attenuation is extremely rapid with increasing depth. In shallow water, however, the depth may not be great enough to eliminate completely the pressure effects of waves passing over a bottomed hydrophone; in such instances, the rough surface can become the dominant source of the low-frequency background of a pressure-sensitive hydrophone.

Seismic Disturbances Because the earth is in a constant state of seismic activity, earth unrest is probably an important cause of low-frequency noise in the sea. One very strong and almost continuous form of seismicity is *microseisms* having a nearly regular periodicity of about $\frac{1}{7}$ Hz and a vertical amplitude on land of the order of 10^{-4} cm. If such amplitudes exist in the deep seabed, and if the disturbance is assumed to be sinusoidal, the pressure that would be produced in the sea above can be found from the expression

$$p = 2\pi f \rho c a$$

where f = frequency
ρ = density of water
c = velocity of water
a = displacement amplitude

With $f = \frac{1}{7}$ Hz and $a = 10^{-4}$ cm, the resulting pressure amplitude turns out to be 10 dynes/cm^2, or 20 db above 1 dyne/cm^2. This is roughly of the same level as that measured as the ambient-noise pressures at frequencies below 1 Hz. It may therefore be surmised that microseismic disturbances, and perhaps earth seismicity in general, are likely sources of sea noise at very low frequencies. In addition, intermittent seismic sources such as individual earthquakes and distant volcanic eruptions (1) are undoubtedly transient contributors to the low-frequency background of the deep sea.

Oceanic Turbulence The role of turbulence in the sea as a source of ambient noise has been considered by Wenz (2). Conceivably, turbulence, in the form of irregular random water currents of large or small scale, is capable of creating a noise background in several ways. First, such currents may shake or rattle the hydrophone and its mounting and so produce a form of self-noise, rather than a part of the ambient noise of the sea. Second, the pressure changes associated with the turbulence may be radiated to a distance, and so appear as part of the background at places distant from the turbulence itself. However, the radiated noise of ocean turbulence is not likely to be of significance because of its quadrupole origin and its consequently rapid falloff with distance. The third, and most important, acoustic effect of turbulence is the turbulent pressure changes created inside the turbulent region. The turbulence gives rise to varying dynamic pressures that are picked up by a pressure-sensitive hydrophone located in the turbulent region. The magnitude of such pressures can be estimated, as was done by Wenz, from a guess as to the magnitude of the turbulent currents in the deep sea. If the turbulent component of the flow is written as u, then the associated dynamic pressure is ρu^2, where ρ is the fluid density. If u is taken to be 5 percent of the steady (d.c.) component of the current, then in a 1-knot steady current, the turbulent component would be 0.05 knot or 2.5 cm/sec, and the dynamic pressure of the turbulent motion would then be, with $\rho = 1$, equal to 6.3 dynes/cm^2, or +16 db above 1 dyne/cm^2. Figure 7.3 shows estimates of the spectrum of turbulent pressures for three values of steady current speed $\bar{u}$, as derived by Wenz from theoretical and experimental relations of turbulent flows. The magnitudes and the slope of these estimated spectra of ambient oceanic turbulence for $\bar{u} = 2$ cm/sec are in remarkable agreement with observed noise spectra in the decade 1 to 10 Hz. As a consequence, it may be inferred that, although no direct observational evidence has come to light, the turbulence of deep ocean currents is a likely cause of low-frequency noise.

Ship Traffic Various pieces of evidence suggest that distant ship traffic is a dominant source of noise at frequencies around 100 Hz. First, ambient noise is known to arrive at a deep hydrophone in a preferentially horizontal direction at frequencies near 100 Hz and to be independent

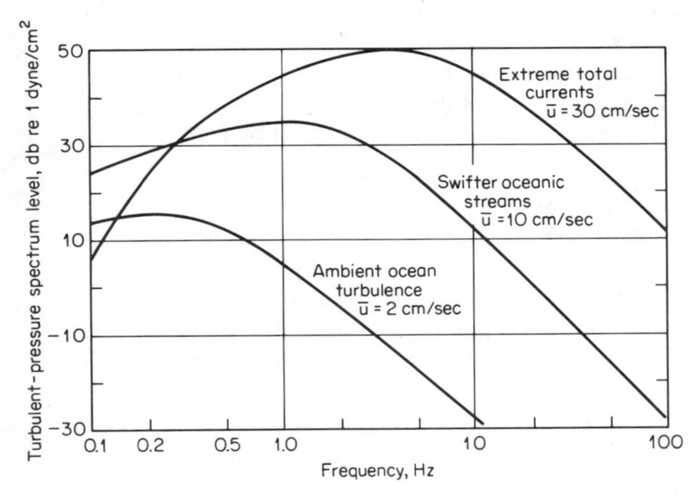

FIG. 7.3. Spectra of the pressures produced by oceanic turbulence, derived theoretically by Wenz. Ref. (2).

of wind and weather. Second, there is a "plateau," or a flattening, in observed ambient-noise spectra in this frequency region that coincides remarkably closely with the maximum in the radiated-noise spectra of ships. Finally, and most important, ambient-noise measurements in areas of high ship activity are higher and less wind-dependent at frequencies from 50 to 500 Hz than they are in areas where shipping is sparse.

The importance of ship traffic as a source of noise may be appreciated from the estimate that there are, on the average, 1,100 ships under way in the North Atlantic Ocean at any one moment of time. Quantitatively, Wenz (2) has made calculations using reasonable estimates of ship traffic and propagation conditions that yield noise spectra similar in shape and level to those measured in the sea.

All of this evidence suggests that distant ship traffic is a principal source of noise in the decade 50 to 500 Hz. Such traffic may occur at distances of 1,000 miles or more from the measurement hydrophone. The competitor of shipping as a noise source is *distant storms*, in which a part of the kinetic energy of the wind is coupled to the sea as underwater sound.

Surface Waves At still higher frequencies, ambient noise is governed by the roughness of the sea surface. During World War II many observations (3) of deep-water ambient noise between 500 Hz and 25 kHz indicated a direct connection between sea state or wind force and the level of ambient noise. Based on these observations, the well-known *Knudsen spectra* were obtained, having sea state or wind force as a parameter. Later data indicated a better correlation of noise with wind speed than with sea state, probably because of the difficulty of estimating the latter. The noise level has been shown to be correlated with the *local* wind speed over the measurement hydrophone.

Although it is clear that the sea surface must generate the major portion of the ambient noise in this frequency range, the process by which it does so is still uncertain. Breaking whitecaps and spray must create some underwater noise at sea states where they exist. Yet, that these cannot be the sole sources of rough-sea noise is shown by the fact that ambient-noise levels increase rapidly from sea state 0 to sea state 2, when whitecaps and spray are absent. Marsh (4) has suggested that wave motion alone near the surface is capable of generating sound that is radiated to a distance and has obtained theoretically computed noise spectra that agree remarkably well in spectral slope and wind-speed variation with observed data. Another possible process is in the wind, which, blowing over the rough surface that it itself produces, generates turbulent pressures, or flow noise, that are coupled to the sea and appear as ambient noise below. But whatever the physical process may be, it is clear that the rough sea surface, not too far from the location of the measurement hydrophone, is the dominant noise source at frequencies between 1 and 30 kHz.

Thermal Noise In 1952, Mellen (5) showed theoretically that thermal noise of the molecules of the sea places a limit on hydrophone sensitivity at high frequencies. Reasoning that since the average energy per degree of freedom is kT (k = Boltzmann's constant, T = absolute temperature) and since the number of degrees of freedom in a large volume of sea is the same as the number of compressional modes possible in that volume, Mellen was able to compute the equivalent plane-wave pressure for thermal noise in water. For a nondirectional hydrophone which is perfectly efficient in converting acoustic to electric energy, the equivalent noise spectrum level is, for ordinary temperatures,

$$\mathrm{NL} = -115 + 20 \log f$$

where f is the frequency in kilohertz. For example, at 100 kHz the limiting spectrum level is $-115 + 40$, or -75 db. For a hydrophone of directivity index DI and efficiency E, expressed in decibels, the equivalent thermal-noise spectrum level becomes

$$\mathrm{NL} = -115 + 20 \log f - \mathrm{DI} - E$$

This noise is the same as the Nyquist noise developed in the radiation resistance of the hydrophone in water, and has been measured experimentally by Ezrow (6). The spectrum level of this kind of acoustic background in underwater hydrophones rises with frequency at the rate of 6 db/octave ($+20 \log f$) and places a threshold on the minimum observable pressure levels in the sea.

7.2 Deep-water Spectra

Figure 7.4 is an example of the spectrum of ambient noise as it might, conceivably, be observed at one deep-sea location. The spectrum is composed of segments of different slope having a different behavior under different conditions. As stated above, this complexity is interpreted as being due to a multiplicity of noise sources over the entire frequency range. A number of regions, or frequency bands, in the spectrum may be identified and associated with the principal sources described in the previous section. Five frequency bands are shown in Fig. 7.4. Band I, lying below 1 Hz, is a largely unknown portion of the spectrum at the present time, in which the noise is likely to be of hydrostatic origin (tides and waves), or originates in the earth as seismic unrest. Band II is characterized by a spectral slope of -8 to -10 db/octave (about

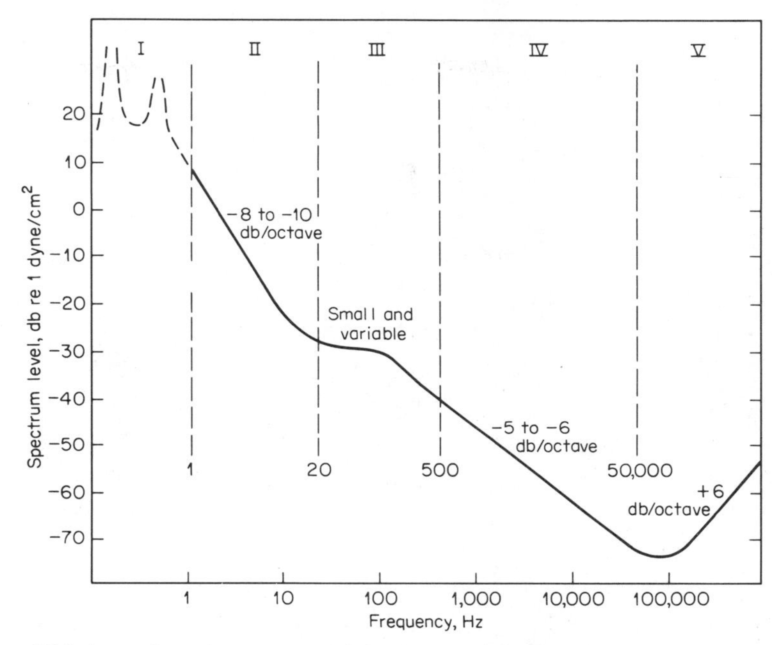

FIG. 7.4. ***Sample spectrum of deep-sea noise showing five frequency bands of differing spectral slopes. The slopes are given in decibels per octave of frequency.***

−30 db/decade), with only a slight wind-speed dependence in deep water, and in which the most probable source of noise in deep water appears to be oceanic turbulence. In Band III, the ambient-noise spectrum flattens out into a "plateau," shown by nearly every measurement, and the noise appears to be dominated by distant ship traffic. Band IV contains the Knudsen spectra having a slope of −5 to −6 db/octave (about −17 db/decade) in which the noise originates at the sea surface at points not very far from the point of measurement. Band V is overwhelmed by thermal noise originating in the molecular motion of the sea and is uniquely characterized by a positive (rising) spectrum having a slope of +6 db/octave.

For prediction purposes, average representative ambient-noise spectra for different conditions are required. Such average working curves are shown in Fig. 7.5 for different conditions of shipping and wind speed. The ambient-noise spectrum at any location at any time is approximated by selecting the appropriate shipping and wind curves and fairing them together at intermediate frequencies where more than one source is important. The "heavy-shipping" curve is used for locations near the

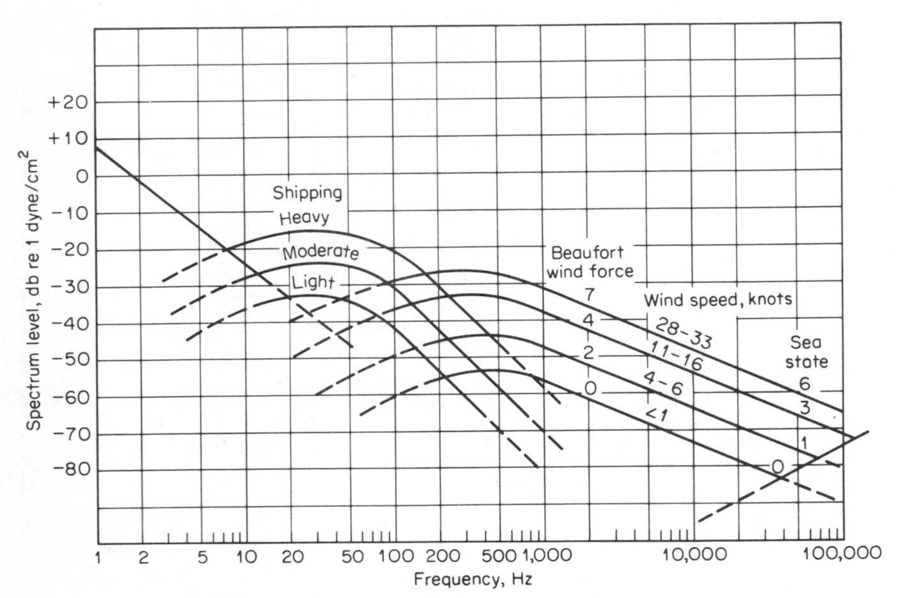

FIG. 7.5. Average deep-water ambient-noise spectra.

shipping lanes of the North Atlantic; the "light-shipping" curve is appropriate for locations remote from ship traffic.

7.3 Shallow-water Ambient Noise

In contrast to the relatively well-defined levels of deep-water ambient noise, the ambient levels in coastal water and in bays and harbors are subject to wide variations. In such locations, the sources of shallow-water noise are highly variable, both from time to time and from place to place. As a consequence, only a rough indication can be given of the levels that might be found in bays and harbors and at offshore coastal locations.

Bays and Harbors A great many measurements of noise inside different bays and harbors were made during World War II. These measurements, motivated by the needs of acoustic mines and harbor-protection sonars, were made in New York Harbor, lower Chesapeake Bay, San Diego Bay, and other bays and harbors. Some examples of the spectra resulting is an from these measurements are given in Fig. 7.6. Included in the figure average line *CC* showing the average of many determinations between 20 and 200 Hz and a shaded area giving the location of other measurements at subsonic frequencies.

In addition to the sources contributing to deep-water noise, other noise sources are important in bays and harbors. At such locations the man-made noises of industrial activity and the noises produced by marine life and by the turbulence of tidal currents all conspire to create a noisy ambient environment.

Coastal Waters In coastal waters, such as on the continental shelves, wind speed again appears to determine the noise level over a wide frequency range. This has been demonstrated by measurements of Piggott (7) over a 1-year period on the Scotian shelf in water about 150 ft in depth. This work showed a dependence of noise level on wind speed at all frequencies between 10 and 3,000 Hz. The increase of level with wind speed was found to be 7.2 db per wind speed doubled, or an increase of intensity slightly greater than the square of the wind speed. The various processes by which the wind conceivably generates noise—hydrostatic effects of wind-generated waves, whitecaps, and direct sound radiation from the rough sea surface—all must play a part in determining the level of the noise in offshore locations.

Figure 7.7 shows the spectra of Piggott for several wind speeds, together with two older average spectra measured in open water 700 ft deep 5 miles off Fort Lauderdale, Florida. It will be seen by comparison with the Knudsen spectra in deep water (Fig. 7.5) that the noise levels in

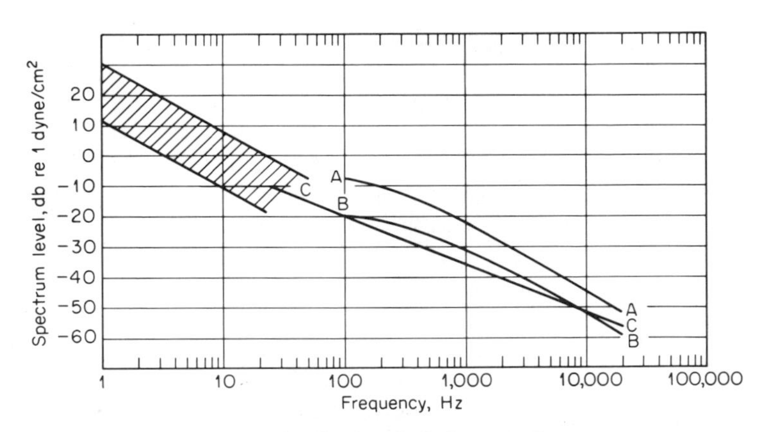

FIG. 7.6. A summary of noise levels in bays and harbors. World War II data. AA: A high noise location: entrance to New York Harbor in daytime. BB: An average noise location: upper Long Island Sound. CC: Average of many World War II measurements. Shaded area: Subsonic background measurements. [*Largely from Knudsen, Alford, and Emling.* (*Ref. 3*).]

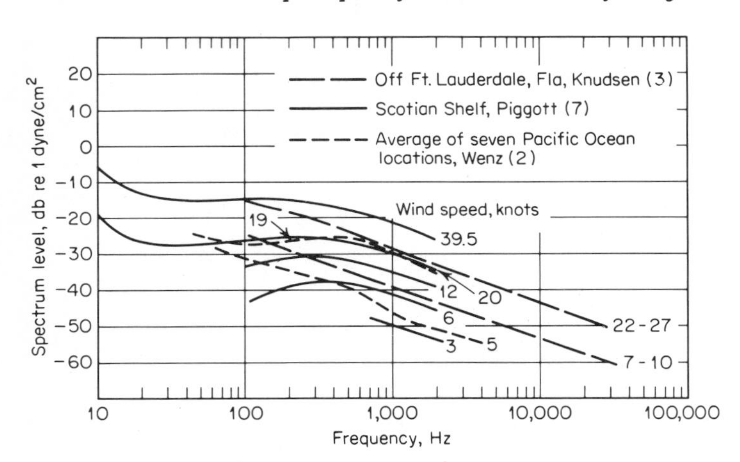

FIG. 7.7. Noise spectra at coastal locations, with wind speed as a parameter.

coastal water are 5 to 10 db higher than in deep water far from shore at frequencies greater than about 500 Hz.

7.4 Variability of Ambient Noise

Like most underwater sound parameters, ambient noise is eminently characterized by variability. Much of this variability arises from changes in the dominant sources of noise, such as changing wind speed and amount of shipping. To the extent possible with present knowledge, this variability has been accounted for in spectra such as those of Figs. 7.5 and 7.7. But a residual variability remains that creates an uncertainty of perhaps 5 to 10 db between an estimate based on average values and an amount that might be measured at a given location over a short period of time. Transient sound sources of short or extended duration, such as those of biological origin, are prime offenders. Another source of variability is changing sound-transmission conditions, which affect the transmission of sound from distant sources. For example, Walkinshaw (8) found from a 4-year period of observation with a bottomed hydrophone at Bermuda and at the Bahama Islands that the average ambient level was 7 db higher in winter than in summer; this difference was attributed to better sound transmission during the winter season.

Other examples of the variability of ambient noise are the peculiar periodic variations observed by Wenz (9) in data collected at six locations spaced over 45° of longitude in the Pacific Ocean. The analysis of data in the frequency band 20 to 100 Hz revealed periodicities of 12 and 24 hr, with a maximum of noise at midnight and noon, local time, at all locations, independent of longitude. Although the change in level was small (normally 1.5 to 5 db), it was undoubtedly real, but the origin of these changes in noise, which are apparently free from biological activity and synchronized to the mean solar day at diverse locations, remains a mystery.

Both tidal and human activity are sources of the variability of the ambient noise in bays and harbors. In Narragansett Bay, for example, the turbulence produced by tidal currents was believed (10) to be the source of observed changes in 25-Hz noise. Industrial activity has obvious periodicities that are likely to be reflected in the observed ambient levels in and near busy harbors. At a location 15 miles off the mouth of San Diego Bay, for example, some 12 db more noise was observed (11) by day than by night, and 10 db more during the day on weekdays than on weekends—changes directly associated with the tempo of the shipping and industrial life of the harbor.

7.5 Intermittent Sources of Ambient Noise

By intermittent noise sources is meant those that do not persist over periods of hours or days, but are of transient occurrence. Although some forms of biological noise cannot be called truly intermittent at some locations, but are, rather, steady and characteristic parts of the ambient background, such noises are included in this category of intermittent noises for convenience.

Biological Sounds The sounds produced by biological organisms in the sea are many and varied, and have been extensively studied. Only a brief description can be given here. Additional information can be found in a summary paper by Backus (12) and in a book (13) devoted to the subject. Although the different sounds are extremely diverse, ranging from the calls of porpoises to the frying noise of a mass of soniferous shrimp, only three groups of marine animals are known to make sound: certain kinds of shellfish (Crustacea), certain kinds of true fish, and the marine mammals (Cetacea), such as whales, dolphins, and porpoises. Among the Crustacea, the most important are snapping shrimp, which are ubiquitous inhabitants of shallow tropical and semitropical waters having a bottom of rock, shell, or weed that offers the animals some concealment. These animals make noise by snapping their claws together, as one snaps thumb and forefinger, and thereby produce a broad spectrum of noise between 500 Hz and 20 kHz. Among the fish should be mentioned the croakers of Chesapeake Bay and other East Coast locations, which make intermittent series of tapping noises like that of a woodpecker by means of the contraction of drumming muscles attached to the air bladder; the same principle is involved in beating a drum. The Cetacea include whales and porpoises, which create noise by blowing air through the larynx. Porpoises, for example, produce frequency-modulated whistles that are associated with certain behavior patterns of these animals. Many peculiar chirps, grunts, boops, groans, yelps, and barks heard in the sea are of biological origin.

Although the sounds of marine organisms have been much studied by marine biologists, very little quantitative data are available to the engineer on the spectra and levels to expect in the natural environments. The bulk of available data dates from World War II, when measurements were made on a variety of biological sounds at different locations. Some of the spectra obtained are given in Figs. 7.8 and 7.9.

Although many species of fish make noise, unfortunately no species of any commercial importance appears to do so. As a result, passive sonars are useless for fish finding. For other sonar applications, the noise of fish and other forms of life in the sea is an annoying, though readily identified, hindrance to sonar operation and forms an erratic and, in large part, unpredictable part of the ambient background in which the sonars must operate.

Twenty-cycle Pulses Some peculiar transient sounds have been heard with low-frequency bottomed hydrophones over widespread areas of the Atlantic and Pacific Oceans. These have been called *20-cycle pulses* (14, 37), and consist of trains of nearly sinusoidal 20-Hz pulses of about 1 sec in duration lasting for several minutes. The repetition rate is remarkably regular, with one pulse approximately every 10 sec. The pulse trains, lasting anywhere from 6 to 25 min, are separated by silent

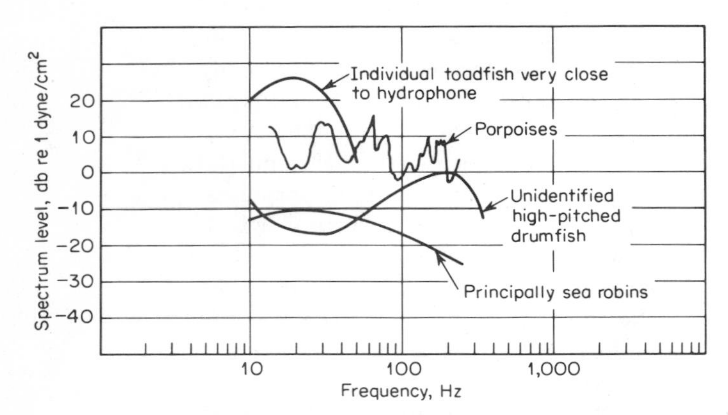

FIG. 7.8. Sample spectra of the noise made by marine animals as observed at sea. (Ref. 3.)

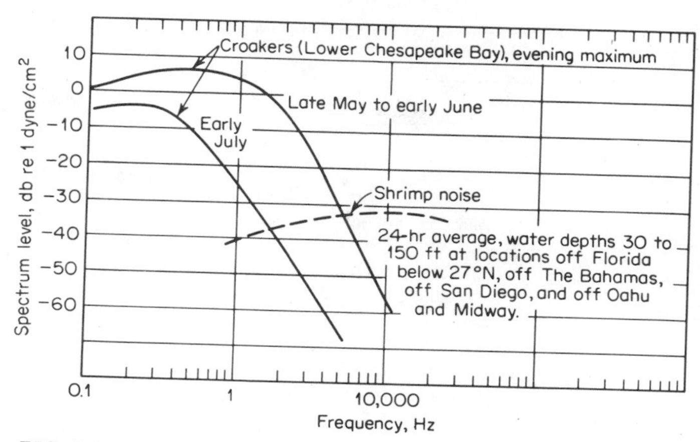

FIG. 7.9. Observed spectra of the noise of croakers and snapping shrimp measured in Chesapeake Bay and at various shallow-water locations in low latitudes. (*Ref. 3.*)

periods 2 to 3 min long, and repeat for many hours. The apparent sources of these sounds have been tracked by triangulation using two or more hydrophones and have been found to move in a random manner with speeds typically 2 or 3 knots. The total radiated acoustic power of a single pulse has been estimated to lie between 1 and 25 watts. Because of this relatively high power, the pulses were found to be detectable to ranges of 35 miles in the shallow water of the continental shelf. The cause of these noises is believed by marine biologists to be a species of whale, although the mechanism of production of sound of such high acoustic intensity and low frequency is uncertain. Heartbeats and some sort of respiratory mechanism have been suggested (38).

Rain Falling rain may be expected to increase ambient-noise levels to an extent depending on the rate of rainfall and perhaps on the area over which the rain is falling. For example, an increase of almost 30 db in the 5- to 10-kHz portion of the spectrum has been noted in a "heavy" rain (15), and at 19.5 kHz in a steady, though not torrential, rain in sea state 2, the levels increased to those corresponding to sea state 6—an increase of 10 db (16). Spectra of the ambient background during rain have been published by Heindsman, Smith, and Arneson (17) from measurements made in 120 ft of water near the eastern end of Long Island Sound. Figure 7.10 shows three spectra of ambient noise in rain of different intensity taken from this paper, together with the no-rain spectrum estimated from Fig. 7.7 for the existing wind speed of 20 to 40 knots. It will be observed that the spectrum of the noise of heavy

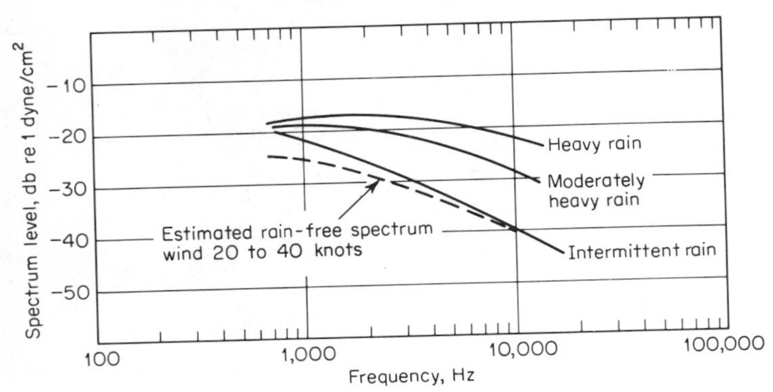

FIG. 7.10. Ambient-noise spectra in rain, observed in Long Island Sound. (*Ref. 17.*) *The dashed curve is a spectrum in the absence of rain for the wind speed applying for the measured data.*

rain is nearly "white" between 1 and 10 kHz, with a noise increase of 18 db in a "heavy" rain at 10 kHz over the no-rain spectrum level. The work of Franz (18), in which the noise produced by individual water droplets falling in air upon a water surface was studied experimentally and theoretically, has led to an estimate of spectra of rain noise that should be expected at sea in terms of rate of rainfall (Fig. 7.11).

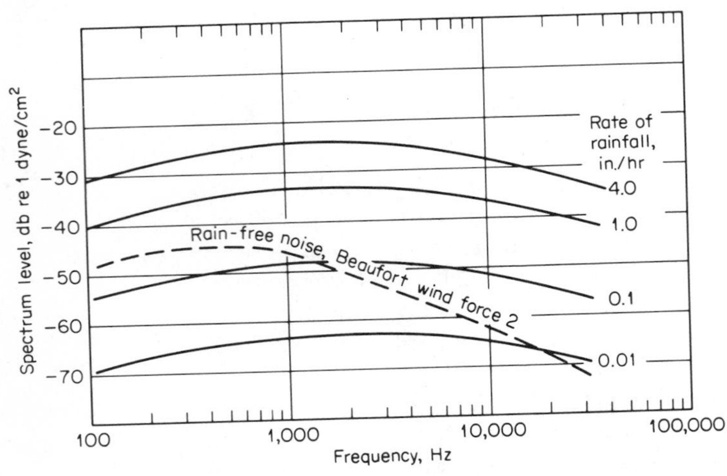

FIG. 7.11. Theoretical spectra of rain noise for different rates of rainfall. Dashed curve is for deep-sea noise at wind force 2 without rain. (*Ref. 18.*)

The agreement in shape and general magnitude between these basically derived spectra and those observed by Heindsman, Smith, and Arneson in the field is noteworthy. In connection with noise production by falling droplets, it may be noted in passing that the efficiency of a falling water droplet as a source of sound was found by Franz to be $2M^3$, where the efficiency is defined as the ratio of total radiated underwater acoustic energy to the kinetic energy of the droplet, and where M (a kind of Mach number) is the ratio of the velocity of the falling droplet in air to the velocity of sound in water.

7.6 Effect of Depth

When the sea surface is the principal source of noise, one might imagine that a deep hydrophone would be more quiet than a shallow hydrophone. However, analysis shows that in the absence of absorption and refraction, the ambient level should be independent of depth when the noise sources are distributed along a plane surface of infinite extent.

Referring to Fig. 7.12, consider what is received by a hydrophone located at point P at depth h beneath a densely packed plane distribution of random-noise sources in a medium with straight-line propagation paths. Let each square yard of surface radiate noise having the pattern

$$I(\theta) = I_0 \cos^m \theta$$

where $10 \log I_0$ is the *source level* of the unit surface area. Consider an infinitesimally thin ring of area dA of such sources located at angle

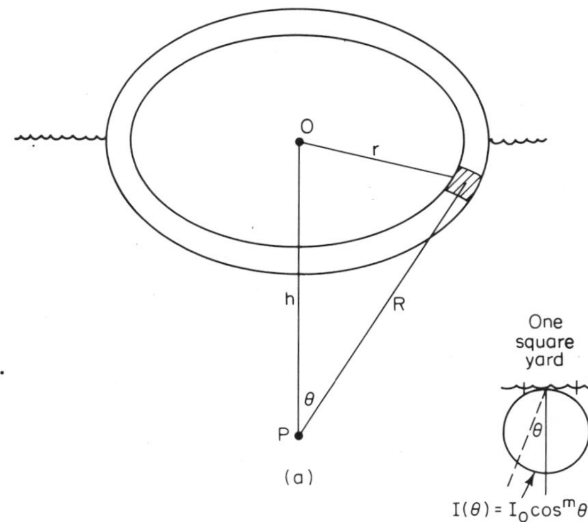

FIG. 7.12. Geometry of ambient noise produced by surface-distributed sources. (a) The noise intensity at P is the sum of noise intensities generated in a succession of rings along the surface. (b) A unit area of sea surface radiates noise with a $\cos^m \theta$ *beam pattern.*

θ around P at a horizontal range r from the point O vertically above P in water of infinite depth. Then, since $dA = 2\pi r\,dr$ and $\cos\theta = h/r$, the intensity received from the ring of randomly phased sources will be

$$dI = 2\pi r I(\theta)\,\frac{10^{(0.1)\alpha R}}{R^2}\,dr = 2\pi r I_0 \left(\frac{h}{R}\right)^m \frac{10^{(0.1)\alpha R}}{R^2}\,dr$$

where the term $10^{(0.1)\alpha R}/R^2$ represents spherical spreading plus absorption at the rate of α db/yd.

The total intensity at depth h will therefore be

$$I_h = 2\pi I_0 \int_h^\infty \left(\frac{h}{R}\right)^m \frac{10^{(0.1)\alpha R}}{R}\,dR$$

This integral has been evaluated (19) for different values of the product αh. The result is shown in Fig. 7.13, which gives the reduction in noise level with depth in terms of the quantity αh (the absorption loss in decibels over distance h) for values of $m = 0$, 1, and 2. As an example, at 50 kHz with $\alpha = 0.014$ db/yd (14 db/kyd), the lowering of a hydrophone from 3 ft ($\alpha h = 0.014$) to 300 ft ($\alpha h = 1.4$) would result in an ambient level lower by 7½ db if $m = 0$, by 4 db if $m = 1$, and by 2½ db if $m = 2$. In water of finite depth, reflection from the bottom would reduce these values.

In the absence of absorption ($\alpha = 0$), the preceding integral evaluates to

$$I_h = \frac{2\pi I_0}{m}$$

Hence the intensity produced by a surface distribution of noise sources in a medium free of absorption and refraction would be a constant,

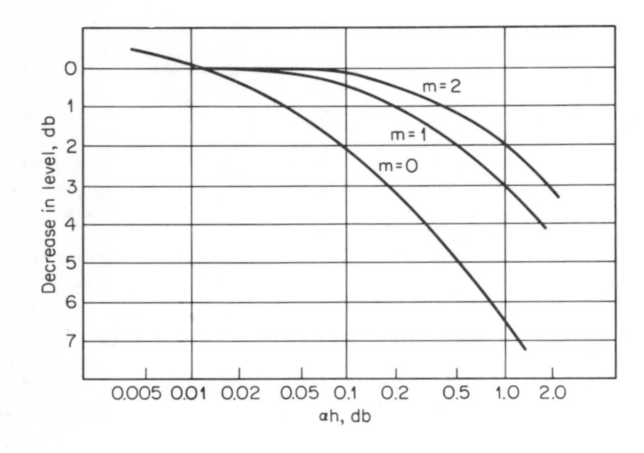

FIG. 7.13. Reduction of ambient noise with depth below an infinite distribution of sources radiating like $\cos^m\theta$. *The product* αh *is the loss in decibels due to absorption in the distance equal to the hydrophone depth.* α *is the absorption coefficient in decibels per yard if the depth* h *is in yards.*

independent of depth. The ambient-noise level everywhere would be higher by 10 log $2\pi = 8$ db if $m = 1$, and 10 log $\pi = 5$ db if $m = 2$, than the source level for ambient noise of 1 yd^2 of sea surface.

Direct measurements of the depth dependence of noise can be made only with special instrumentation and through the use of a hydrophone whose calibration is known as a function of pressure and temperature. Both a slowly sinking hydrophone connected to a ship by a very light, flexible cable (20), and the manned bathyscaphe "Trieste" (21), have been used for this purpose.

In the latter work done in the Mediterranean at frequencies from 10 to 240 Hz, although no decrease of noise was found with depth in sea state zero, a decrease of about 15 db was found at the highest frequency in sea state 2 between a shallow depth and the bottom at 3,000 m. This depth effect was found to become less at lower frequencies. Although the magnitude of the effect appears extremely high, these findings are consistent with the view that high-frequency high-sea-state noise is of surface origin. In the work with a slowly sinking hydrophone, no depth dependence was found in the range 20 to 295 Hz down to 4,000 ft.

Very close to the ocean surface, however, profound effects occur. If the hydrophone is less than one-quarter wavelength (acoustically) beneath the surface, pressure release takes place, and a pressure-sensitive hydrophone will observe a lower ambient level. On the other hand, the hydrostatic effect of the ocean waves becomes important at depths just below the surface. Another effect of the close proximity of the sea surface is the occurrence of a more "spiky" character of the noise, as will be mentioned in the following section.

Considerable data on the level of ambient noise in deep water has been obtained in recent years with deep-bottomed hydrophones (2, 39). These measurements indicate spectrum levels at frequencies above 500 Hz some 5 to 10 db lower than those predicted by the "classic" Knudsen curves derived from measurements made at shallower depths.

7.7 Amplitude Distribution

Ambient noise has been found by probability-density analyses (22) of data in one deep- and two shallow-water areas to have a Gaussian amplitude distribution at moderate depths. This is consistent with the view that the noise originates through a great many sources of random amplitude and phase. Near the sea surface, however, as with a hydrophone at a depth of a few feet below the surface, ambient noise has been observed to be more spiky than Gaussian, and to contain individual spikes or crashes of sound originating at close-by sources such as breaking wavelets. Similarly, under an ice cover, the noise is at times spiky because of the creation of tensile cracks in a continuous cover and because of the rubbing of ice masses together in a noncontinuous ice sheet.

7.8 Noise in Ice-covered Waters

Under an ice cover, ambient noise is notably different in character and level from the noise under ice-free conditions. When the ice is not continuous, as in a brashy ice pack, noise levels 5 to 10 db higher than those measured at the same sea state in ice-free waters have been observed. On the other hand, with a continuous shore-fast ice cover under rising temperatures, very low noise levels occur. Such levels demand extreme care in electronic design to achieve the necessary low self-noise levels for their measurement.

In ice-covered waters, the level and character of the noise is highly variable and depends upon ice conditions, wind speed, snow cover, and air-temperature changes. With a solid shore-fast ice cover, Milne and Ganton (23) found that the noise is spiky or impulsive when the air temperature decreases and tensile cracks in the ice are formed. Under rising temperatures, the spiky character of the noise disappears and a

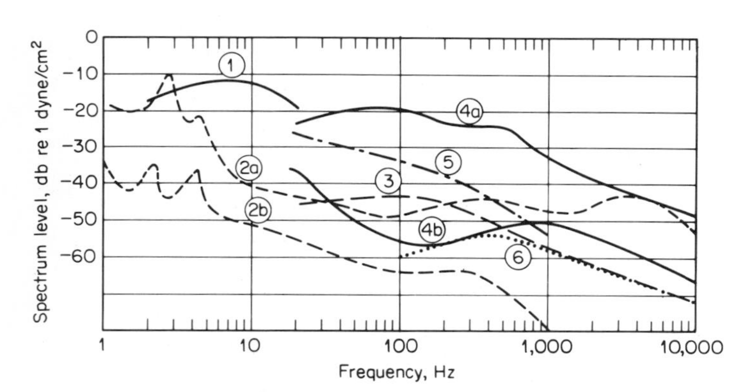

FIG. 7.14. Spectra of ambient noise observed under ice. (1) 80 to 90 in. ice, no cracking noises, Barrow Strait, April, 1959. (Ref. 26.) (2) Old polar ice with frozen leads and pressure ridges, April, 1961. (2a) Noisy periods, (2b) quiet periods. (Ref. 23.) (3) 70 percent 1-year ice 3 to 5 ft thick, 30 percent polar floes. September, 1961. (Ref. 23.) (4) Shore-fast ice like that of (2), February, 1963. Cracking noises prevalent (4a), absent (4b). (Ref. 23.) (5) Average over a 2-week period in Beaufort Basin. (Ref. 27.) (6) Ice-free deep water, wind force 0, Fig. 7.5.

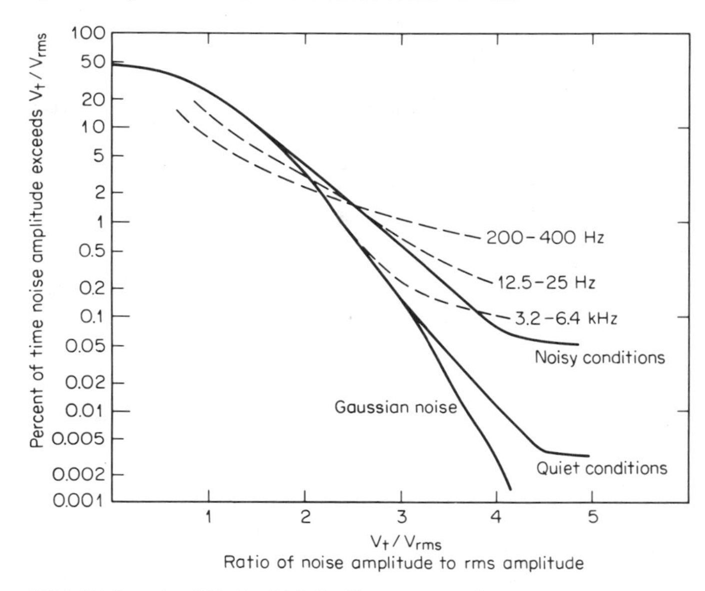

FIG. 7.15. ***Amplitude-distribution curves of ambient noise under an ice cover. Dashed curves, highly impulsive noise under decreasing temperature.*** (***Ref. 23.***) ***Solid curves, broadband noise under "quiet" and "noisy" conditions.*** (***Ref. 27.***)

Gaussian amplitude distribution reappears. Another source of noise under an ice sheet is the wind, which appears to make noise through its own turbulence and the motion of drifting granular snow impinging on the rough sea ice. Wind noise is more prominent under a noncontinuous ice cover than it is under a continuous ice cover, when, according to the observations of Macpherson (25), the noise levels become independent of wind speed. A final source of noise that should be mentioned is the bumping and scraping together of ice flows in a noncontinuous ice cover.

At a frequency equal to the reciprocal of the round-trip travel time between surface and bottom and at its harmonics, line components have been identified (25) in the ambient-noise spectrum in shallow water under an ice cover. These tonal components in the noise were attributed to the existence of low-order propagation modes in the shallow-water duct. Similar line components in ice-free ambient noise in shallow water appear not to have been reported.

It is not possible at the present time, as Milne and Ganton state, to predict the spectra and amplitude distribution of the noise to be expected under an ice cover. Very low levels, well below the lowest Knudsen curve for sea state zero, have been found under a continuous ice sheet with calm winds and rising temperatures; on the other hand, levels some 40 db higher have been observed when the ice had been actively cracking under falling air temperatures. A compilation of various published noise spectra measured in the Arctic is given in Fig. 7.14. The non-Gaussian spiky amplitude distribution of ice-cover noise, as measured in the Arctic by Milne (23, 26) and by Greene and Buck (27) is shown in Fig. 7.15.

7.9 Directional Characteristics of Deep-water Ambient Noise

Until recently, very little was known about the directionality of ambient noise in deep water. In recent years, however, the advent of modern processing techniques and observations with deep-bottom-mounted hydrophone arrays have revealed a reasonably consistent picture of the directional nature of the ambient noise at a bottomed hydrophone in deep water.

A relatively large number of independent determinations have been made. Anderson (28) and Becken (29) used a volume array with simultaneous preformed beams hung at depth from a surface ship; Forster (30) and Watson (31) used a bottom-mounted "fixed acoustic buoy" consisting of a vertical buoyed array in which narrow beams were formed electronically. VonWinkle (32), Fox (33), and Axelrod, Schoomer and VonWinkle (39) all employed a 300-ft long 40-element vertical array, located south of Bermuda, in which successive vertical beams were formed and steered electronically on shore; Cron, Hassel, and Keltonic (34) and Linnette and Thompson (35) correlated pairs of vertical hydrophones together and interpreted the correlograms in terms of a directional radiation function of the noise.

Figure 7.16*a* illustrates the findings of Axelrod, Schoomer, and VonWinkle (39) at frequencies of 112 and 1,414 Hz. These are polar plots of the ambient intensity per unit solid angle $N(\theta)$ arriving at a bottomed hydrophone as a function of vertical angle θ. At 112 Hz, more noise appears to arrive at the hydrophone from the horizontal than from the vertical, with a difference that *diminishes* with increasing wind force. At 1,414 Hz, the reverse is the case, with more noise arriving from overhead than horizontally, with a difference that *increases* with increasing wind force. This behavior is illustrated in Fig. 7.16*b*, where the ordinate is $N(\theta)$ normalized to an isotropic pattern by the factor 4π, the number of steradians in a unit sphere, and divided by the measured omnidirectional intensity N_0. When plotted in this way, the curves show the deviation from isotropicity of the noise.

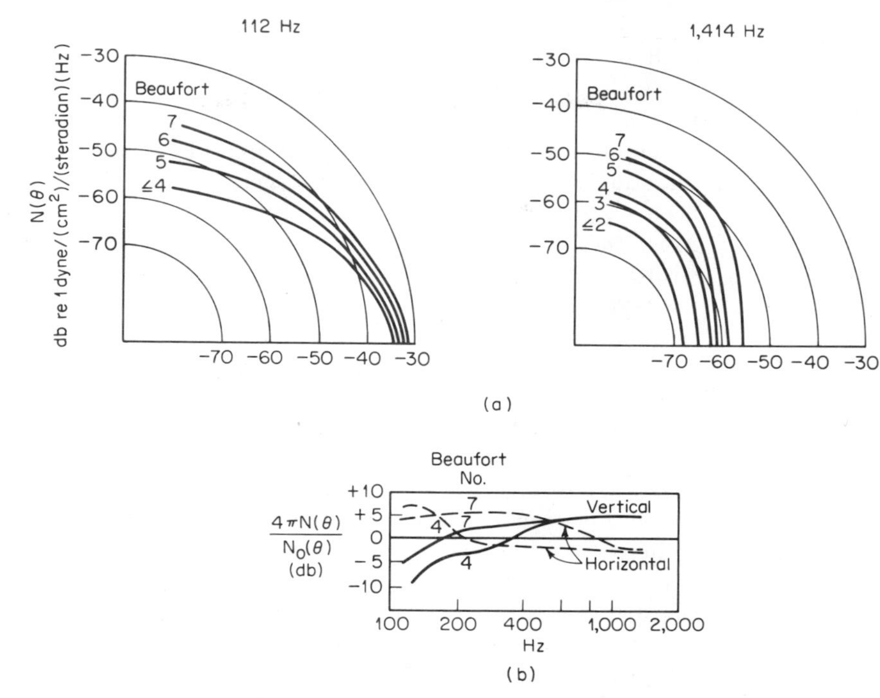

FIG. 7.16. *(a) Distribution of ambient noise in the vertical at a bottomed hydrophone at two frequencies. (Ref. 39.) (b) Curves of the ratio $4\pi N(\theta)/N_0(\theta)$ for noise arriving from the horizontal and the vertical at two wind speeds. (Ref. 39.)*

This directional behavior is consistent with the view that low-frequency noise originates at great distances and arrives at the hydrophone via primarily horizontal paths, whereas high-frequency noise originates at the sea surface more nearly overhead.

The polar plots of Fig. 7.16*a* may be used to find the ambient-noise output of any directional hydrophone having a beam pattern described by the function $B(\theta,\varphi)$, where θ and φ are angles in polar coordinates. If the ambient-noise intensity per steradian is $N(\theta)$, the ambient-noise power or voltage appearing at the terminals of the hydrophone will be

$$k \int_0^{4\pi} N(\theta)B(\theta,\varphi)\, d\Omega = k \int_0^{2\pi} \int_{-\pi}^{\pi} N(\theta)B(\theta,\varphi) \cos\theta \, d\theta \, d\varphi$$

where $d\Omega$ = infinitesimal solid angle
k = proportionality factor between sound intensity and power or voltage at hydrophone terminals

The fact that at frequencies greater than 1 kHz the noise comes dominantly from the vertical suggests that the sea surface, by whatever process it produces noise, radiates preferentially downward and that each small area of the sea surface has a beam pattern of its own in the vertical. Most investigators (17, 28, 29, 32) of sea-surface sound radiation have suggested a function of the form $I(\theta) = I_0 \cos^m \theta$, where I_0 is the intensity radiated by a small area of sea surface in the downward direction ($\theta = 0$), and m is an integer. Values of $m = 1$, 2, or 3 have been obtained, depending on conditions and method of measurements. Most measurements center roughly about $m = 2$, a value consistent with the hypothesis of a dipole source formed by the actual source and its image in the sea surface.

7.10 Spatial Coherence of Ambient Noise

The design of hydrophone arrays to minimize the pickup of ambient-noise centers basically around the spatial coherence of the noise between separated receivers. By spatial coherence is meant the time-averaged crosscorrelation coefficient of the noise observed with hydrophones a distance apart in the sea. It is shown in Sec. 3.8 that the mean-square noise output of an array can be expressed in terms of the spatial correlation between all pairs of hydrophones of the array.

These functions have been worked out by Cron and Sherman (36) for pairs of hydrophones (1) in an isotropic noise field and (2) in the noise field produced by sources distributed along an infinite plane surface and having a radiation pattern of the $\cos^m \theta$ type mentioned above. For isotropic noise at a single frequency, the correlation coefficient, or normalized time-averaged product of the output of two hydrophones spaced a distance d apart, can be easily shown to be

$$\rho(d) = \frac{\sin kd}{kd}$$

where $k = 2\pi d/\lambda$. For a surface-noise distribution of sources with $m = 2$, that is, for $I(\theta) = I_0 \cos^2 \theta$, Fig. 7.17, taken from the work of Cron and Sherman, shows the theoretical correlation between hydrophones separated horizontally and vertically, as well as for isotropic single-frequency noise. Somewhat different spatial correlation curves apply for band-limited noise and for values of m different from 2. Observations on the coherence of noise made with a bottom-mounted vertical array have shown (34) excellent agreement with theoretical predictions for $m = 2$ at frequencies between 400 and 1,000 Hz.

Time-delay correlograms show the correlation coefficient between two hydrophone outputs as a function of the time delay between them. For deep-sea ambient noise, time-delay correlograms have been obtained by Arase and Arase (40) and by Urick (41). Some examples are shown in

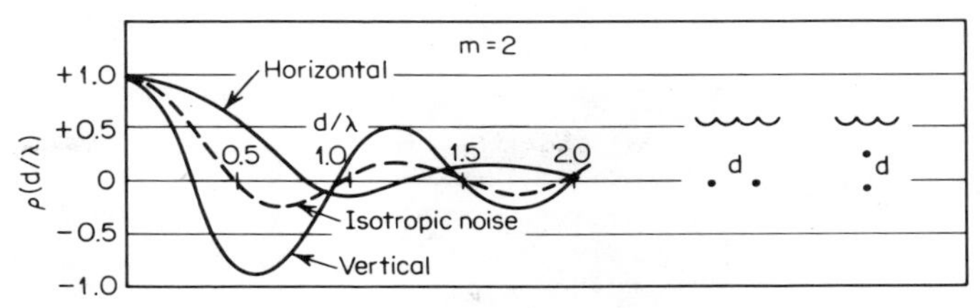

FIG. 7.17. Spatial-correlation curves for isotropic noise at a single frequency and for a surface distribution of sources radiating like $\cos^2 \theta$, *as received with hydrophones spaced a distance* d *apart horizontally and vertically.* $\rho(d/\lambda)$ *is the correlation coefficient as a function of the ratio of spacing to wavelength.* (*Ref. 36.*)

Fig. 7.18 for different pairs of vertically separated hydrophones near the bottom in deep water off Bermuda at two different wind speeds. The correlograms are seen to be different in the two cases. At the lower wind speed, the correlograms have a peak that remains high and centered near zero delay as the hydrophone separation is increased; at the higher wind speed, the correlogram peak becomes weaker and shifts to the right with increasing hydrophone separation. This behavior is consistent with the view that the noise originates in distant shipping or storms at low wind speeds, and becomes dominated by sea-surface noise at high wind speeds. Values of noise correlation coefficient measured or read from such correlograms are useful for the design of noise-discriminating arrays steered in different vertical directions (Sec. 3.8).

7.11 Summary

The characteristics of the spectrum of ambient noise in deep water indicate the existence of a number of different causes of the noise in different portions of the spectrum (Fig. 7.4). These include tides, waves, and earth seismicity at the low-frequency end, distant shipping and rough sea surface at moderate frequencies, and molecular thermal motion of the sea at very high frequencies. By means of the component spectra of Fig. 7.5, the level of the deep-sea ambient at any frequency can be estimated with a moderate degree of assurance. By contrast, the magnitude of the noise in bays and harbors (Fig. 7.6) and in coastal locations (Fig. 7.7) is not nearly so predictable, as is indicated by the scatter of the reported data. Intermittent sources of the ambient background include biological sounds (Figs. 7.8 and 7.9), rain (Figs. 7.10 and 7.11), and the once-mysterious twenty-cycle pulses. Under an ice cover, the

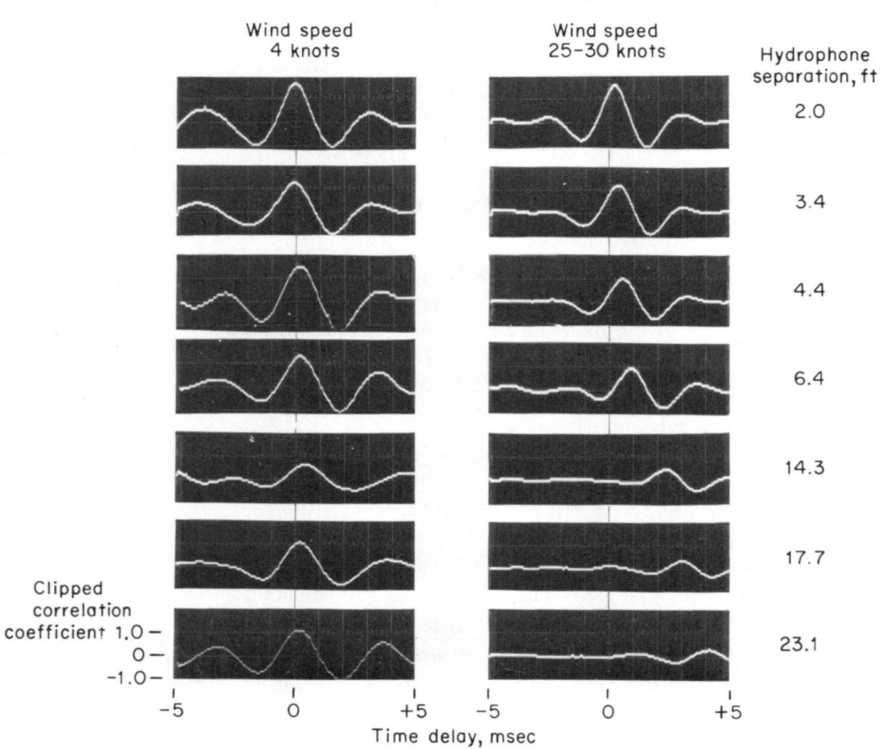

FIG. 7.18. ***Time-delay correlograms of ambient noise in the 200- to 400-kHz octave received by two vertically separated hydrophones spaced at different distances apart. Horizontal scale is relative time delay; a positive delay means that the upper hydrophone of the vertical pair is delayed electronically relative to the lower.*** (***Ref. 41.***)

ambient background (Figs. 7.14 and 7.15) has some peculiar properties of its own, such as the transient sounds of thermal cracks in the ice and the rubbing and bumping of ice blocks together. In recent years, the directional nature (Fig. 7.16) and spatial coherence (Fig. 7.17) of deep-sea ambient noise has been investigated, in large part through the use of bottom-mounted hydrophone arrays. These studies have shown that in deep water the noise is not isotropic, but has a directionality and a coherence in space consistent with presently accepted ideas about the origin of the noise in deep open water.

REFERENCES

1. Snodgrass, J. M., and A. F. Richards: Observations of Underwater Volcanic Acoustics, *Trans. Am. Geophys. Union*, **37**:97 (1956).

2. Wenz, G. M.: Acoustic Ambient Noise in the Ocean: Spectra and Sources, *J. Acoust. Soc. Am.*, **34**:1936 (1962).
3. Knudsen, V. O., R. S. Alford, and J. W. Emling: Underwater Ambient Noise, *J. Marine Res.*, **7**:410 (1948).
4. Marsh, H. W.: Origin of the Knudsen Spectra, *J. Acoust. Soc. Am.*, **35**:409 (1963).
5. Mellen, R. H.: Thermal-noise Limit in the Detection of Underwater Acoustic Signals, *J. Acoust. Soc. Am.*, **24**:478 (1952).
6. Ezrow, D. H.: Measurement of the Thermal Noise Spectrum of Water, *J. Acoust. Soc. Am.*, **34**:550 (1962).
7. Piggott, C. L.: Ambient Sea Noise at Low Frequencies in Shallow Water of the Scotian Shelf, *J. Acoust. Soc. Am.*, **36**:2152 (1965).
8. Wilkenshaw, H. M.: Low-frequency Spectrum of Deep Ocean Ambient Noise (abstract), *J. Acoust. Soc. Am.*, **32**:1497 (1960).
9. Wenz, G. M.: Some Periodic Variations in Low-frequency Acoustic Ambient Noise Levels in the Ocean, *J. Acoust. Soc. Am.*, **33**:64 (1961).
10. Willis, J., and F. T. Dietz: Some Characteristics of 25 cps Shallow-water Ambient Noise, *J. Acoust. Soc. Am.*, **37**:125 (1965).
11. Wenz, G. M.: Informal communication, *U.S. Navy Electron. Lab. Rept.* 338, 1962.
12 Backus, R. H.: Informal communication, Office of Naval Research, 1958.
13. Tavolga, W. N. (ed.): "Marine Bioacoustics," Pergamon Press, New York, 1964.
14. Walker, R. A.: Some Intense Low Frequency Underwater Sounds of Wide Geographic Distribution, Apparently of Biological Origin, *J. Acoust. Soc. Am.*, **35**:1816 (1963).
15. Teer, C. A.: Informal report, (British) Underwater Detection Establishment, 1949.
16. Richard, J. D.: Underwater Ambient Noise in the Straits of Florida and Approaches, *Univ. Miami Marine Lab. Rept.* 56-12, June, 1956.
17. Heindsman, T. E., R. H. Smith, and A. D. Arneson: Effect of Rain upon Underwater Noise Levels, *J. Acoust. Soc. Am.*, **27**:378 (1955).
18. Franz, G. J.: Splashes as Sources of Sound in Liquids, *J. Acoust. Soc. Am.*, **31**:1080 (1959).
19. Urick, R. J.: Some Directional Properties of Deep-water Ambient Noise, *Naval Res. Lab. Rept.* 3796, 1951.
20. Berman, A., and A. J. Saur: Ambient Noise as a Function of Depth (abstract), *J. Acoust. Soc. Am.*, **32**:915 (1960). Also, Hudson Laboratories Contribution No. 77.
21. Lomask, M., and R. Frassetto: Acoustic Measurements in Deep Water Using the Bathyscaphe, *J. Acoust. Soc. Am.*, **32**:1028 (1960).
22. Calderon, M. A.: Probability Density Analysis of Ocean Ambient and Ship Noise, *NEL Rept.* 1248, November, 1964.
23. Milne, A. R., and J. H. Ganton: Ambient Noise under Arctic Sea Ice, *J. Acoust. Soc. Am.*, **36**:855 (1964). Also, *Pacific Naval Lab. Can. Rept.* GS-1, January, 1965.
24. Milne, A. R., and S. R. Clark: Resonances in Seismic Noise under Arctic Sea Ice, *Bull. Seismological Soc. Am.*, **54**:1797 (1964).
25. Macpherson, J. D.: Some Under-ice Acoustic Ambient Noise Measurements, *J. Acoust. Soc. Am.*, **34**:1149 (1962).
26. Milne, A. R.: Shallow Water under Ice Acoustics in Barrow Strait, *J. Acoust. Soc. Am.*, **32**:1007 (1960).
27. Greene, C. R., and B. M. Buck: Arctic Ocean Ambient Noise, *J. Acoust. Soc. Am.*, **36**:1218 (1964).
28. Anderson, V. C.: Arrays for the Investigation of Ambient Noise in the Ocean, *J. Acoust. Soc. Am.*, **30**:470 (1958).
29. Becken, B. A.: Directional Distribution of Ambient Noise in the Ocean, *Scripps Inst. Oceanog. Rept.* 61-4, 1961.

30. Forster, C. A.: Ambient Sea Noise Directivity (abstract), *J. Acoust. Soc. Am.*, **34**:1986 (1962).
31. Watson, J. G.: Deep Ocean Noise Model Determined by a Deep Line Array (abstract), *J. Acoust. Soc. Am.*, **34**:1986 (1962).
32. VonWinkle, W. A.: Vertical Directionality of Deep Ocean Noise, *U.S. Navy Underwater Sound Lab. Rept.* 600, 1963.
33. Fox, G. R.: Ambient Noise Directivity Measurements, *J. Acoust. Soc. Am.*, **36**:1537 (1964).
34. Cron, B. F., B. C. Hassel, and F. J. Keltonic: Comparison of Theoretical and Experimental Values of Spatial Correlation, *U.S. Navy Underwater Sound Lab. Rept.* 596, 1963.
35. Linnette, H. M., and R. J. Thompson: Directivity Study of the Noise Field in the Ocean Employing a Correlative Dipole, *J. Acoust. Soc. Am.*, **36**:1788 (October 1964).
36. Cron, B. F., and C. H. Sherman: Spatial-correlation Functions for Various Noise Models, *J. Acoust. Soc. Am.*, **34**:1732 (1962).
37. Patterson, B., and G. R. Hamilton: Repetitive 20 cps Biological Hydroacoustic Signals at Bermuda, paper in "Marine Bioacoustics," p. 125, Pergamon Press, New York, 1964.
38. Schevill, W. E., W. A. Watkins, and R. H. Backus: The 20 Cycle Signals and Balaenoptera (Fin Whales), paper in "Marine Bioacoustics," p. 147, Pergamon Press, New York, 1964.
39. Axelrod, E. H., B. A. Schoomer, and W. A. VonWinkle: Vertical Directionality of Ambient Noise in the Deep Ocean at a Site near Bermuda, *J. Acoust. Soc. Am.*, **37**:77 (1965).
40. Arase, E. M., and T. Arase: Correlation of Ambient Sea Noise, *J. Acoust. Soc. Am.*, **40**:205 (1966).
41. Urick, R. J.: Correlative Properties of Ambient Noise at Bermuda, *J. Acoust. Soc. Am.*, **40**:1108 (1966).

EIGHT

Scattering in the Sea: Reverberation Level

The sea contains, within itself and on its boundaries, inhomogeneities of many different kinds, ranging in size from the tiny particles of dust that cause the deep sea to be blue, to schools of fish within the volume of the sea, and to pinnacles and sea mounts on the seabed. These inhomogeneities form discontinuities in the physical properties of the medium and thereby intercept and reradiate a portion of the acoustic energy incident upon them. This reradiation of sound is called *scattering*, and the sum total of the scattering contributions from all the scatterers is called *reverberation*. It is heard as a long, slowly decaying, quivering tonal blast following the ping of an active sonar system, and is particularly obnoxious in systems of high power and/or low directivity. Since it often forms the primary limitation on system performance, a necessary part of the design process of new active sonars is to make an estimate of the reverberation level to be encountered under the conditions in which the system will be used. In this chapter some useful formulas will be given for this purpose, together with a description of the characteristics and sources of the reverberation observed in the sea.

8.1 Types of Reverberation

The reverberation-producing scatterers in the sea are of three basically different classes. One type of scatterer occurs in the volume, or body, of the sea and produces *volume reverberation*. Examples of scatterers producing volume reverberation are the marine life and inanimate matter distributed in the sea and the inhomogeneous

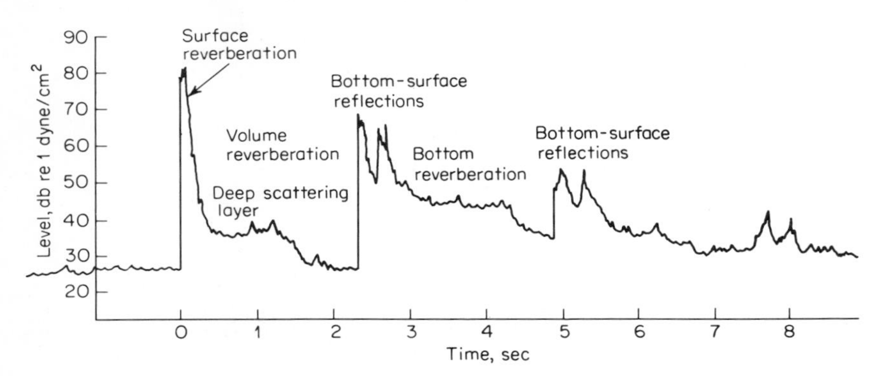

***FIG. 8.1.** Reverberation following a 2-lb explosive charge detonating at 800 ft in water 6,500 ft deep, as observed with a nearby hydrophone at a depth of 135 ft. Filter band 1 to 2 kHz.*

structure of the sea itself. *Sea-surface reverberation* is produced by scatterers located on or near the sea surface, and *bottom reverberation* originates at scatterers on or near the sea bottom. The last two forms of reverberation may be analytically considered together as *surface reverberation*, since a two-dimensional distribution of scatterers is involved. Figure 8.1 is an example of the reverberation that is observed to follow an explosive charge in deep water and is received on a shallow, nearby nondirectional hydrophone. In this example, the reverberation occurring immediately after the direct blast from the charge is surface reverberation, which quickly dies away into volume reverberation originating in large part in the deep scattering layer. The remainder of the return is caused by the sea bottom, which, together with the sea surface, produces a long and irregular reverberation tail caused by multiple reflection and scattering at the two boundaries.

8.2 The Scattering-strength Parameter

The fundamental ratio upon which reverberation depends is called *scattering strength.* It is the ratio, in decibel units, of the intensity of the sound scattered by a unit area or volume, referred to a distance of 1 yd, to the incident plane-wave intensity. In symbols, if I_{scat} is the intensity of the sound scattered by an area of 1 yd^2 or a volume of 1 yd^3, when the intensity is measured at a greater distance and reduced to 1 yd, and if I_{inc} is the intensity of the incident plane wave, then the scattering strength is defined to be

$$S_{s,v} = 10 \log \frac{I_{scat}}{I_{inc}}$$

The parameter scattering strength may be conceptually visualized as illustrated in Fig. 8.2*a* and *b* for volume reverberation and for surface reverberation, respectively. The direction of scattering is normally *back* toward the source, in which case the quantity S is more specifically termed *backscattering* strength. The point of reference for S is the point P distant 1 yd from the unit volume or area in the backward direction. In "bistatic" sonars, having a widely separated source and receiver, other directions will be of interest. From their manner of definition, it follows that scattering strength and target strength are analogous sonar parameters.

In radar (1) it is customary to use the *backscattering cross section* of a unit area or volume as the basic parameter involving scattering. This unit was adopted for sonar during World War II and was given the symbols m_v and m_s. It is the ratio of the scattered *power*, referred to 1 yd, to the *intensity* incident on a unit volume or area. Since

$$\text{Power} = \text{intensity} \times \text{area}$$

and if it is assumed arbitrarily that the scattering is uniformly distributed over a sphere and a hemisphere, respectively, it follows that

$$S_v = 10 \log \frac{m_v}{4\pi}$$

$$S_s = 10 \log \frac{m_s}{2\pi}$$

where m_v and m_s are the backscattering cross sections of a unit volume and area. These cross sections may be viewed as the areas, which, when multiplied by the intensity, equal the power removed from the incident

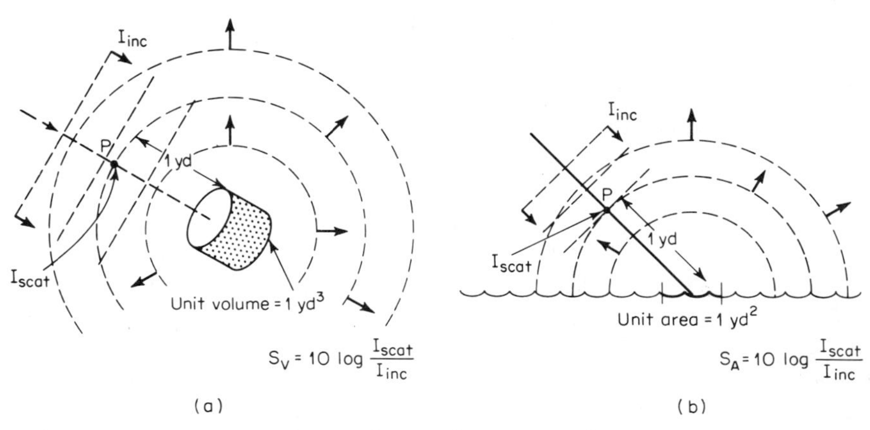

FIG. 8.2. Conceptual definitions of scattering strength for volume and surface scattering.

wave by a unit volume or area and reradiated in all directions. However, the parameter scattering strength, first introduced in 1954 (2), is rapidly replacing m_v and m_s in most sonar usage.

8.3 Equivalent Plane-wave Reverberation Level

In the sonar equations as previously written (Chap. 2), the term RL refers to the *equivalent plane-wave reverberation level.* This is the level of the axially incident plane wave which produces the same hydrophone voltage across the hydrophone terminals as that produced by the received reverberation. RL is therefore directly analogous to the echo level EL, with which it may be compared to obtain the echo-to-reverberation ratio across the hydrophone terminals.

In seeking an explicit, rather than a purely formal, general expression for RL, we need certain simplifying assumptions. Although these may seem to restrict the results to purely idealized situations, the resulting expressions for reverberation have been found to be useful for design and prediction purposes under many actual conditions. The assumptions necessary are the following:

1. Straight-line propagation paths, with all sources of attenuation other than spherical spreading neglected. The effect of absorption of sound in the sea on the reverberation level can, in any case, be readily allowed for.
2. A random, homogeneous distribution of scatterers throughout the area or volume producing reverberation at any one instant of time.
3. A density of scatterers so large that a large number of scatterers occur in an elemental volume dV or area dA.
4. A pulse duration short enough for propagation effects over the range extension of the elemental volume or area to be neglected.
5. An absence of multiple scattering, that is, the reverberation produced by reverberation is negligible.

8.4 Volume-reverberation Theory

Consider, in Fig. 8.3*a*, a directional projector in the ideal medium implied by the above assumptions, containing in its volume a large number of uniformly distributed scatterers. Let the beam pattern of the projector be denoted by $b(\theta,\varphi)$, and let the axial intensity at unit distance be I_0 (the source level SL $= 10 \log I_0$). By the definition of the beam-pattern function, the intensity at 1 yd in the θ, φ direction is $I_0 b(\theta,\varphi)$. At a distance r in this direction, let there be a small volume dV of volume scatterers. The incident intensity at dV will be, by assump-

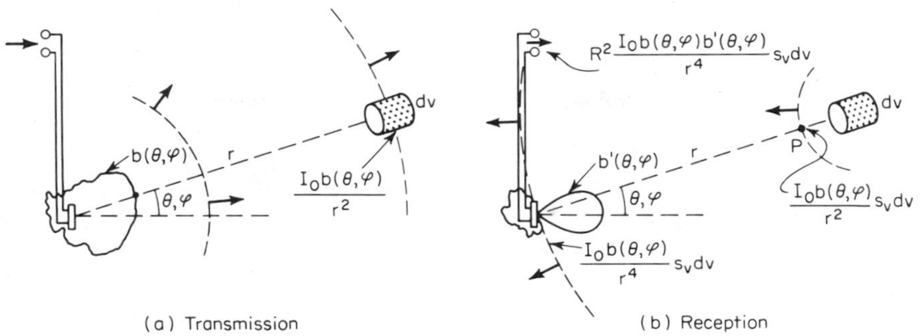

(a) Transmission (b) Reception

FIG. 8.3. Geometric view of volume scattering.

tion (1) above, $I_0 b(\theta,\varphi)/r^2$. The intensity of the sound backscattered by dV, at point P distant 1 yd back toward the source (Fig. 8.3*b*), will be $[I_0 b(\theta,\varphi)/r^2]\, s_v\, dV$, where s_v is the ratio of the intensity of the backscattering produced by a unit volume, at a distance of 1 yd from the volume, to the intensity of the incident sound wave. As defined in the previous section, the quantity 10 log s_v is the backscattering strength for volume reverberation and is denoted by the symbol S_v. S_v is dependent on the type and density of scatterers that give rise to the reverberation, as will be discussed thoroughly later on. Back in the neighborhood of the source, the reverberation contributed by dV will have the intensity $(I_0/r^4)b(\theta,\varphi)s_v\, dV$ and will produce a mean-squared voltage output of $R^2(I_0/r^4)b(\theta,\varphi)b'(\theta,\varphi)s_v\, dV$ at the terminals of a hydrophone having a receiving beam pattern $b'(\theta,\varphi)$ and voltage response R. (In simple sonars, the same transducer is used for projection and reception, and the patterns b and b' are identical; in many modern sonars, however, the projector and receiver beam patterns are different.) By assumption (3), the elemental volumes dV can be made so small that their total contribution can be summed up by integration, and by assumption (2), the coefficient s_v is a constant and can be removed from under the integral sign. The total reverberation hydrophone output will accordingly become $(R^2 I_0/r^4)s_v \int_v b(\theta,\varphi)b'(\theta,\varphi)\, dV$. Defining now the *equivalent plane-wave reverberation level RL* as the intensity, in decibel units, of an axially incident plane wave producing the same hydrophone output as the observed reverberation, we obtain

$$\mathrm{RL} = 10 \log \left(\frac{I_0}{r^4} s_v \int_v bb'\, dV \right)$$

where the θ, φ symbols have been dropped for economy in writing.

To proceed from here, we must now have a closer look at the elemental volumes dV. As shown in Fig. 8.4, let us take dV to be an infinitesimal

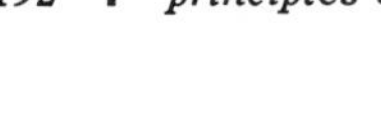

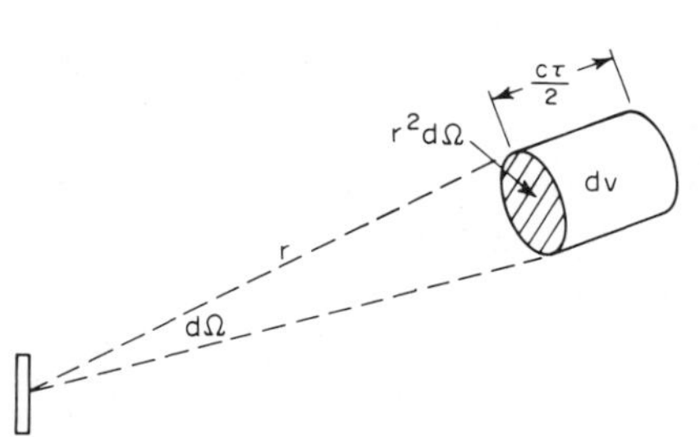

FIG. 8.4. Selection of the elemental volume for volume reverberation.

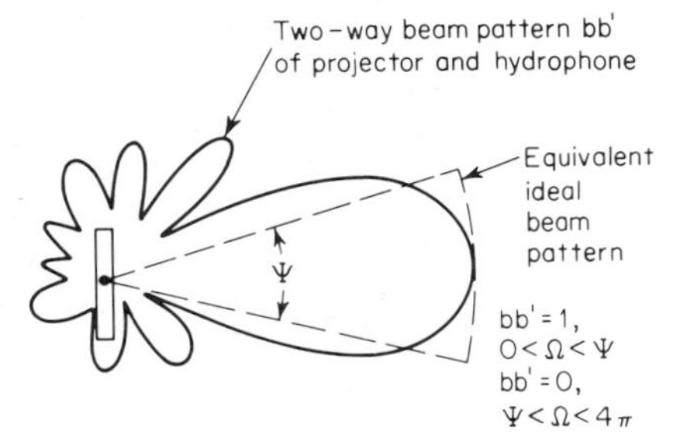

FIG. 8.5. Actual and equivalent beam patterns.

cylinder of finite length with ends normal to the incident direction. The area of the end face of this infinitesimal cylinder dV may be written $r^2\,d\Omega$, where $d\Omega$ is the elemental solid angle subtended by dV at the source. The extension in range of dV is such that, with a pulsed sonar, the scattering produced by all portions of dV arrives back near the source at the same instant of time. This requirement means that if one imagines a sound pulse to be incident upon the scatterers within dV, the scattering of the "front" end of the pulse by the "rear" scatterers in dV will arrive back at the source at the same instant as the scattering of the "rear" of the pulse by the "front" scatterers in dV. The extension in range must accordingly be $c\tau/2$, where τ is the pulse duration, and c the velocity of sound. Hence the elemental volume becomes

$$dV = r^2 \frac{c\tau}{2}\,d\Omega$$

and

$$\mathrm{RL} = 10 \log \left(\frac{I_0}{r^4} r^2 \frac{c\tau}{2} s_v \int bb'\,d\Omega \right)$$

where, by assumption (4), the range extension $c\tau/2$ is small compared to the range r.

The integral $\int bb'\,d\Omega$ can be interpreted as the equivalent beam width of the projector-hydrophone combination for volume reverberation. Let the actual two-way beam pattern bb' be replaced by an ideal pattern of unit relative response within a solid angle Ψ, and zero relative response beyond Ψ, as shown in Fig. 8.5. Then the width Ψ of this equivalent ideal beam is given by

$$\int_0^{4\pi} bb'\,d\Omega = \int_0^{\Psi} 1 \times 1\,d\Omega = \Psi$$

The solid angle Ψ may therefore be regarded as the opening angle of the ideal beam pattern, having a flat response within Ψ and none beyond, which, for reverberation, is equivalent to the actual two-way beam pattern.

In terms of Ψ, the expression for the equivalent plane-wave reverberation level becomes

$$\mathrm{RL}_v = 10 \log \left(\frac{I_0}{r^4} s_v \frac{c\tau}{2} \Psi r^2 \right)$$

which may be written as

$$\mathrm{RL}_v = \mathrm{SL} - 40 \log r + s_v + 10 \log V$$

$$V = \frac{c\tau}{2} \Psi r^2$$

The quantity V may be termed the *reverberating volume*, and is the volume of scatterers returning reverberation at any one instant of time.

In this geometric view, the reverberation is caused by those scatterers lying within the equivalent beam at the range of interest and contained in the volume V, determined by the equivalent beam width Ψ of the projector-hydrophone combination, the pulse duration τ, and the range r. Expressions for Ψ for some simple transducers are given in Table 8.1 in terms of the transducer dimensions and the width, in degrees,

TABLE 8.1 Equivalent Two-way Beam Widths in Logarithmic Units, Based on Ref. 65

Array	10 log Ψ db re 1 steradian	10 log Φ db re 1 radian
Integral expression	$10 \log \int_0^{2\pi} \int_{-\pi/2}^{\pi/2} b(\theta,\varphi)b'(\theta,\varphi) \times \cos\theta \, d\theta \, d\varphi$	$10 \log \int_0^{2\pi} b(0,\varphi)b'(0,\varphi) \cdot d\varphi$
Circular plane array, in an infinite baffle of radius $a > 2\lambda$	$20 \log \left(\frac{\lambda}{2\pi a}\right) + 7.7$ or $20 \log y - 31.6$	$10 \log \frac{\lambda}{2\pi a} + 6.9$ or $10 \log y - 12.8$
Rectangular array in an infinite baffle, side a horizontal, b vertical, with $a, b \gg \lambda$	$10 \log \frac{\lambda^2}{4\pi ab} + 7.4$ or $10 \log y_a y_b - 31.6$	$10 \log \frac{\lambda}{2\pi a} + 9.2$ or $10 \log y_a - 12.6$
Horizontal line of length $l > \lambda$	$10 \log \frac{\lambda}{2\pi l} + 9.2$ or $10 \log y - 12.8$	$10 \log \frac{\lambda}{2\pi l} + 9.2$ or $10 \log y - 12.8$
Nondirectional (point) transducer	$10 \log 4\pi = 11.0$	$10 \log 2\pi = 8.0$

y is the half angle, in degrees, between the two directions of the two-way, or product, beam pattern in which the response is 6 db down from the axial response. That is, y is the angle from the axis of the two-way beam pattern such that $b(y)b'(y) = 0.25$. For the rectangular array, y_a and y_b are the corresponding angles in planes parallel to the sides a and b.

of the two-way beam pattern. For other projector-hydrophone combinations, Ψ can be found by integration of the pattern function bb', as indicated by the integral expressions of Table 8.1.

For a sonar having a nondirectional projector and hydrophone, $b = b' = 1$, and

$$\Psi = \int bb'\, d\Omega = 4\pi$$

The ratio of the reverberation received by such a sonar to that received by a directional sonar of the same source level is $4\pi/\int bb'\, d\Omega$. During World War II the quantity $10 \log (4\pi/\int bb'\, d\Omega)$ was termed (3) the *directivity index for volume reverberation*, in analogy with its counterpart for isotropic noise, and given the symbol J_v.* In this alternative formulation the additional 4π is absorbed into the term s_v, and the expression for reverberation level becomes

$$\text{RL} = \text{SL} - 40 \log r - J_v + 10 \log m_v + 10 \log \left(\frac{c\tau}{2} r^2 \right)$$

where $m_v = 4\pi s_v$ is the backscattering cross section of a unit volume.

A geometric, rather than a purely mathematical, view of reverberation is valuable in practical sonar problems. The magnitude and dimensions of the reverberating volume, its location in the sea, and the kind of scatterers it might contain are all useful aids for intelligent performance predictions of active sonar equipments.

8.5 Surface-reverberation Theory

By surface reverberation is meant reverberation produced by scatterers distributed over a nearly plane surface, rather than throughout a volume. Most notably, such scattering surfaces are the surface and the bottom of the sea.

The derivation of an expression for the equivalent plane-wave level of surface reverberation proceeds as that for volume reverberation. The result is

$$\text{RL}_s = 10 \log \left(\frac{I_0}{r^4} s_s \int b(\theta,\varphi) b'(\theta,\varphi)\, dA \right)$$

where dA is an elemental area of the scattering surface and $10 \log s_s = S_s$ is the *scattering strength for surface reverberation*. If dA is taken as a portion of a circular annulus in the plane of the scatterers, with center

* Originally both J_v and DI were defined so as to be *negative*. Current practice is to define DI and J_v as *positive* quantities.

directly above or below the transducer, we may write

$$dA = \frac{c\tau}{2} r \, d\varphi$$

where $d\varphi$ is the subtended plane angle of dA at the center of the annulus. The reverberation level becomes

$$\mathrm{RL} = \frac{I_0}{r^4} s_s \frac{c\tau}{2} r \int_0^{2\pi} b(\theta,\varphi)b'(\theta,\varphi) \, d\varphi$$

This integral is difficult to evaluate analytically at high angles between the transducer beam and the scattering surface. Some useful graphs for this case have been given in a paper by Urick and Hoover (4). In the more usual sonar case, where the axis of the beam is only slightly inclined toward the scattering surface, the scattering surface nearly corresponds to the $\varphi = 0$ plane of the transducer beam pattern, so that

$$\mathrm{RL}_s = 10 \log \left(\frac{I_0}{r^4} s_s \frac{c\tau}{2} r \int_0^{2\pi} b(0,\varphi)b'(0,\varphi) \, d\varphi \right)$$

The plane angle of the equivalent ideal beam is such that

$$\int_0^{2\pi} b(0,\varphi)b'(0,\varphi) \, d\varphi = \int_0^{\Phi} 1 \times 1 \, d\varphi = \Phi$$

and, therefore,

$$\mathrm{RL}_s = 10 \log \left(\frac{I_0}{r^4} s_s \frac{c\tau}{2} r\Phi \right)$$

or

$$\mathrm{RL}_s = \mathrm{SL} - 40 \log r + S_s + 10 \log A$$

$$A = \frac{c\tau}{2} \Phi r$$

Expressions for Φ are included in Table 8.1. The area A is the area of surface of scattering strength S_s lying within the ideal beam width Φ which produces the same reverberation as that actually observed. In Fig. 8.6, the reverberating area A is illustrated for a downward-pointing transducer receiving bottom reverberation. The point P, directly beneath, is the center of a circular annulus of which only the portion A, of width Φr and range extension $c\tau/2$, returns reverberation at any one instant of time. S_s, in general, varies with the angle θ of the incident sound beam upon the scattering surface; for a transducer at a constant distance above or below the scattering surface, S_s will therefore vary with range or time after emission of the pulse. As in the case of volume reverberation, it is important that the design engineer be able to visualize

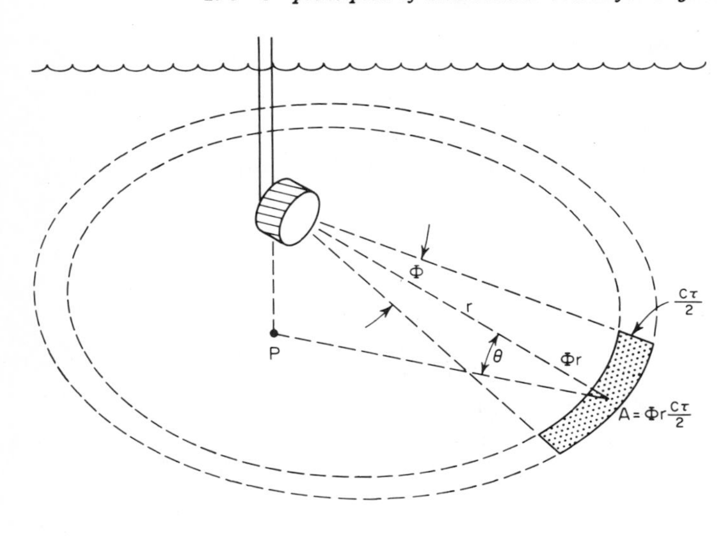

FIG. 8.6. The reverberating area for surface reverberation. The transducer is visualized as a downward-pointing circular plane array receiving bottom reverberation.

where in the sea the sound emitted by his active sonar is going and what scatterers it is likely to encounter.

8.6 Target Strength and Scattering Strength

As mentioned above, the scattering strength for volume or surface reverberation is defined as 10 times the logarithm to the base 10 of the ratio of the scattered intensity produced by 1 yd^3 of ocean or 1 yd^2 of scattering surface, respectively, to the incident intensity. The scattered intensity is referred to a distance of 1 yd from the unit volume or surface in the direction back toward the source of sound, unless some other direction is specified. When so defined, the scattering strength of the sea and its boundaries and the target strength of sonar targets are essentially identical parameters. Scattering strength is the target strength, for scattering, of a unit volume or area.

Accordingly, in reverberation backgrounds, it becomes easy, in simple situations, to find the echo-to-reverberation ratio for a target of known target strength. The level of the echo from the target is (Chap. 2)

$$\mathrm{EL} = \mathrm{SL} - 2\mathrm{TL}_E + \mathrm{TS}$$

where SL = source level
TL_E = transmission loss for echo
TS = target strength

The reverberation level, or the level of the axial plane wave producing the same mean-squared hydrophone voltage as the reverberation, is

$$RL = SL - 2TL_R + S_{s,v} + 10 \log A,V$$

where TL_R is the transmission loss for reverberation, and $S_{s,v}$ and A,V are the scattering strength and the area or volume of the reverberating area or volume, respectively. If, as is usually the case, the scattering volume or area surrounds the target so that the target lies in the midst of the reverberation-producing area or volume, TL_R and TL_E are the same, and the echo-to-reverberation ratio becomes

$$EL - RL = TS - (S_{s,v} + 10 \log A,V)$$

The echo-to-reverberation ratio is thus simply the difference of two "strengths," one for the echo and the other for the scattering, without reference to source or medium characteristics. The target must, however, lie in the vicinity of the reverberation-producing scatterers in order for the two transmission losses to be the same and in order to avoid beam-pattern complications. As an example, let a target of target strength +10 db lie on a sea bottom of scattering strength −30 db. The echo-to-reverberation ratio is desired for conditions of range, pulse duration, and equivalent beam width such that 10 log A equals 15 db. For these conditions, the echo-to-reverberation ratio becomes simply 10 + 30 − 15 = +25 db. At other ranges and times, the echo-to-reverberation ratio will be different because both the angle of incidence of the sound beam upon the bottom and the size of the reverberating area will vary with range or time after emission of the sonar pulse.

8.7 Surface Scattering by a Layer of Volume Scatterers

Some kinds of scatterers in the sea, such as the deep scattering layer or a layer of air bubbles just below the sea surface, lie in layers of finite thickness rather than being diffusely distributed throughout an irregular volume. Such *layered reverberation* is best considered as a form of surface reverberation. If the average volume scattering strength of the layer is S_v, the corresponding surface scattering strength is

$$S_s = S_v + 10 \log H$$

where H is the layer thickness in yards. In general, if the volume scattering strength is not constant within the layer,

$$S_s = 10 \log \int_0^H s_v(h)\, dh$$

where $s_v(h)$ is the volume scattering strength at a depth h within the layer, expressed in linear units ($S_v = 10 \log s_v$).

8.8 Reverberation Level for Short Transients

Hitherto we have considered the reverberation from a sonar pulse having a constant average intensity I_0 during a fixed definite duration τ. Although these conditions apply for many sonars, they do not apply for short transients of indefinite waveform and duration.

For such waveforms, it can be shown that the product $I_0\tau$ occurring in the reverberation equations can be replaced by the integral $\int_0^\infty i(t)\, dt$, in which $i(t)$ is the instantaneous intensity of the source at 1 yd, and the integration is taken for the entire duration of the transient. A restriction to short transients is necessary in order for the range r and scattering strength S in the reverberation equations to be taken as constants. The integral, which reduces to $I_0\tau$ for flat-topped pulses of duration τ, is the energy flux density of the source at a distance of 1 yd, in units of the energy flux density of a plane wave having an rms pressure of 1 dyne/cm^2 taken for a period of 1 sec. This unit amounts to 10 log ($1/\rho c$), or -52 db below 1 erg/cm^2. A source of energy flux density E in these units at 1 yd can be said to have an energy-flux-density source level SL_E defined by

$$\text{SL}_E = 10 \log E$$

The reverberation equations become

$$\text{RL}_v = \text{SL}_E - 20 \log r + S_s + 10 \log \Psi + 10 \log \frac{c}{2}$$

$$\text{RL}_s = \text{SL}_E - 30 \log r + S_v + 10 \log \Phi + 10 \log \frac{c}{2}$$

The quantity 10 log ($c/2$), where c is the velocity of sound, is a kind of conversion factor necessary to equate intensity level, on the left, to an energy-density level SL_E, on the right; with 1 yd as the unit of distance, c must be taken in yards per second so that 10 log ($c/2$) $= 29$ db. For nondirectional sonars, $\Psi = 4\pi$, and $\Phi = 2\pi$.

These expressions for the reverberation level in terms of the energy-density of a short transient were first obtained by Urick (5) and have been independently derived and used for the measurement of sea-surface scattering strengths with explosive sources by Chapman and Harris (6). For explosives, the quantity SL_E has been established for certain conditions by the work of Weston (7), Stockhausen (8), and Christian and Blaik (9) (Sec. 4.3).

8.9 Air Bubbles in Water

Occurrence Air bubbles, or more generally gas bubbles, occur in various forms in the sea. They occur immediately below the surface of the sea, where they are produced by the breaking of waves and are carried by turbulence beneath the surface. They occur within certain biological organisms, as noted above, as, for example, the swim bladders of fish. They are generated in the wakes of ships, where they are known to persist for long periods of time. Free air bubbles in the sea are quite small, since the larger bubbles tend to rise quickly to the surface. They form only a very small percentage, by volume, of the sea in which they occur. Nevertheless, because air has markedly different density and compressibility than seawater, and because of the resonant characteristics of bubbles, the suspended air content of seawater has a profound effect upon underwater sound. Some of these effects are considered in the following sections.

Physical Processes When a sound wave strikes an air bubble, the bubble partakes, to an extent, in the compression and rarefactions of the incident sound wave. This response of the bubble to the excitation of the incident sound wave depends upon the frequency of the sound wave and the size of the bubble. At a certain frequency, a resonant response occurs when the inertial properties of the system—the mass of the bubble plus a portion of the surrounding water—become matched to the compressibility of the system. At resonance, a maximum oscillation of the bubble size develops, and a maximum amount of energy is extracted from the incident sound wave. A portion of this energy is scattered in all directions by the pulsating bubble, and the remainder is converted to heat—a process brought about by heat conduction inside the bubble, the viscosity of the surrounding water, and the surface tension at the bubble surface. The oscillating bubble may therefore be viewed as intercepting a portion of the incident sound wave characterized by the *extinction cross section* of the bubble and reradiating it as scattered sound in all directions, as well as converting it to heat. These processes depend upon the frequency of oscillation, the size of the bubble, and the physical and thermodynamic properties of the two mediums involved.

The theory of the resonant behavior of air bubbles in water was worked out in World War II, and may be found in one of the Summary Technical

Reports of the NDRC (10), as well as in the book by Albers (11). This theory will not be repeated here, but only the results will be given, along with an occasional reference to subsequent literature.

Sound Velocity in a Bubbly Medium When air is *dissolved* in water, measurements show (12) that the effect on the velocity of sound is completely negligible, even when the water is completely saturated with air. When, however, air is *suspended* as tiny bubbles, its effect is profound; a minute amount of air substantially reduces the velocity of sound in the bubbly fluid. When the bubbles are all much smaller than the resonant size, the velocity of sound has been found experimentally (13) to be given by the simple mixture theory. By this theory (14), the compressibility and the density of air and water, which determine the sound velocity, are weighted by the proportionate amount of the two mediums present in a given volume of mixture. If the compressibilities of mixture, air, and water are denoted by κ, κ_a, and κ_w, respectively, and if the corresponding densities are ρ, ρ_a, and ρ_w, then the sound velocity of a mixture of air and water containing a fraction β of air by volume is, by the mixture theory, for small β,

$$v = \left(\frac{1}{\rho\kappa}\right)^{1/2} = \left\{\frac{1}{[\beta\rho_a + (1-\beta)\rho_w][\beta\kappa_a + (1-\beta)\kappa_w]}\right\}^{1/2}$$
$$= \left[\frac{1}{\rho_w\kappa_w(1+\beta\kappa_a/\kappa_w)}\right]^{1/2} = v_w\left[\frac{1}{1+\beta\kappa_a/\kappa_w}\right]^{1/2}$$
$$= v_w\left(\frac{1}{1+2.5\times 10^4\beta}\right)^{1/2}$$

since $\kappa_a \gg \kappa_w$, $\rho_a \ll \rho_w$, and $\beta \ll 1$. The formula indicates that a volume fraction of air of only 0.01 percent ($\beta = 10^{-4}$) reduces the sound velocity to 53 percent of the velocity v_w of air-free water.

By contrast, at high frequencies well beyond the resonant frequency of the smallest bubble present in the mixture, the effect of suspended air content is negligible. In this region of bubble size, large compared to the resonant size, the two velocities are related by (15)

$$c^{-2} = c_0^{-2} - \frac{4\pi na}{(2\pi f)^2}$$

where a = bubble radius

n = number of bubbles in a unit volume

The second term on the right is small compared to the first for small bubble concentrations.

In the vicinity of resonance, large changes in sound velocity take place. The velocity in a bubbly mixture over a wide frequency range surrounding the resonant frequency is illustrated by Fig. 8.7, which shows the smoothed velocity curve obtained experimentally by Fox, Curley, and Larson (16) in a cloud of uniform air bubbles having an average diameter of 0.011 cm and a volume concentration β of 2×10^{-4}. The velocity passes through its value for bubble-free water at the resonant frequency.

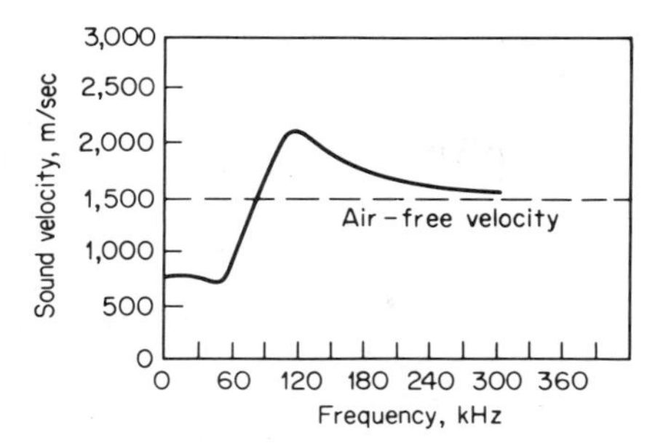

FIG. 8.7. Curve showing the measured sound velocity in a cloud of bubbles of uniform size (diameter 0.011 cm) over a wide range of frequency. (Ref. 16.)

Resonant Frequency At a certain frequency, a bubble of a given size is in resonance with the exciting sound wave and a maximum extinction cross section occurs. This frequency is (Ref. 10, eq. 18)

$$f_r = \frac{1}{2\pi a}\sqrt{\frac{3\gamma P_0}{\rho}}$$

where γ = ratio of specific heats of the gas
P_0 = hydrostatic pressure
ρ = density of the water

For air ($\gamma = 1.4$) at zero depth in the sea ($P_0 = 1$ atm) this becomes

$$f_r = \frac{326}{a}$$

where f_r = resonant frequency, Hz
a = bubble radius, cm

At a depth of d ft in water, the corresponding expression is

$$f_r = \frac{326}{a}\sqrt{1 + 0.03d}$$

The value of $k_r a = 2\pi a/\lambda_r$ for a resonant bubble at zero depth is 0.0136.

Damping Constant A measure of the above-mentioned dissipation processes is the damping constant δ associated with bubble pulsation. δ is the reciprocal of the Q of the bubble oscillation and is dimensionless. In terms of free oscillations, δ is the reciprocal of the number of cycles required for the amplitude of motion to decrease to $e^{-\pi}$ of its original value; for forced oscillations, δ is defined as

$$\delta = \frac{f_2 - f_1}{f_r}$$

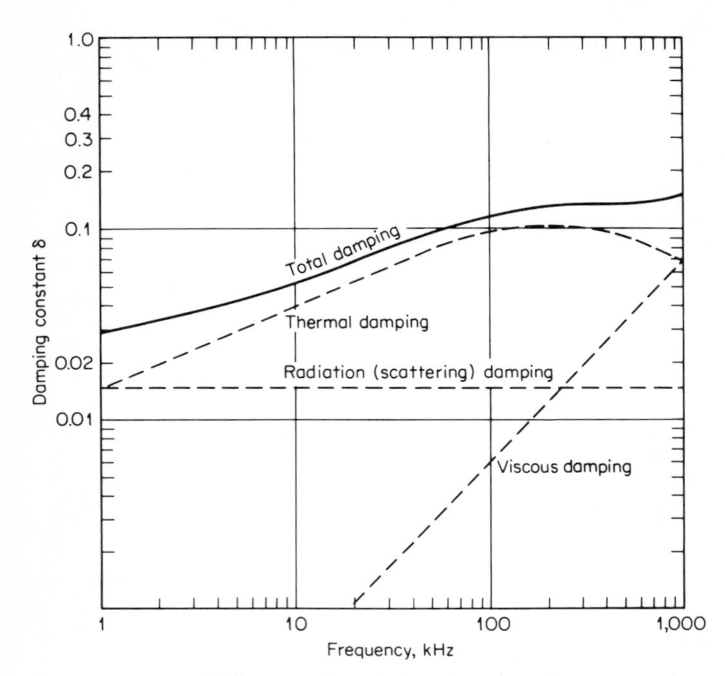

FIG. 8.8. Theoretical damping of resonant air bubbles in water. [*After Devin* (*Ref. 17*).]

where f_2 and f_1 are the frequencies above and below resonance at which the square of the amplitude diminishes to one-half of its value at the resonant frequency f_r (half-power points).

The damping constant δ of air bubbles in water has been determined by various investigators using different methods, with none-too-concordant results. A survey made by Devin (17) of the theory and observations of oscillatory damping of air bubbles in water shows that the damping constant predicted by hydrodynamic and thermodynamic theory is valid. Figure 8.8 shows the theoretical damping constant δ for resonant air bubbles, together with the contributions to the total damping produced by (1) heat conduction during bubble pulsations, (2) radiation of sound by the pulsating bubble, and (3) the viscosity of the surrounding water.

Extinction, Scattering, and Absorption Cross Sections In terms of the damping constant δ, the extinction cross section at frequency f of a bubble of radius a resonant at frequency f_r is (Ref. 10, eq. 43)

$$\sigma_e = \frac{4\pi a^2 \delta / k_r a}{(f_r^2/f^2 - 1)^2 + \delta^2}$$

where $k_r a = 2\pi a/\lambda_r = 2\pi f_r a/c_0$. The absorption cross section is (Ref. 10, eq. 45)

$$\sigma_a = \frac{4\pi a^2(\delta/k_r a - 1)}{(f_r^2/f^2 - 1)^2 + \delta^2}$$

and the *scattering cross section* is (Ref. 10, eq. 34)

$$\sigma_s = \frac{4\pi a^2}{(f_r^2/f^2 - 1)^2 + \delta^2}$$

so that

$$\sigma_e = \sigma_a + \sigma_s$$

These cross sections have a maximum at the resonant frequency and fall off with frequency away from resonance in the same manner as the response curve of a tuned electric circuit containing a resistive loss.

Target Strength and Scattering Strength The target strength of a single bubble becomes

$$\text{TS} = 10 \log \frac{\sigma_s}{4\pi} = 10 \log \frac{a^2}{(f_r^2/f^2 - 1)^2 + \delta^2}$$

and the scattering strength of a unit volume (1 yd^3) of water containing n bubbles/yd^3, each of scattering cross section σ_s, is (multiple scattering neglected)

$$S_v = 10 \log \frac{n\sigma_s}{4\pi} = 10 \log \frac{na^2}{(f_r^2/f^2 - 1)^2 + \delta^2}$$

where the units of the radius a must be taken in yards.

Attenuation Coefficient The attenuation of a plane wave propagating through water containing bubbles is expressed by an attenuation coefficient defined by

$$\alpha = 10 \log \frac{I_1}{I_2}$$

where I_1 and I_2 are the intensities of the wave at two points located a unit distance apart in the direction of propagation. If there are n resonant bubbles/yd^3, each with extinction cross section σ_e yd^2, the attenuation coefficient is (Ref. 10, eq. 53)

$$\alpha = 4.34\, n\sigma_e \qquad \text{db/yd}$$

Distributions of Bubbles When, as is usual, the bubbles are not of uniform size, the effective cross section must be obtained by integration over the range of sizes of the bubbles. If the number of bubbles per cubic yard with radii between a and $a + da$ is denoted by $n(a)\, da$, the effective scattering cross section is

$$\sigma = \int_0^\infty n(a)\sigma(a)\, da$$

where $\sigma(a)$ = extinction, absorption, or scattering cross section
$n(a)$ = size-distribution function of bubble population

Because only bubbles of near-resonant size will make a large contribution to σ, and because $n(a)$ does not vary rapidly with a, the integral can be easily evaluated by approximate methods for a given bubble-distribution function.

8.10 Volume Reverberation

At the time when the first analytic studies in reverberation were made, it was realized that under certain conditions the reverberation observed in sonars must come from the body of the sea. These conditions involved downward refraction in deep water when the sound beam of a horizontally pointing sonar was bent downward deep into the body of the sea. It was also observed that the volume scatterers, whatever they may be, were not uniformly distributed in depth, but tended to be concentrated in a diffuse layer called, at the time, the ECR layer, in recognition of its three discoverers, C. F. Eyring, R. J. Christensen, and R. W. Raitt (18). The present name for this layer is the deep scattering layer, abbreviated DSL, which is now known to be more complex than previously realized.

The scatterers responsible for volume scattering appear undoubtedly to be biological in nature, that is, they are a part of the marine life existing in the sea. Nonbiological sources such as dust and sand particles, thermal microstructure (19), and turbulence of natural or man-made origin, such as ships' wakes, can be shown by analysis to be usually insignificant contributors to the scattering strength observed at sea.

Although the source of volume scattering in the sea has been definitely established as biological, the organisms responsible for it have not been clearly identified. Many studies, such as that of Johnson, Backus, Hersey, and Owen (20) using short pulses of sound, deep-towed net hauls, and underwater photography, have been made of the organisms most likely responsible for the volume scattering of the deep scattering layer. Shrimplike euphausids, squid, and copepods have been postulated; and fish—more particularly their gas-filled swim bladders (21)—are the most likely source of the frequency-selective acoustic behavior of the layer at lower frequencies. Undoubtedly, the layer is a complex aggregate of different biological organisms, and therefore may be expected to possess a scattering strength varying with frequency, location, season, and even time of day.

Variation with Depth The variation of volume scattering strength with depth has been studied by using depth-sounding sonars pointing vertically downward and measuring the return as a function of time (and therefore depth) before the onset of the bottom return. An example of relative reverberation level as a function of depth, as observed in World War II studies, is shown in Fig. 8.9. When the relative reverberation level is converted to scattering strength, the observations show that S_v generally decreases with depth, except for greater values within the deep scattering layer. Figure 8.10*a* and *b* illustrates measurements of S_v at different depths in two Pacific Ocean areas at a frequency of 24 kHz. From many individual measurements, the average smoothed decrease of S_v with depth may be estimated to be about 5 db/1,000 ft, with increases of 5 to 15 db within the DSL. In the abyssal depths of the sea, a lesser decrease may be expected, whereas at shallow depths, a greater decrease of scattering coefficient is likely just beneath the sea surface. This variation with depth is consistent with the general distribution of biological organisms with depth in the sea.

The DSL is well known to exhibit a diurnal migration in depth and to be at a greater depth by day than by night, with a rapid depth change near sunrise and sunset. This diurnal migration extends in depth over many hundreds of feet and tends to be just sufficient (22) to keep the intensity of light illumination constant at the depth of the layer.

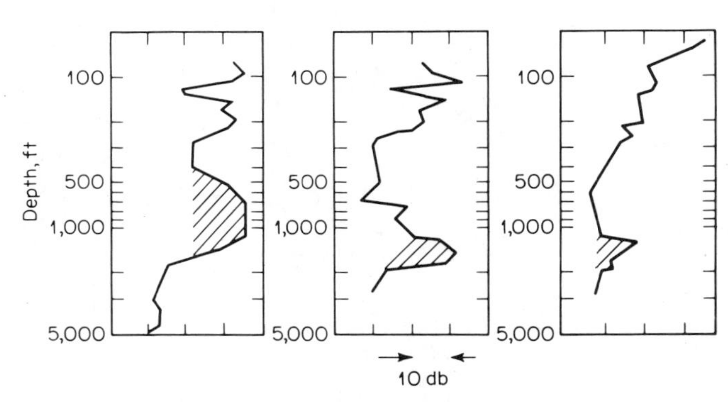

FIG. 8.9. Observed depth variation of reverberation observed with a downward-pointing transducer. Frequency 10 kHz. Relative levels only; logarithmic depth scale. (Ref. 3, Fig. 7.) The crosshatching shows the deep scattering layer.

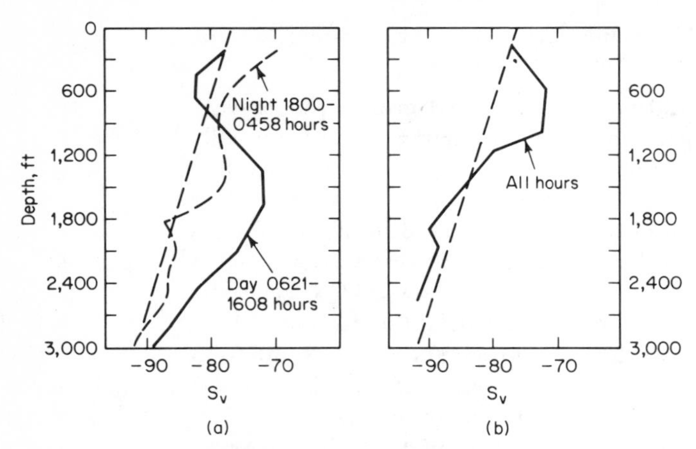

FIG. 8.10. Volume scattering strength at 24 kHz versus depth as measured in two Pacific Ocean areas: (a) Guadalupe Island area, latitude 29°N; (b) Queen Charlotte Island area, latitude 51°N. Dashed lines are estimated minimum values. (United States Navy Electronics Laboratory data quoted in Ref. 65.)

Variation with Frequency At frequencies in excess of 10 kHz, S_v has been found to rise with frequency at the rate of about 3 to 5 db/octave. Examples of observed data are shown in Fig. 8.11. The increase with frequency is far less than would be expected from the fourth-power law (12 db/octave) of Rayleigh scattering, which applies only for fixed, rigid obstacles smaller in size than one wavelength. The scattering most likely originates in scatterers of different sizes, many of which are of the same dimensions as the wavelength, and all of which are deformable and free to move in the sound field.

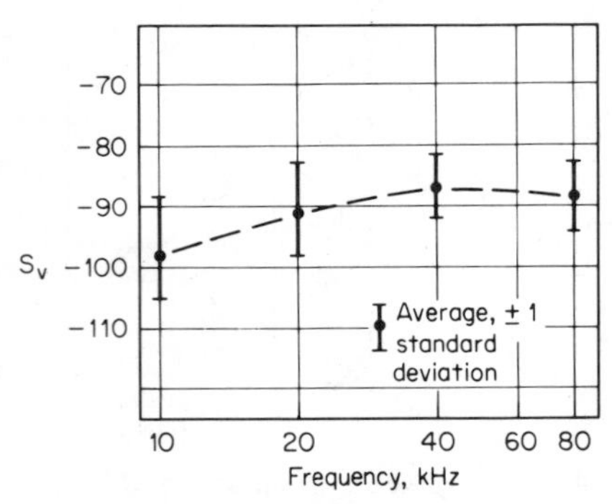

FIG. 8.11. Variation of volume scattering strength with frequency between 10 and 80 kHz, averaged from data at nine locations at depths less than 250 ft. The low values of S_v are unexplained. (Ref. 3, Fig. 13.)

At lower frequencies, observations with broadband short explosive pulses have revealed the existence of numerous and highly variable peaks in the volume reverberation spectra, each corresponding to a different depth in the sea. Different frequencies are found to peak at different depths—

thereby indicating a multilayered structure within the DSL. This was most clearly indicated by the observations of Hersey, Backus, and Hellwig (23) using explosive sound sources at three locations in the western North Atlantic. At all locations, scattering layers with different resonant frequencies between 5 and 20 kHz were found. Most, but not all, layers showed a diurnal vertical migration, accompanied by a change in frequency. The depth, intensity, and resonant frequencies varied greatly from locality to locality.

Subsequent studies by Marshall and Chapman (24) using explosives confirmed this complex frequency-depth structure of the layer. For example, at one location northwest of Bermuda, a frequency analysis of the explosive reverberation showed a peak frequency of 5.4 kHz by day that shifted suddenly at sunset to a frequency of 4.7 kHz. The depth of the resonant layer (corresponding to the time of arrival of the onset of the peak frequency) changed from 2,200 ft by day to 1,600 ft by night, with a layer thickness of about 500 ft. The frequency change was interpreted as consistent with the behavior with pressure of a resonant air bubble of constant volume—suggesting the swim bladders of bathypelagic fish.

In summary, at frequencies of several kilohertz, the spectrum of volume reverberation originating at depths above about 3,000 ft is likely to contain one or more broad peaks of different frequency at different depths, locations, and times of day. Such frequency peaks apparently do not occur below about 1 kHz, since the lowest frequency thus far observed (25) is 1.6 kHz.

Deep-scattering-layer Characteristics Since the DSL will often be the dominant source of volume reverberation in broad-beamed or downward-tilted sonars, its acoustic characteristics will be of interest to the sonar engineer. These may be summarized as follows:

1. The DSL occurs in all oceans and is a regular acoustic and biological feature of the world's oceans.

2. The DSL ascends at sunset, descends at sunrise, and remains at constant depth during daytime and nighttime hours.

3. The depth of the DSL can be expected to lie between 600 and 3,000 ft, with a representative depth of 1,300 ft and thickness of 300 ft.

4. At frequencies near 24 kHz, the volume scattering strength within the layer is -70 to -80 db, but is variable from place to place within the layer. At lower frequencies, S_v within the layer is variable and unpredictable.

5. S_v is frequency selective at frequencies between 1.6 and 12 kHz, with different frequencies occurring at different depths. The DSL is therefore multilayered and contains layers of different resonant frequency

at different depths, not all of which participate in the diurnal depth migration.

6. Layers of scatterers occur in shallow water as well. Most notable are schools of fish feeding at a particular depth. In the North Sea, a layer of scatterers, presumably biological, has been observed (26) at a depth of 15 fathoms, just at the top of a strong thermocline.

Wakes A target frequently encountered at sea, especially under conditions when ship maneuvers are involved, is the wake left behind by a ship or submarine. Because a ship's wake is an extended, rather than a discrete, target, its echo has some of the characteristics of reverberation, even though it may appear as a clear-cut, though extended, underwater target.

The acoustic characteristics of wakes were much studied during World War II, and a thorough discussion of the subject may be found in one of the Summary Technical Reports of the NDRC (10).

The wake of a ship is a length of bubbly turbulent water created by the ship's propeller. It extends an indefinite distance astern and gradually dissipates as the air bubbles rise to the surface or dissolve. Its width is initially that of the wake-laying ship, but gradually increases with distance astern. The thickness of the wake of a surface ship is initially about twice the draft and varies with distance astern in an uncertain manner.

The acoustic backscattering of wakes is described by a parameter termed *wake strength*, defined as the scattering strength *per yard of wake length*.* Thus, wake strength is defined per unit length of wake in a manner analogous to volume scattering strength, per unit volume and area, respectively. The level EL of the echo from a wake of strength W is accordingly given by a pair of equations similar to those for surface reverberation:

$$\mathrm{EL} = \mathrm{SL} - 40 \log r + W + 10 \log L$$
$$L = \Phi r$$

where the term 40 log r implies spherical spreading in the absence of absorption, and the echoing length L of the wake is the product of the equivalent plane-angle beam width Φ, and r is the range in yards. This formulation applies only for long pulse durations such that the dimensions of the wake returning sound at any one instant of time are determined by the beam width and range alone.

Some typical values of W, measured at distances a few hundred yards astern of the wake-producing vessel, are given in Table 8.2. W was found

* W was originally defined (10) per *foot* of wake. A reference length of 1 yd is used here for dimensional conformity with scattering strength. Values of W quoted in Ref. 10 have accordingly been increased by 10 log 3 = 5 db.

TABLE 8.2 Wake Strengths W for Various Ships and Submarines (Ref. 10, Tables 4 and 6)

Ship	W, db	Frequency, kHz	Speed, knots	Depth, ft
Surface ships:				
CVE's and AP's	−9	24		
DD's and DE's	−11	24		
Laboratory yachts ("Scripps" and "Jasper")	−15	24		
Small boats	−19	24		
Submarines				
USS-S23 (SS-128)	−13	60	9.5	Surfaced
	−21	60	6	90
USS-S34 (SS-139)	−8	45	9.5	Surfaced
	−18	45	6	90
USS "Tilefish" (SS-307)	−8	45	9.5	Surfaced
	−15	45	6	90
USS-S-18 (SS-123)	−28	45	6	45

to decrease with the "age" of the wake (in effect, the distance astern) at the rate of about 1 db/min and to increase with the size of the vessel (for surface ships). No clear variation with frequency over the range 15 to 60 kHz was evident.

8.11 Sea-surface Reverberation

Because of its roughness and the possibility of the occurrence of entrapped air bubbles just beneath it, the sea surface is a profound scatterer of sound. This scattering has long been noted to be responsible for the reverberation received in horizontally pointing sonars under conditions such that the sound beam is not carried down into the depths of the sea by refraction. It is analogous in many ways to the "sea clutter" of radar.

Sea-surface scattering strengths have been measured by nondirectional (mostly explosive) sources and receivers and by directional sonars in which a sound beam is pointed upward to intercept the sea surface at a desired angle. The scattering strength of the sea surface has been found to vary with angle, frequency, and the roughness of the surface, measured by the speed of the wind above it.

Variation with Angle and Wind Speed The first systematic postwar investigation of sea-surface backscattering appears to have been published by Urick and Hoover (4) through measurements of the sea return at 60 kHz made with a tiltable directional transducer hung beneath the surface from a float. The results of this study are shown in Fig. 8.12. S_s is seen to increase with wind speed at low grazing angles, but to decrease with wind speed at high angles near normal incidence, with a "crossover" near 80°. This peculiar behavior was attributed (27) to different processes responsible for the backscattering over different ranges of angle. At low angles (<30°), the scattering was attributed to a layer of air bubbles just beneath the sea surface, for which observational evidence was presented; at intermediate angles (30 to 70°), scattering was invoked as the dominant process, and at high angles (70 to 90°), the return was postulated to originate as reflection from normally inclined wave facets acting as tiny acoustic mirrors. Scattering curves at 60 kHz similar to those of Fig. 8.12, but without the flattening at low angles attributed

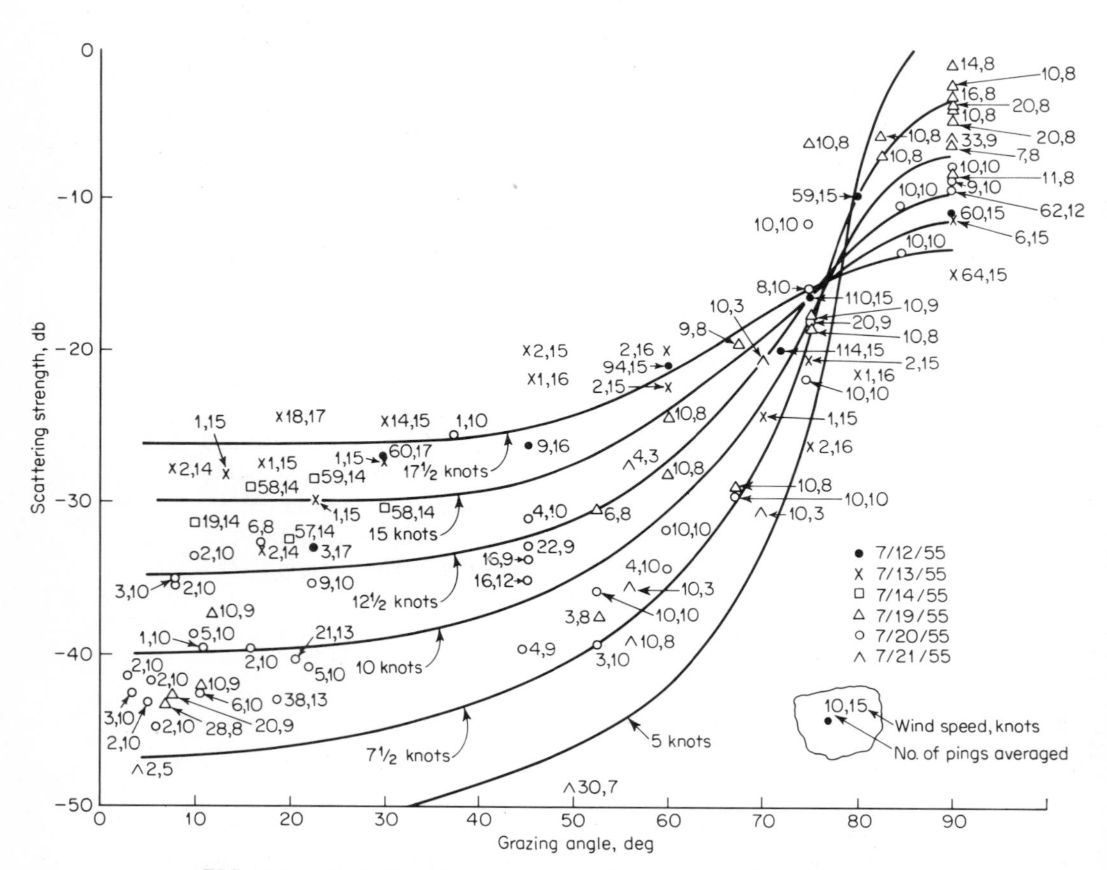

FIG. 8.12. Variation of sea-surface scattering strength at 60 kHz with angle at different wind speeds off Key West, Florida. Plotted points show the number of pulses averaged and the wind speed for individual determinations. (*Ref. 4.*)

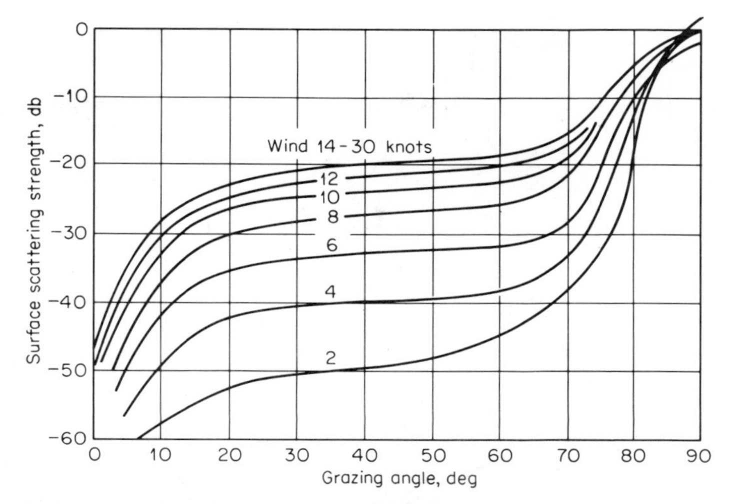

FIG. 8.13. ***Variation of sea-surface scattering strength at 60 kHz with angle at different wind speeds in Dabob Bay, Puget Sound, Washington.*** (***Ref. 28.***)

to a bubbly layer, have been derived by Garrison, Murphy, and Potter (28) from measurements in Dabob Bay, Puget Sound, Washington. These curves are shown in Fig. 8.13.

At lower frequencies, explosive sound sources together with a nearby nondirectional hydrophone have been used to obtain the scattering strength of the sea surface. In this experimental arrangement, illustrated in Fig. 8.14, the time after detonation is related to the angle between the sea surface and the incident direction, an angle which becomes progressively smaller with increasing time. Fig. 8.15 shows curves for a number of octave bands, with wind speed (in knots) as a parameter, as reported by Chapman and Scott (29) from observations over a 62-hr period at an open-ocean location in the

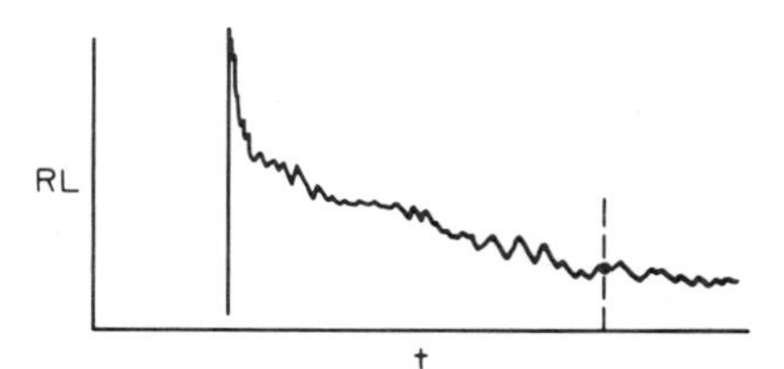

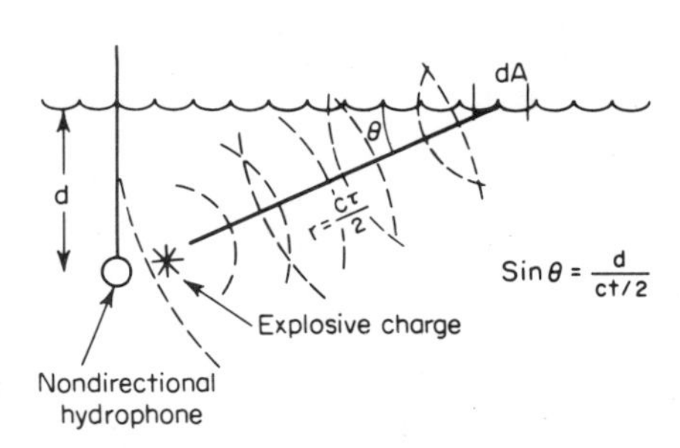

FIG. 8.14. ***Measurement of sea-surface scattering strength at different angles with a nondirectional hydrophone and an explosive charge.***

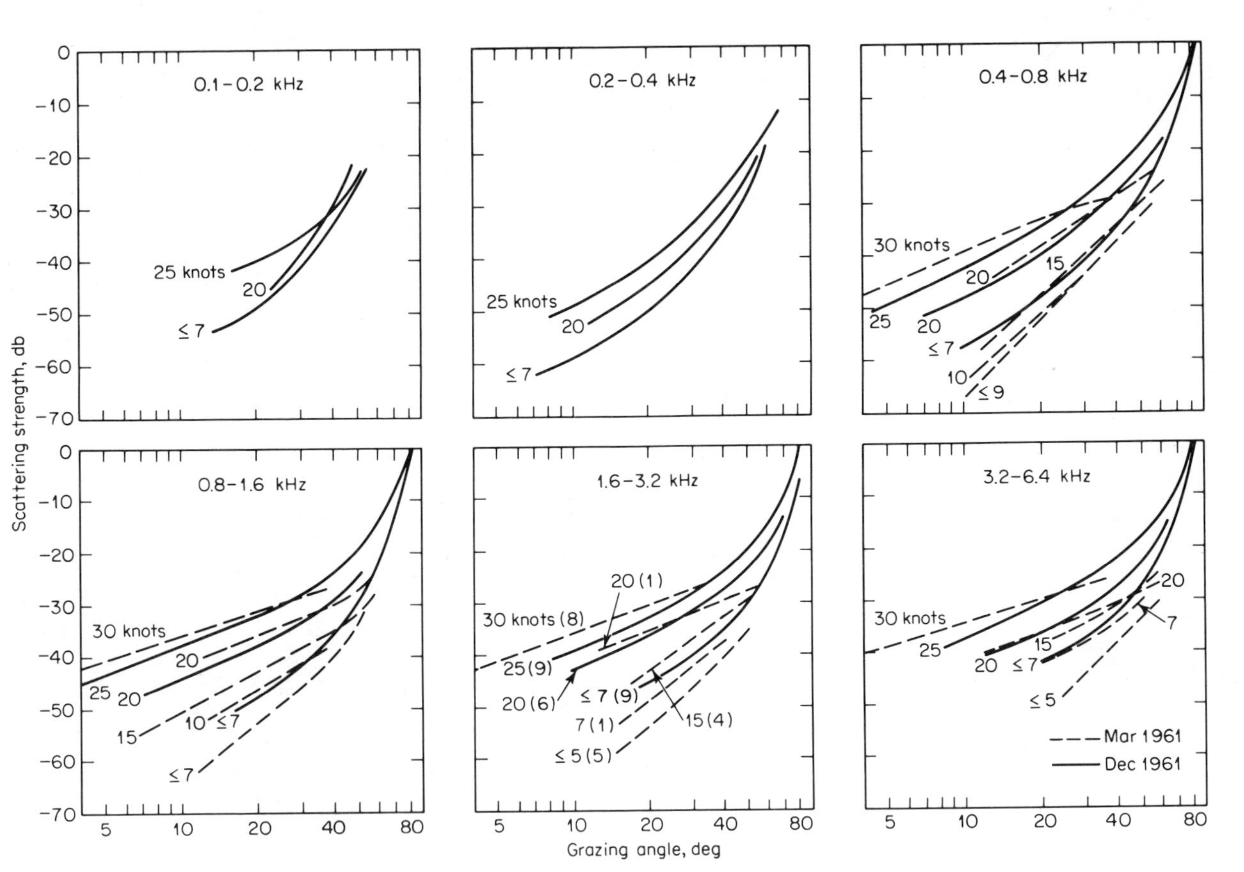

FIG. 8.15. Low-frequency sea-surface scattering strengths at different wind speeds in octave bands between 0.1 and 6.4 kHz. (Ref. 29.)

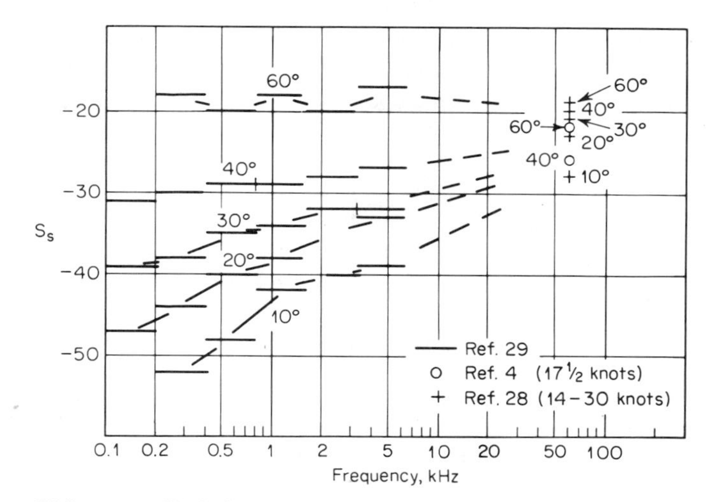

FIG. 8.16. Variation with frequency of the surface scattering strength at a wind speed of 20 knots for various grazing angles.

North Atlantic. Similar curves, but over a lesser range of frequency and grazing angles, had previously been published by Chapman and Harris (6).

As at 60 kHz, the low-frequency scattering strength of the sea surface increases with wind speed and angle over most of the angular range. Most measurements, such as those of Richter (30), made with an explosive source and a nondirectional hydrophone, show S_s to be independent of angle at low angles and low wind speeds—somewhat like the measurements of Fig. 8.12, taken at 60 kHz with a directional transducer. More recently, it has become clear that under these conditions the deep scattering layer dominates the reverberation return and the reverberation received is not properly interpreted as arising at the sea surface.

Variation with Frequency A composite plot of a number of determinations of S_s over a wide frequency range is given in Fig. 8.16, showing measurements by Chapman and Scott (29), Urick and Hoover (4), and Garrison, Murphy, and Potter (28). It is clear that at low angles a strong frequency dependence exists, amounting to about 3 db/octave. This dependence vanishes at angles approaching normal incidence. This behavior is consistent with present understanding of the origin of sea-surface scattering, as will be explained in the section to follow.

8.12 Theories and Causes of Sea-surface Scattering

The theory of the scattering of sound by rough surfaces has received considerable attention, and a relatively abundant literature is available.

An introduction to this literature may be found in the references quoted in a paper by Miles (31); an extensive theoretical background also exists for the rough-surface scattering of electromagnetic waves (32). Unfortunately, much of this literature is of interest only to the theoretician, and only two of the many theoretical discussions of the subject have been carried to the point where a comparison with observed data of sound scattering by the sea surface is possible.

One of these is the theory of Eckart (33) based on the Helmholtz solution of the wave equation in the form of a surface integral over the scattering surface. The result has been reduced to practical form by Chapman and Scott (29):

$$S_s = -10 \log 8\pi\alpha^2 + 2.17\alpha^{-2} \tan^2 \theta$$

where θ = grazing angle, deg
α^2 = mean-square slope of surface waves

From measurements of the sun's glitter, Cox and Munk (34) obtained the following empirical relationship between mean-square wave slope and wind speed:

$$\alpha^2 = 0.003 + 5.12 \times 10^{-3}W$$

where W is the speed of the wind in meters per second. Good agreement was found by Chapman and Scott between this theory and their field data, shown previously in Fig. 8.15, at angles greater than 60°. No frequency dependence is predicted by the Eckart relationship, and none is evident in the field data at angles of 60° and beyond.

Another theoretical result has been derived by Marsh (35, 37) and by Marsh, Schulkin, and Kneale (36). Here, the result for scattering back toward the source of sound is

$$S_s = 10 \log \left[\frac{\tan^4 \theta}{32} \frac{\omega^5 A^2(\omega)}{g^2} \right]$$

where θ = grazing angle
g = acceleration of gravity
$A^2(\omega)$ = "power" spectrum of sea-surface elevation at angular frequency ω

The sea surface is postulated to act as a diffraction grating; the scattering is produced by only those waves or wavelets of the sea-surface roughness which "match" the incident sound, that is, those which produce reinforcement in the sound scattered back toward the source. This process will be considered somewhat more fully in the section on bottom scattering. If, following Marsh, the sea-surface spectrum is taken to be that postulated by Burling (38), namely,

$$A^2(\omega) = 7.4 \times 10^{-3} g^2 \omega^{-5}$$

a particularly simple result is obtained. It is

$$S_s = -36 + 40 \log \tan \theta$$

This result was compared by Marsh (37) with measured explosive-source sea-surface scattering strengths in sea states ranging from 0 to 4. The comparison is shown in Fig. 8.17. The measured data points fell in the area shown shaded in the figure, with no apparent dependence on frequency or sea roughness. Although some degree of agreement is evident, it should be noted that the Marsh theory predicts no dependence of scattering on either frequency or wind speed, in contradiction to the bulk of measured data on sea-surface scattering, and fails, like the Eckart theory, to give meaningful results in the vicinity of normal incidence ($\theta = 90°$).

It is almost intuitively apparent that a rough surface must become effectively smooth when the wavelength with which the surface is "viewed" becomes large enough or when the grazing angle becomes small enough. That is, the backscattering must vanish at long wave-

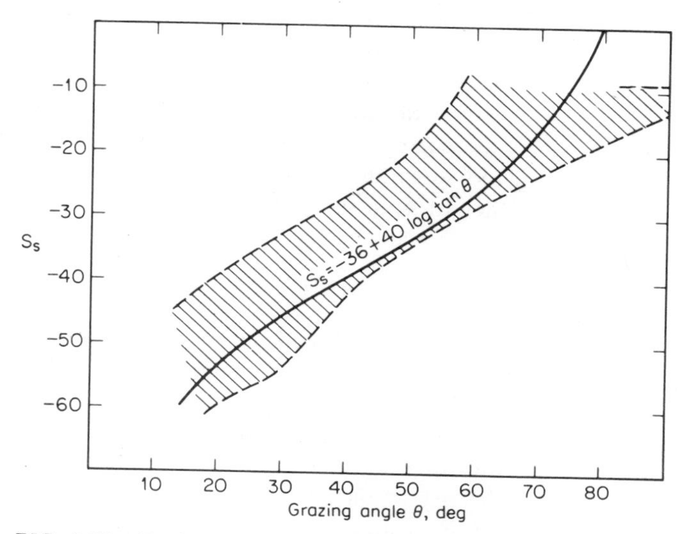

FIG. 8.17. Backscattering strengths (shaded area) measured at five locations in the Norwegian Sea, compared with theoretical curve based on the sea roughness spectrum. [*After Marsh (Ref. 37).*]

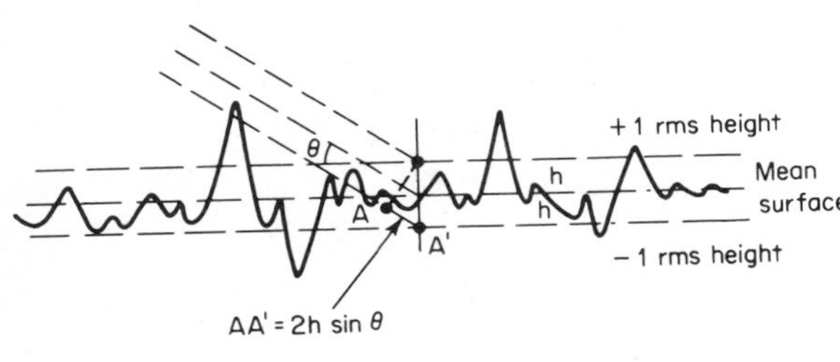

FIG. 8.18. Relation between roughness height, wavelength, and angle as a backscattering parameter.

lengths and small grazing angles. In Fig. 8.18, the backscattering must depend on the ratio of distance $2AA'$ (the average path difference for backscattering between the peaks and troughs of the rough surface) to wavelength. This led Schulkin and Shaffer (39) to interpret essentially all of the available published data on the sea-surface backscattering coefficient in terms of the ratio $2h(\sin\theta)/\lambda$, where h is the mean peak-trough roughness, λ the wavelength, and θ the grazing angle. In terms of the frequency f, this ratio becomes proportional to $fh \sin\theta$. The summary expression obtained was

$$S_s = 10 \log (fh \sin\theta)^{0.99} - 45.3$$

It was found that the data could be fitted by this expression with a standard deviation of 5 db over a range of $fh \sin\theta$ between 4 and 200. However, this empirical expression must not be used near normal incidence, nor at angles so small that some source of scattering other than the sea surface may dominate the observed reverberation. As a practical aid to using this expression, a relationship between crest-to-trough wave height h, in feet, and wind speed V, in knots, was given by Schulkin and Shaffer as

$$h = 0.0026V^{5/2}$$

From a physical point of view, the peculiar behavior of the scattering coefficient with angle suggests that different scattering processes operate in different regions of grazing angle in somewhat the same way that the peculiar spectra of ambient noise suggest different origins of the noise in different spectral regions. As mentioned above, three processes have been postulated (27) to account for the S-shape characteristic of curves of S_s versus angle. In Region I of Fig. 8.19, the backscattering is conjectured to be due to a layer of scatterers (probably air bubbles whipped into the sea at moderate and high sea states) just below the sea surface itself or, when a nondirectional system is involved, to scatterers in the deep scattering layer. In this angular region, the characteristics of the bubbly near-surface layer required to produce the observed backscattering strengths have been considered by Clay and Medwin (64). In Region II, the dominant process is one of scattering by roughnesses comparable in size to the acoustic wavelength, whereas in Region III, the process is essentially one of reflection (that is, coherent scattering) by wave facets inclined at right angles (for backscattering) to the direction of the incident sound. It may be noted that, from a scattering point of view, some interesting analogs exist, both as to processes and magnitude of the scattering, in the behavior of the sea surface in scattering electromagnetic waves incident from above and in scattering sound waves incident from below (40).

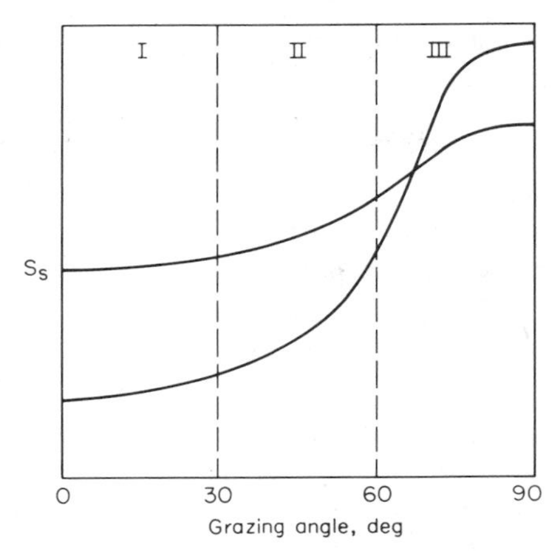

FIG. 8.19. Three regions of different sea-surface backscattering processes.

8.13 Sea-bottom Reverberation

Like the sea surface, the bottom is an effective reflector and scatterer of sound and acts to redistribute in the ocean above it a portion of the sound incident upon it. This redistribution is illustrated in Fig. 8.20, which shows the "beam pattern" of the sea bottom for sound incident on it. In this diagram the radius of the pattern in any direction is proportional to the intensity of the sound scattered or reflected in that direction. Pattern A might be characteristic of smooth reflecting bottoms, while pattern B might be typical of rough scattering bottoms, for which the intensity is less in the specular direction OR and correspondingly greater in other directions. Patterns such as these have been much studied in connection with the glossiness of surfaces for light (41), but have not yet been obtained for sound scattering from the sea bottom. In the specular direction OR, the *reflection*

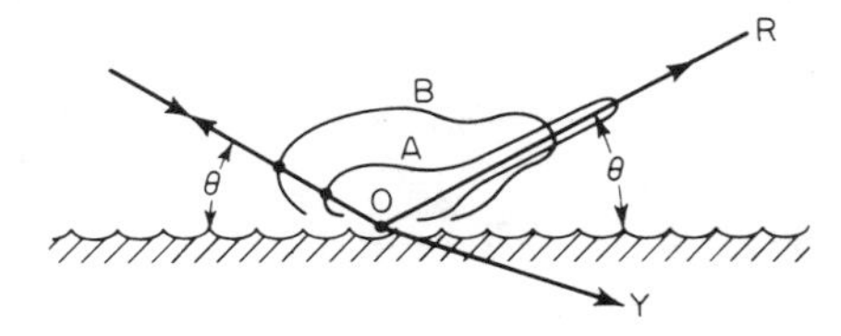

FIG. 8.20. "Beam pattern" for scattering of sound incident at angle θ. The radius of the pattern is proportional to the intensity, per unit solid angle, of sound scattered or reflected in any direction. OR is the specular direction. The arrow Y indicates that some sound enters the bottom as compressional and shear waves.

coefficient has been considered as a part of the subject of sound propagation in the sea (Sec. 5.8).

Scattering occurs three dimensionally, as well as in the plane containing the source and the normal to the surface. In one study (42) this "side scattering" was found to be uniform; a sand bottom at very low grazing angles was found to be an isotropic scatterer of 24-kHz sound in horizontal directions.

Variation with Angle and Bottom Type It was observed at an early date that the reverberation from the seabed was greater over rocky bottoms than over mud bottoms. It has since become customary to relate bottom scattering strength to the type of bottom, such as mud, silt, sand, boulders, rock, even though it is realized that the size of the particles comprising a sedimentary bottom is only an indirect indicator of acoustic scattering.

A relatively large number of mostly discordant measurements have been reported in the literature. As for the sea surface, the scattering strength of a particular bottom is found, in practice, by measuring the backscattering it produces and converting to scattering strength by the appropriate form of the basic equation for reverberation. Measurements have been made at high frequencies with tiltable directional transducers, in which the angle at which the sound beam strikes the bottom is the angle of tilt, corrected if necessary for refractive bending, and at low frequencies with nondirectional sources and receivers, where the angle is determined by the time after emission of the ping or the explosive pulse.

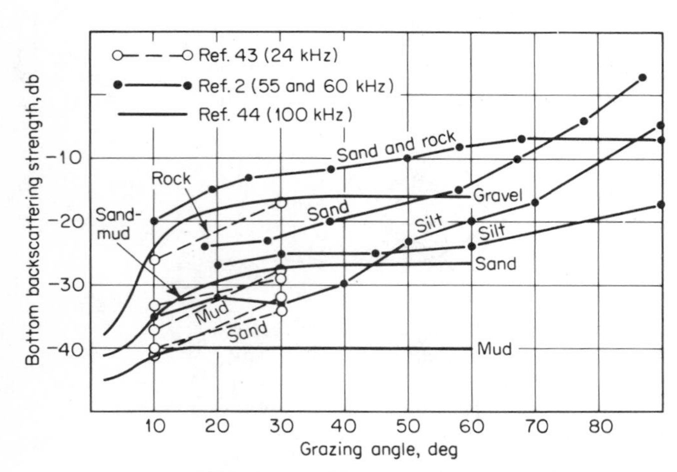

FIG. 8.21. Measured backscattering strength of the seabed at various coastal locations.

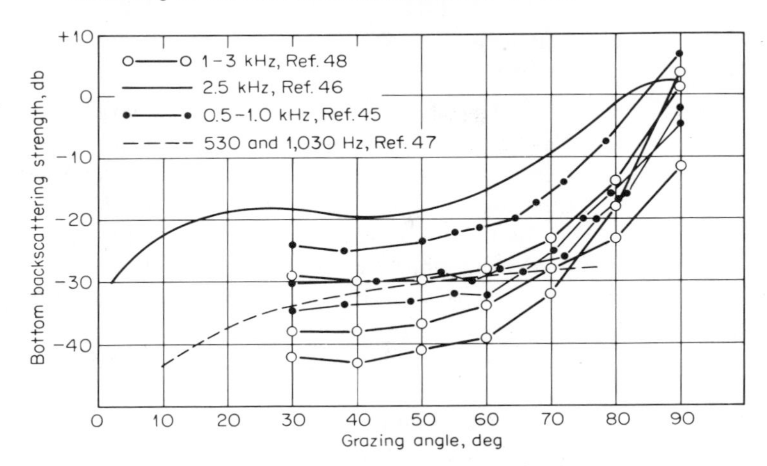

FIG. 8-22. Low-frequency deep-water bottom backscattering strength as reported in four sets of measurements.

Figure 8.21 is a compilation of high-frequency measurements in the range 24 to 100 kHz for a variety of different bottom types, as reported by different observers. Most of these observations were made in shallow coastal locations. Included are average values obtained during World War II (Ref. 43, p. 317), examples of individual determinations in Narragansett Bay, Rhode Island (2), and average curves obtained from measurements in 16 different coastal locations (44).

Figure 8.22 is a similar compilation, but for low-frequency deep-water measurements, all made in water depths in excess of 350 fathoms. In most cases, these data were obtained with nondirectional transducers, with explosive sources being used for two sets of measurements. The bottom at such deep-water locations is the fine mud of the deep abyssal sea. Included here are data reported by Urick and Saling (45) at three locations at the outer edge of the continental slope in water 2,100 to 2,800 fathoms deep, by Patterson (46) in water 12,000 ft deep, by Mackenzie (47) in 2,000-fathom water near San Diego, and by Burstein and Keane (48) at three locations in depths of 350, 1,500 and 2,050 fathoms.

The curve attributed to Mackenzie in this figure can be represented by the function $S_s = -28 + 10 \log \sin^2 \theta$; this is a Lambert's law relationship (see below) which was found to fit much of the then-extant data reasonably well at moderate grazing angles.

It is evident from these figures that the engineer will have considerable difficulty in selecting a value of S_s for use in a practical problem. Bottom composition, as expressed conveniently by particle-size designators like "mud" and "sand," gives only an approximate indication of the backscattering strength at a given grazing angle. Some property of the seabed other than the size of its sedimentary particles apparently must govern the backscattered return.

Variation with Frequency A comparison of the preceding figure for low frequencies with the mud and sand curves of Fig. 8.21 at higher frequencies would indicate an absence of a strong frequency dependence of the bottom backscattering coefficient. This lack of frequency dependence was pointed out by Mackenzie (47), who, by comparison with older data, found coefficients at 530 and 1,030 Hz essentially identical to previously reported values at 10 kHz and above.

However, a closer look at the data would suggest a definite frequency variation at frequencies beyond 10 kHz for the smoother (mud and sand) bottoms. Figure 8.23 shows the data of Urick (2) at a grazing angle of 30° and of McKinney and Anderson (44) at an angle of 10° extending over a wide frequency range. The sand and silt bottoms show a rise in scattering strength with frequency at the rate of about 3 db/octave (or a variation as the 1.0 power of the frequency), whereas little or no frequency dependence appears to exist for the "rock," "sand and rock," and "silt and shell" bottoms. This behavior may be attributed to a difference in the scale of bottom roughness. Bottoms with roughness

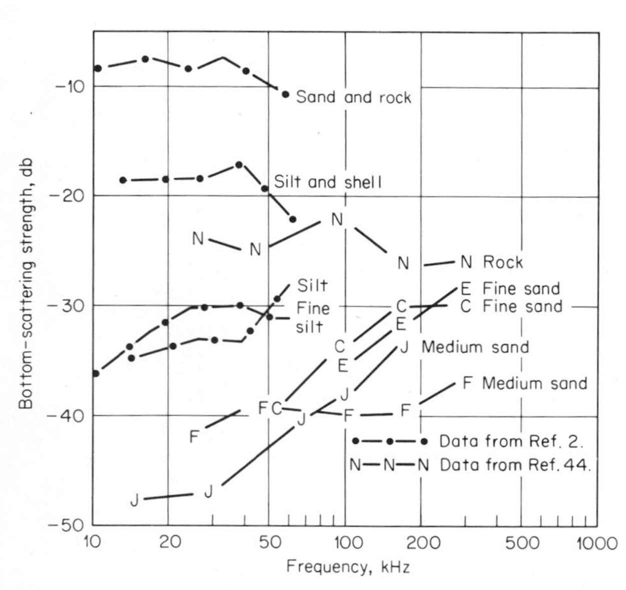

FIG. 8.23. ***Frequency variation of bottom backscattering. Dots, 30° grazing angle (Ref. 2); letters, 10° grazing angle (Ref. 44).***

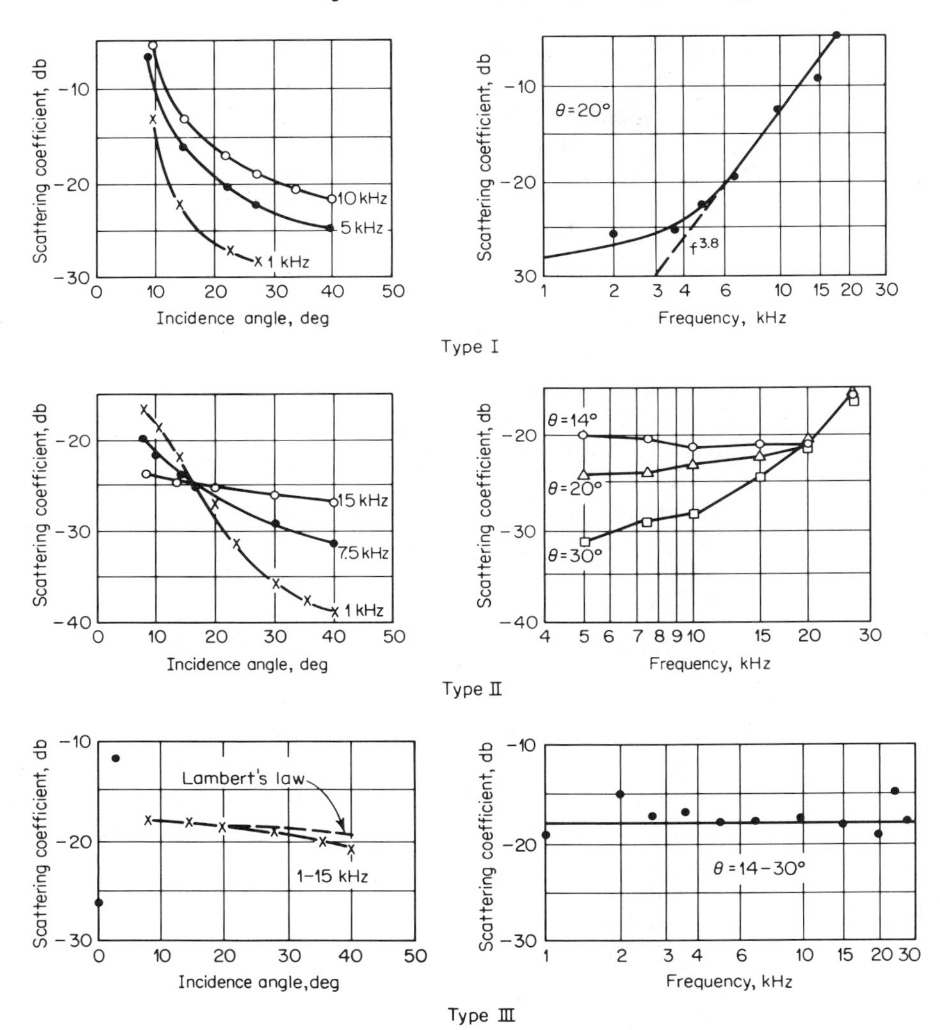

FIG. 8.24. Bottom backscattering strength in deep-water areas of three different bottom types plotted against angle and frequency. In the left-hand figures the angle is the angle of incidence, with 0° normal incidence and 90° grazing incidence. (*Ref. 49.*)

large compared to a wavelength have a backscattering coefficient independent of frequency; bottoms having an appreciable portion of their roughness spectrum at roughnesses small compared to a wavelength have a scattering strength which increases with frequency.

This behavior has been confirmed by Russian observations (49) of deep-sea backscattering at 100 areas of the Atlantic Ocean, using small explosive charges and filtering in the band 1 to 30 kHz. The results, shown in Fig. 8.24, were broken down into three broad areas, corresponding to different degrees of bottom roughness with the results plotted against angle (at a number of fixed frequencies) and against frequency (at a number of fixed angles). Three basic bottom types were recognized. One, here called Type I, is the deep-sea abyssal plain having little roughness, in which the angular variation and frequency variation (as $f^{3.8}$) are high. At the other extreme are areas which we may call Type III, having heavily dissected bottoms with underwater ridges, in which a Lambert's law variation with angle is observed with no frequency dependence over the 1- to 30-kHz range. An intermediate type of bottom may be called Type II, called "hill regions" (in probable reference to the continental slope and the continental rise) in which an intermediate behavior with angle and frequency was observed. These observations were said to be in agreement with theoretical results in which the scattering coefficient is expressed as the sum of two terms, one frequency independent and contributed to by large irregularities, the second strongly frequency dependent (as f^3) and resulting from irregularities small compared to a wavelength.

Causes and Theories of Sea-bottom Backscattering Although particle size, for sedimentary bottoms, serves as a first-cut means of classifying bottoms in terms of acoustic backscattering, the roughness of the sea bottom appears to be the dominant determining characteristic for backscattering. Just as wind speed is an indicator of surface roughness, so is bottom type an indicator of bottom roughness. However, since the sea bottom is partly transparent to sound, in the sense that a portion of the incident acoustic energy is transmitted to the earth below, bottom type must in part determine the partition of energy between the water above and the bottom beneath.

In addition to the roughness, other processes may, in some circumstances, play a significant role in the scattering process. One is the particulate nature of sedimentary bottoms, where each sedimentary particle may itself be imagined to be a scatterer of sound, and where the return of sound from the bottom is produced by a form of volume reverberation within the bottom itself. However, studies in model tanks having perfectly smooth bottoms (50) show far too little backscattering in comparison with natural bottoms; moreover, only a little roughening of the smooth sand suffices to greatly increase the backscattered return. Penetration of sound into the bottom and its subsequent reradiation as backscattering may well be an important process at low frequencies and high grazing angles, where reflection and scattering from subsurface layers and other inhomogeneities may occur. At high frequencies, however, there appears little justification for invoking processes other than the irregular discontinuity of acoustic properties occurring at the bottom itself to explain the principal characteristics of the bottom return.

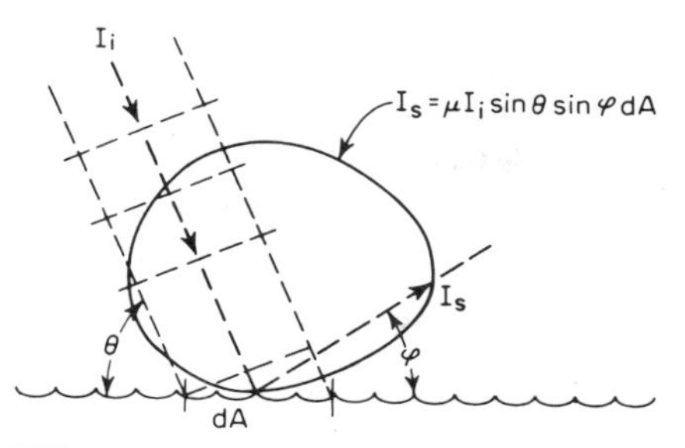

FIG. 8.25. Lambert's law for a scattering surface.

The function called Lambert's law is a type of angular variation which many rough surfaces appear to satisfy for the scattering of both sound and light. In Fig. 8.25, let sound of intensity I_i be incident at angle θ on the small surface area dA. The power intercepted by dA will be $I_i \sin \theta \, dA$. This power is *assumed*, by Lambert's law, to be scattered proportionately to the sine of the angle of scattering, so that the intensity at unit distance in the direction φ will be

$$I_s = \mu I_i \sin \theta \sin \varphi \, dA$$

where μ is a proportionality constant. For a unit area we have, on taking 10 times the logarithm of each side,

$$10 \log \frac{I_s}{I_i} = 10 \log \mu + 10 \log (\sin \theta \sin \varphi)$$

and in the backward direction, for which $\varphi = \pi - \theta$,

$$S_s = 10 \log \mu + 10 \log \sin^2 \theta$$

Thus, by Lambert's law, the backscattering strength must vary as the square of the sine of the grazing angle. If all of the incident acoustic energy is redistributed into the upper medium, with none lost by transmission into the medium below, then it can be shown by integration that $\mu = 1/\pi$; the normal-incidence backscattering strength would therefore be $10 \log (1/\pi) = -5$ db. It should be emphasized that the law rests on a particular assumption as to the redistribution of the scattered energy in space. Although many materials follow Lambert's law closely in scattering light, none does so exactly. Lambert's law applies specifically to the radiation of light by radiant, absorptive materials (51); the "law" should properly be called Lambert's "rule" for scattering. Nevertheless, it is, as we have seen, a good description of the backscattering of sound by very rough bottoms.

Marsh's theory of scattering, previously referred to in connection with the sea surface, is applicable to the sea bottom as well. In terms of the wave numbers k and K, referring to the wavelengths of the incident acoustic wave and of a component of the bottom roughness, respectively, the Marsh formula for a bottom transmitting no energy to the solid

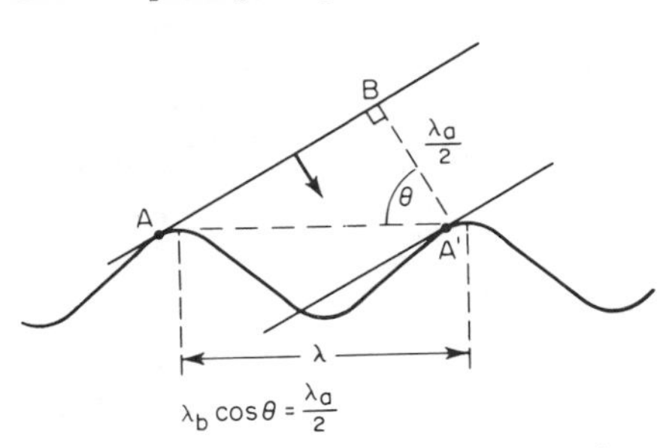

FIG. 8.26. Reinforcement criterion for backscattering.

medium below can be written

$$S_s = 10 \log \left[\frac{\sin^4 \theta}{\pi \cos \theta} k^3 A^2(K) \right]$$

where $A^2(K)$ is the amplitude squared of the irregularity in the bottom contour having wave number K. $A^2(K)$ is the bottom-roughness power spectral density. If λ_a is the wavelength of the incident sound wave, and λ_b the wavelength of a component of bottom roughness, then $k = 2\pi/\lambda_a$, and $K = 2\pi/\lambda_b$. The essential concept of the theory is that the rough bottom acts as a diffraction grating for scattering in the manner first described by Rayleigh (52), wherein the scattering is dominantly produced by that wavelength component of the bottom roughness for which phase reinforcement occurs for scattering back in the direction of the incident sound. In Fig. 8.26, the wavelength component λ_b of the roughness spectrum responsible for backscattering is that for which the scattering from points A and A' are in phase at a large distance. The distance BA' must therefore be one-half wavelength,

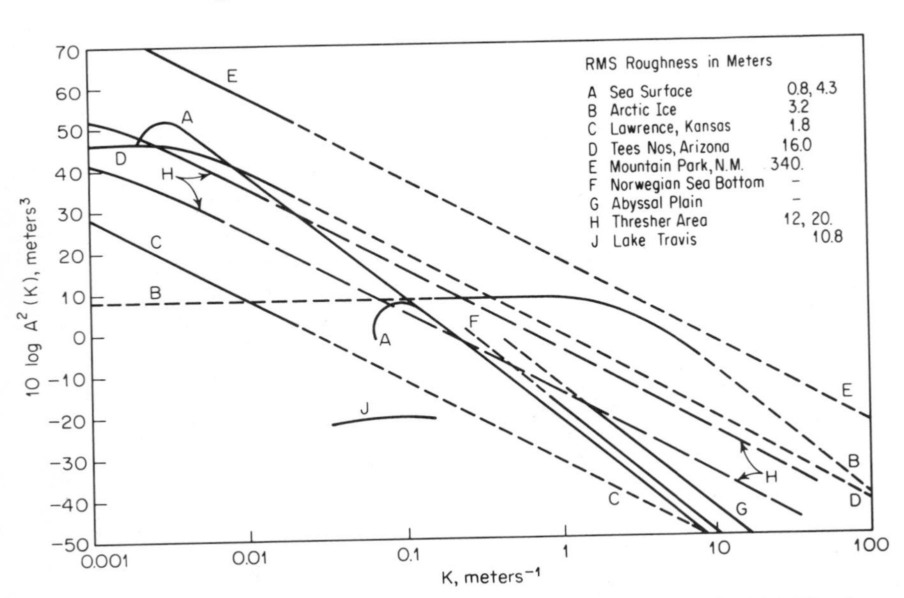

FIG. 8.27. Geophysical elevation spectra. [*Compiled by Marsh* (53).] *Included are spectra of the sea and ice surfaces* (*A, B*), *land areas* (*C, D, E*), *and the sea bottom* (*F, G, H, J*).

so that λ_a and λ_b become related by

$$\lambda_b \cos \theta = \frac{\lambda_a}{2}$$

or

$$K = 2k \cos \theta$$

Unfortunately, little data exist on the roughness power spectrum of the sea bottom. Marsh (53) has compiled the roughness spectra for various land, sea, and bottom surfaces, as shown in Fig. 8.27. It is interesting to observe that bottoms having a "white" spectrum, such as Area J of Fig. 8.27, would, by this theory, have a scattering strength varying as the cube of the frequency; those for which $A^2(K)$ varies as K^{-3}, or as the inverse cube of the frequency, would show a scattering strength independent of frequency.

8.14 Underice Reverberation

The undersurface of an ice cover may be expected to produce strong reverberation when it is rough and has a jagged, irregular underice topography. Figure 8.28 shows the results of measurements of the back-scattering strength of the ice-covered sea for two Arctic locations at different times of year. The spring pack ice measured by Milne (54)

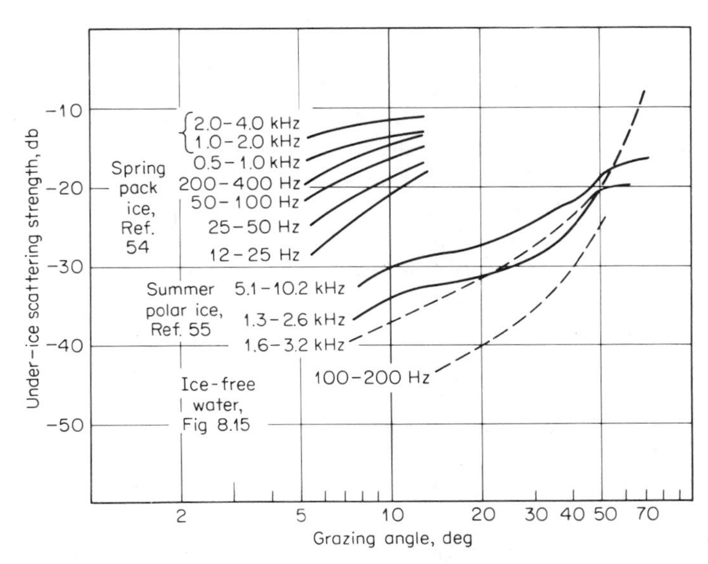

FIG. 8.28. Scattering strengths of an ice cover.

during late April and early May was described as consisting "largely of broken and pressure-packed one-year ice intermixed with broken frozen leads." The summer polar ice, for which much smaller backscattering strengths were obtained by Brown (55) during the month of September, was not described, but was probably less irregular in its bottom contour. Included in Fig. 8.28 are curves of the backscattering of the ice-free sea at a wind speed of 25 knots, copied from Fig. 8.15. Both sets of data for underice scattering, while discordant, show an increase of scattering strength with frequency and grazing angle. Certain other data (56), however, indicate an absence of a frequency variation, and so may be characteristic of a third kind of underice topography. The engineer, accordingly, has only fragmentary data available on which to base a prediction of the reverberation to be expected under an ice cover, although it is clear that backscattering strengths higher than those for the ice-free sea at low and moderate wind speeds must prevail.

8.15 Characteristics of Reverberation

With a pulsed sinusoidal sonar, the reverberation that follows the emitted pulse is heard as an irregular, quivering, slowly decaying tone. Within the smooth decay there appear onsets of increased reverberation whenever the emitted sound intercepts the sea surface and bottom, as well as blobs of reverberation of roughly the same duration as that of the pulse. In this section, some of the characteristics of reverberation, other than its intensity, are briefly described.

Amplitude Distribution Because the backscattered return originates at a large number of scattering obstacles in the sea and on its boundaries, such as the roughness irregularities of the seabed, the instantaneous reverberation amplitude fluctuates from instant to instant as the number and phase relationship of the scatterers change with range or time. Field observations long ago (57) showed that the instantaneous reverberation amplitude is Rayleigh distributed; that is to say, the probability distribution of the amplitude is that found originally by Rayleigh (58) for a large number of sine waves of equal amplitude but random phase. In such a distribution, the probability that the instantaneous intensity, or amplitude squared, is greater than the value I is $\exp(-I/\bar{I})$, where $\bar{I}$ is the average intensity, or mean-square amplitude. This means that the reverberation amplitude is greater than its rms value about 37 percent of the time and less than its rms value 63 percent of the time, with a marked tendency for the reverberation to be below the rms value at any given time. The Rayleigh distributed property of reverberation has been repeatedly confirmed in investigations since World War II (59, 60).

Coherence By *coherence* is meant the autocorrelation and crosscorrelation properties of reverberation. The reverberation observed with a pulsed sonar is well known to have an envelope containing blobs of approximately the same duration as that of the emitted pulse. This blobby character of reverberation is illustrated in Fig. 8.29. It is evidenced statistically by the finding (57) that the autocorrelation

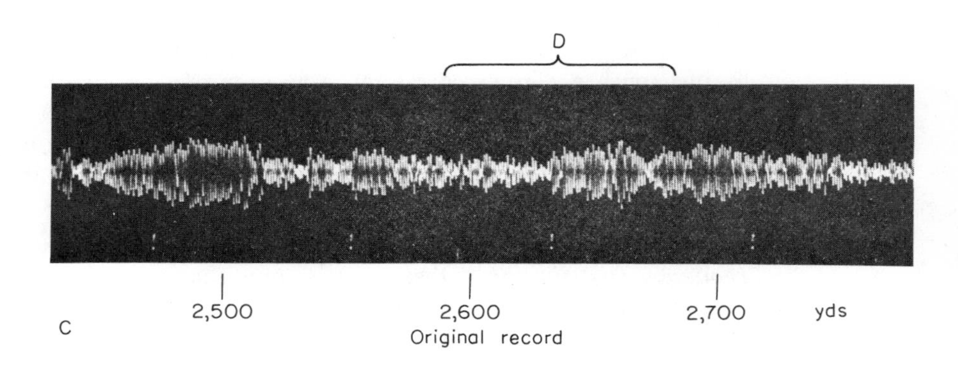

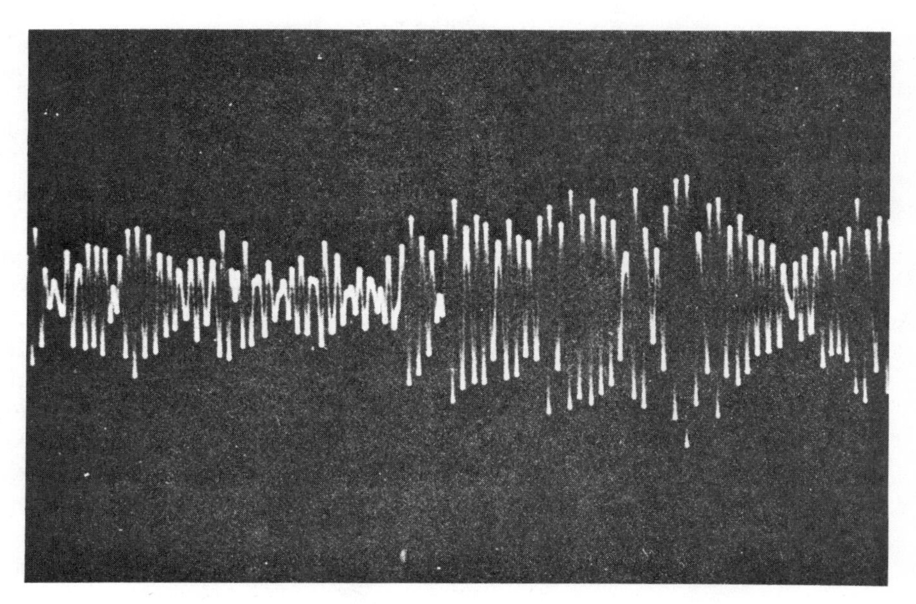

FIG. 8.29. A portion of an original recording of reverberation and an enlarged portion, illustrating the blobby, irregular envelope of pulsed sinusoidal reverberation. (Ref. 57, Fig. 12.)

coefficient of the reverberation envelope decreases to nearly zero in a time interval equal to a pulse duration.

The coherence of the reverberation received at two vertically separated hydrophones has also been studied (25). It was found that bottom reverberation at the times near the onset of the bottom return is highly coherent and that that from the deep scattering layer is much less so, with a crosscorrelation coefficient diminishing for both sources of reverberation with increasing hydrophone separation and increasing frequency. As an example, the peak correlation coefficient, read from correlograms obtained with a clipper correlator, was found to be about 0.8 for the bottom return near normal incidence, but only 0.3 or less for surface or deep-scattering-layer reverberation at a vertical separation of hydrophones of 10 ft and in the frequency band 1 to 2 kHz. Basic information of this sort is of interest for the design of hydrophone arrays to suppress reverberation (Sec. 3.8).

Frequency Distribution Observations show that the reverberation that follows a sinusoidal sonar pulse does not in general lie entirely at the frequency of the pulse, but is both shifted in frequency and spread out into a frequency band. The shift in center frequency is simply the Doppler shift caused by the velocity of the platform on which the sonar is mounted, together with any uniform velocity of the reverberation-producing scatterers themselves. The other frequency characteristic of reverberation, frequency spread, is due to several causes. One is the finite duration of the sonar pulse itself, which, if it is t sec long, has a frequency spread of $1/t$ hertz about the center frequency. Another is the fact that on a moving ship the reverberation arriving from different directions experiences a different Doppler shift because of the differing components of motion in different directions. For example, with a transducer of an equivalent plane beam width Φ facing in the forward direction on a vessel under way with velocity v, the reverberation emitted and arriving on the axis of the beam would have a Doppler shift of $(2v/c)f$ hertz, whereas that on the edges of the beam would be shifted by $(2v/c)f \cos(\varphi/2)$ (Fig. 8.30). The reverberation band is therefore $(2vf/c)[1 - \cos(\varphi/2)]$ hertz wide. A third factor causing a band spread of reverberation is the Doppler spread caused by the motion of the scatterers themselves.

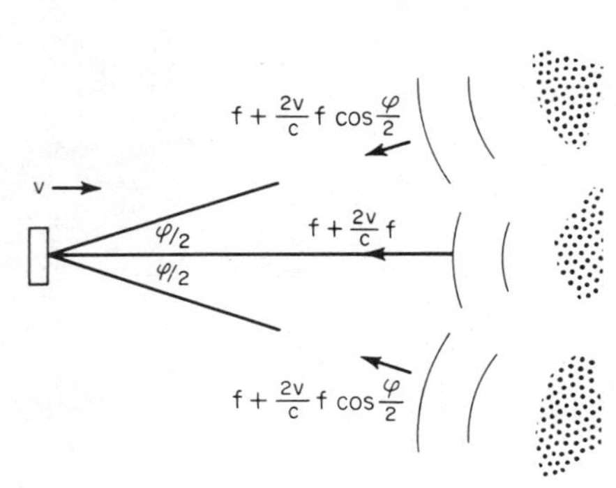

FIG. 8.30. Frequency spread of reverberation in a transducer moving in its axial direction with velocity v. The velocity of sound is c.

Although the first two of these causes are readily calculated and can be allowed for in sonar design, the motion of the reverberation-producing scatterers is a largely unknown oceanographic characteristic of the sea.

Early work during World War II was done with a frequency analyzer called a period meter (57), but the interpretation of the results obtained in terms of the spectrum of reverberation has remained obscure. Later work has been done in a laboratory tank (61) and at sea (62). The latter investigation concerned the frequency shift and the frequency spread of sea-surface reverberation. At 85 kHz and in a 20-knot wind, the mean reverberation frequency was shifted downward by 25 Hz when pointing downwind, suggesting a mean downwind scatterer velocity of 35 cm/sec or a ratio of 0.035 of the wind speed. At the same time, the reverberation was spread over a band 25 Hz wide between the -3 db points above and below the mean frequency, so as to suggest a range of scatterer velocities of 0 to 70 cm/sec downwind. Such effects of scatterer motion would be absent for bottom reverberation, but would be expected to be relatively large for the mobile biological scatterers causing volume reverberation.

Both the mean frequency and the spread of the frequency of reverberation are of direct concern for the design of filters to suppress reverberation.

8.16 Reverberation Prediction

In predicting the reverberation to be expected in a particular echo-ranging sonar, it is necessary, as mentioned above, to visualize what kind of reverberation is apt to be encountered at different times after emission of the sonar pulse and where it is coming from. For this purpose, a rough ray diagram is a necessity, for it indicates not only the likely sources of reverberation, but also the grazing angles involved in the encounters of the sonar beam with the surface and bottom.

In simple situations, such as at short ranges where only one form of reverberation is involved, a reverberation computation proceeds in a straightforward manner. After the appropriate form of the reverberation-level equation is selected, depending on the type of reverberation to be expected, the equivalent beam width for the transducer used can be found from Table 8.1 and a value of scattering strength can be selected for the kind of reverberation to be expected from the various curves given above. At different times and ranges, different sources of reverberation and different values of scattering strength, depending on grazing angle, will be involved.

Concerning a prediction of sonar reverberation in such a manner, particular attention should be given to the assumptions underlying the reverberation-level equations. In particular, the transmission loss to and

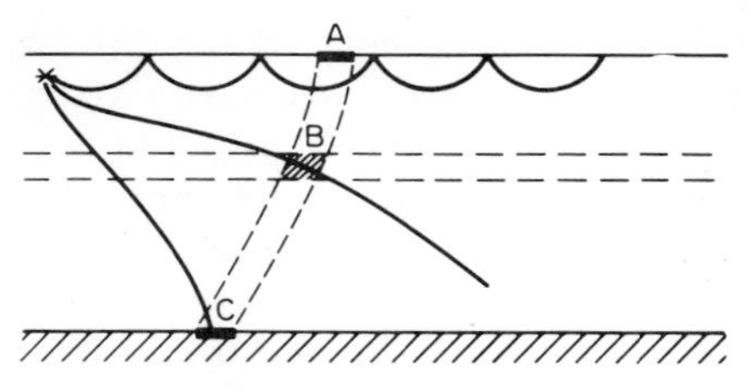

FIG. 8.31. Multiple sources of reverberation at long ranges.

from the scatterers was taken to be that of spherical spreading, and the effects of shadow zones and convergence zones in affecting the propagation of sound to the scattering region were neglected. Hence, the equations must be modified where necessary to take account of conditions peculiar to a prediction problem.

At long ranges, a reverberation computation becomes increasingly difficult because of the increasing number of paths and sources of reverberation. Referring to Fig. 8.31, for example, at some one instant of time, reverberation may be received from the surface at A, the deep-scattering layer at B, and the bottom at C. The contribution of each source must be evaluated separately and combined to give the level of reverberation at the instant of time being considered. At long ranges in shallow water, a reverberation prediction requires special techniques and computation on a digital computer such as those used by Mackenzie (63).

Under conditions which approximate those for which the expressions for RL were derived, a valid estimate of the expected reverberation can readily be made. As an example, we compute RL for an echo-ranging sonar operating at 50 kHz, with a source level of 120 db and a pulse duration of 1 msec. The transducer is assumed to be a line transducer 1 ft long located 100 ft above a mud bottom. The bottom-reverberation level is desired for the instant of time corresponding to a diagonal range to the bottom of 200 yd. At 50 kHz, the wavelength is 0.1 ft. By Table 8.1, $10 \log \Phi = 10 \log (0.1/2\pi) + 9.2 = -8.8$. The equivalent beam width is therefore 0.13 rad, or 7.5°. The reverberating area is therefore

$$A = \frac{c\tau}{2} \Phi r = (0.83)(0.13)(200) = 22 \text{ yd}^2$$

The grazing angle at the bottom is 9.5°. From Fig. 8.21, a value of S_s of -39 db is selected. The reverberation level is therefore

$$\begin{aligned} \text{RL} &= \text{SL} - 40 \log r + S_s + 10 \log A \\ &= 120 - 92 - 40 + 13 \\ &= +1 \text{ db re 1 dyne/cm}^2 \end{aligned}$$

REFERENCES

1. Skolnik, M. I.: "Introduction to Radar Systems," chap. 12, McGraw-Hill Book Company, New York, 1962.

2. Urick, R. J.: Backscattering of Sound from a Harbor Bottom, *J. Acoust. Soc. Am.*, **26**:231 (1954).
3. Physics of Sound in the Sea: Part II, Reverberation, *Natl. Defense Res. Council Div. 6 Sum. Tech. Rept.* 8, 1947.
4. Urick, R. J., and R. M. Hoover: Backscattering of Sound from the Sea Surface: Its Measurement, Causes, and Application to the Prediction of Reverberation Levels, *J. Acoust. Soc. Am.*, **28**:1038 (1956).
5. Urick, R. J.: Generalized Form of the Sonar Equations, *J. Acoust. Soc. Am.*, **34**:547 (1962).
6. Chapman, R. P., and J. H. Harris: Surface Backscattering Strengths Measured with Explosive Sound Sources, *J. Acoust. Soc. Am.*, **34**:1592 (1962).
7. Weston, D. E.: Underwater Explosions as Acoustic Sources, *Proc. Phys. Soc. London*, **76**:233 (1960).
8. Stockhausen, J. H.: Spectrum of the Shock Wave from 1-lb TNT Charges Exploded Underwater, *J. Acoust. Soc. Am.*, **36**:1220 (1964).
9. Christian, E. A., and M. Blaik: Near Surface Measurements of Deep Explosions, *J. Acoust. Soc. Am.*, **38**:57 (1965).
10. Physics of Sound in the Sea: Part IV, Acoustic Properties of Wakes, *Natl. Defense Res. Comm. Div. 6 Sum. Tech. Rept.* 8, pp. 460–474, 1947.
11. Albers, V. M.: "Underwater Acoustics Handbook," 2d ed., chap. 7, The Pennsylvania State University Press, University Park, Pa., 1965.
12. Weissler, A., and V. A. Del Grosso: The Velocity of Sound in Sea Water, *J. Acoust. Soc. Am.*, **23**:219 (1951).
13. Urick, R. J.: A Sound Velocity Method for Determining the Compressibility of Finely Divided Substances, *J. Appl. Phys.*, **18**:983 (1947).
14. Wood, A. B.: "A Textbook of Sound," 2d ed., p. 360, G. Bell & Sons, Ltd., London, 1941.
15. Meyer, R. F., and B. W. Romberg: Acoustic Scattering in the Ocean, *Proj. Trident Tech. Rept.* 1360863, Arthur D. Little, Inc., 1963.
16. Fox, F. E., S. R. Curley, and G. S. Larson: Phase Velocity and Absorption Measurements in Water Containing Bubbles, *J. Acoust. Soc. Am.*, **27**:534 (1955).
17. Devin, C.: Damping of Pulsating Air Bubbles in Water, *J. Acoust. Soc. Am.*, **31**:1654 (1959).
18. Eyring, C. F., R. J. Christensen, and R. W. Raitt: Reverberation in the Sea, *J. Acoust. Soc. Am.*, **20**:462 (1948).
19. Pekeris, C. L.: Note on the Scattering of Radiation in an Inhomogeneous Medium, *Phys. Rev.*, **71**:268 (1947).
20. Johnson, H. R., R. H. Backus, J. B. Hersey, and D. M. Owen: Suspended Echo Sounder and Camera Studies of Mid-water Sound Scatterers, *Deep-sea Res.*, **3**:266 (1956).
21. Hersey, J. B., and R. H. Backus: New Evidence That Migrating Gas Bubbles, Probably the Swim Bladders of Fish? Are Largely Responsible for Scattering Layers on the Continental Rise North of New England, *Deep-sea Res.*, **7**:190 (1954).
22. Moore, H. B.: Scattering Layer Observations, *Atlantis Cruise 151, Sci. Rept.* 3, Woods Hole Oceanographic Institution, Reference 49-2, 1949. Also, The Relation between the Scattering Layer and the Euphausiacea, *Biol. Bull.*, **99**:181 (1950).
23. Hersey, J. B., R. H. Backus, and J. Hellwig: Sound Scattering Spectra of Deep Scattering Layers in the Western North Atlantic Ocean, *Deep-sea Res.*, **8**:196 (1962).
24. Marshall, J. R., and R. P. Chapman: Reverberation from a Deep Scattering Layer Measured with Explosive Sound Sources, *J. Acoust. Soc. Am.*, **36**:164 (1964).
25. Urick, R. J., and G. R. Lund: Vertical Coherence of Explosive Reverberation, *J. Acoust. Soc. Am.*, **36**:2164 (1964).

26. Weston, D. E.: Unpublished report, 1957.
27. Urick, R. J.: The Processes of Sound Scattering at the Ocean Surface and Bottom, *J. Marine Res.*, **15**:134 (1956).
28. Garrison, G. R., S. R. Murphy, and D. S. Potter: Measurements of the Backscattering of Underwater Sound from the Sea Surface, *J. Acoust. Soc. Am.*, **32**:104 (1960).
29. Chapman, R. P., and H. D. Scott: Surface Backscattering Strengths Measured over an Extended Range of Frequencies and Grazing Angles, *J. Acoust. Soc. Am.*, **36**:1735 (1964).
30. Richter, R. M.: Measurements of Backscattering from the Sea Surface, *J. Acoust. Soc. Am.*, **36**:864 (1964).
31. Miles, J. W.: On Non-specular Reflections at a Rough Surface, *J. Acoust. Soc. Am.*, **26**:191 (1954).
32. Beckman, P., and A. Spizzichino: "The Scattering of Electromagnetic Waves from Rough Surfaces," The Macmillan Company, New York, 1963.
33. Eckart, C.: Scattering of Sound from the Sea Surface, *J. Acoust. Soc. Am.*, **25**:566 (1953).
34. Cox, C., and W. Munk: Measurement of the Roughness of the Sea Surface from Photographs of the Sun's Glitter, *J. Opt. Soc. Am.*, **44**:838 (1954).
35. Marsh, H. W.: Exact Solution of Wave Scattering by Irregular Surfaces, *J. Acoust. Soc. Am.*, **33**:330 (1961).
36. Marsh, H. W., M. Schulkin, and S. G. Kneale: Scattering of Underwater Sound by the Sea Surface, *J. Acoust. Soc. Am.*, **33**:334 (1961).
37. Marsh, H. W.: Sound Reflection and Scattering from the Sea Surface, *J. Acoust. Soc. Am.*, **35**:240 (1963).
38. Burling, R. W.: "Wind Generation of Waves on Water," Ph.D. dissertation, Imperial College, University of London, 1955.
39. Schulkin, M., and R. Shaffer: Backscattering of Sound from the Sea Surface, *J. Acoust. Soc. Am.*, **36**:1699 (1964).
40. Hoover, R. M., and R. J. Urick: Sea Clutter in Radar and Sonar, *Inst. Radio Engrs. Conv. Record*, part 9, p. 17, 1957.
41. Harrison, V. G.: "Gloss: Its Definition and Measurement," Tudor Publishing Company, New York, 1945.
42. Urick, R. J.: Side Scattering of Sound in Shallow Water, *J. Acoust. Soc. Am.*, **32**:351 (1960).
43. Physics of Sound in the Sea: Part II, Reverberation, *Natl. Defense Res. Comm. Div. 6 Sum. Tech. Rept.* 8, p. 316, 1947.
44. McKinney, C. M., and C. D. Anderson: Measurements of Backscattering of Sound from the Ocean Bottom, *J. Acoust. Soc. Am.*, **36**:158 (1964).
45. Urick, R. J., and D. S. Saling: Backscattering of Explosive Sound from the Deep-sea Bed, *J. Acoust. Soc. Am.*, **34**:1721 (1962).
46. Patterson, R. B.: Back-scatter of Sound from a Rough Boundary, *J. Acoust. Soc. Am.*, **35**:2010 (1963).
47. Mackenzie, K. V.: Bottom Reverberation for 530 and 1030 cps Sound in Deep Water, *J. Acoust. Soc. Am.*, **33**:1498 (1961).
48. Burstein, A. W., and J. J. Keane: Backscattering of Explosive Sound from Ocean Bottoms, *J. Acoust. Soc. Am.*, **36**:1596 (1964).
49. Jitkovskiy, Yu. Yu., and L. A. Volovova: Sound Scattering from the Ocean Bottom, Paper E67, *Proc. Fifth Intern. Acoust. Congr. Liège Belg.*, 1965.
50. Nolle, A. W., et al.: Acoustical Properties of Water Filled Sands, *J. Acoust. Soc. Am.*, **35**:1394 (1963).
51. Houston, R. S.: "A Treatise on Light," p. 368, Longmans, Green & Co., Inc., New York, 1938.

52. Rayleigh, Lord: "The Theory of Sound," vol. II, pp. 89–96, Dover Publications, Inc., New York, 1945.
53. Marsh, H. W.: "Reflection and Scattering of Sound by the Sea Bottom," paper presented at the 68th meeting of the Acoustical Society of America, October, 1964, Avco Marine Electronics Office, New London, Connecticut.
54. Milne, A. R.: Underwater Backscattering Strengths of Arctic Pack Ice, *J. Acoust. Soc. Am.*, **36**:1551 (1964).
55. Brown, J. R.: Reverberation under Arctic Ice, *J. Acoust. Soc. Am.*, **36**:601 (1964).
56. Mellen, R. H., and H. W. Marsh: Underwater Sound Reverberation in the Arctic Ocean, *J. Acoust. Soc. Am.*, **35**:1645 (1963).
57. Principles of Underwater Sound, *Natl. Defense Res. Comm. Div. 6 Sum. Tech. Rept.*, **7**:90–98 (1947).
58. Rayleigh, Lord: "The Theory of Sound," vol. I, pp. 35–42, Dover Publications, Inc., New York, 1945.
59. Cron, B. F., and W. R. Schumacher: Theoretical and Experimental Study of Underwater Sound Reverberation, *J. Acoust. Soc. Am.*, **33**:881 (1961).
60. Olshevskii, V. V.: Probability Distribution of Sea Reverberation Levels, *Soviet Phys. Acoust.*, **9**:378 (1964).
61. Liebermann, L. W.: Analysis of Rough Surfaces by Scattering, *J. Acoust. Soc. Am.*, **35**:932 (1963).
62. Mellen, R. H.: Doppler Shift of Sonar Backscatter from the Surface, *J. Acoust. Soc. Am.*, **36**:1395 (1964).
63. Mackenzie, K. V.: Long-range Shallow-water Reverberation, *J. Acoust. Soc. Am.*, **34**:62 (1962).
64. Clay, C. S., and H. Medwin: High Frequency Acoustical Reverberation from a Rough Sea Surface, *J. Acoust. Soc. Am.*, **36**:2131 (1964).
65. Barakos, P. A.: Underwater Reverberation as a Factor in ASW Acoustics, *U.S. Navy Underwater Sound Lab. Rept.* 620, September, 1964.
66. The Discrimination of Transducers against Reverberation, *Univ. Calif. War Res. Mem* UT5, file 01.40, NDRC Div. 6.1-5530-968, May, 1943.

NINE

Reflection and Scattering by Sonar Targets: Target Strength

In active sonar the parameter *target strength* refers to the echo returned by an underwater target. Such targets may be objects of military interest, such as submarines and mines, or they may be schools of fish sought by fish-finding sonars. Excluded from the category of "targets" are inhomogeneities in the sea of indefinite extent, such as scattering layers and the ocean surface and bottom, which, because of their indefinite size, return sound in the form of *reverberation* rather than as *echoes*.

In the context of the sonar equations, target strength is defined as 10 times the logarithm to the base 10 of the ratio of the intensity of the sound returned by the target, at a distance of 1 yd from its "acoustic center" in some direction, to the incident intensity from a distant source. In symbols,

$$\mathrm{TS} \equiv 10 \log \left. \frac{I_r}{I_i} \right|_{r=1}$$

where I_r = intensity of return at 1 yd

I_i = incident intensity

Pictorially this is shown in Fig. 9.1, where P is the point at which I_r is imagined to be measured and C is the acoustic center of the target. This fictitious point, inside or outside of the target itself, is the point from which the returned sound appears to originate on the basis of measurements made at a distance. In "monostatic" sonars, having the same, or closely adjacent source and receiver, the point P lies in the direction back toward the source of sound. In "bistatic" sonars, P can lie in any direction relative to the target, and target strength then becomes a function of both the

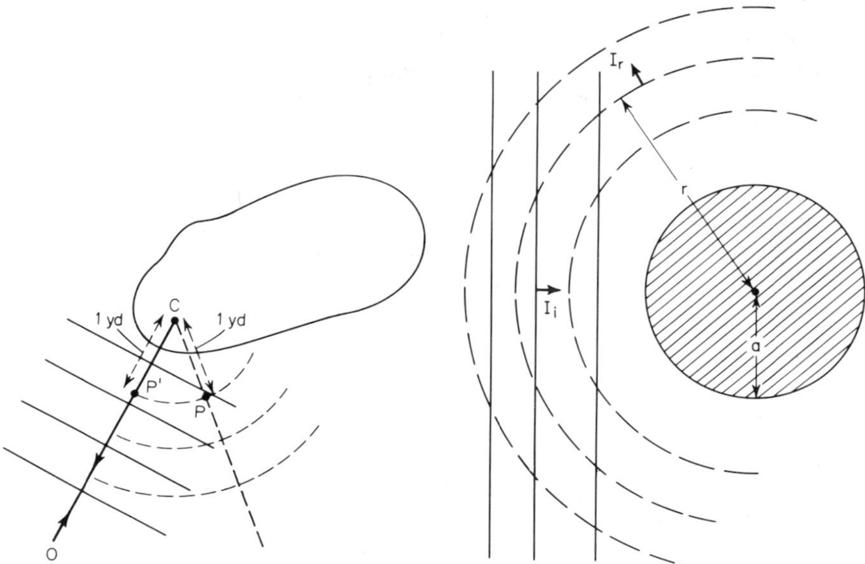

***FIG. 9.1.** Geometry of target strength. A plane wave is incident on the target in the direction OC. Target strength refers to the points P and P′ 1 yd from the acoustic center C.*

***FIG. 9.2.** A sphere as an isotropic reflector of an incident plane wave.*

incident direction and the direction of the receiver, both relative to some axis of symmetry of the target. Because most sonars are "monostatic," we will restrict our discussion to "back reflection" and "backscattering," in which the reference point lies at P' back in the direction of the incident sound.

Special mention should be made of the use of 1 yd as the reference distance for target strength. This arbitrary reference often causes many underwater objects to have *positive* values of target strength. Such positive values should not be interpreted as meaning that more sound is coming back from the target than is incident upon it; rather, they should be regarded as a consequence of the arbitrary reference distance. If, instead of 1 yd, 1 kyd were used, all customary targets would have a negative target strength.

The meaning of target strength can be shown by computing the target strength of a sphere, large compared to a wavelength, on the assumption that the sphere is an isotropic reflector; that is, it distributes its echo equally in all directions. Let a large, perfect, rigid sphere (Fig. 9.2) be insonified by a plane wave of sound of intensity I_i. If the sphere is of radius a, the power intercepted by it from the incident wave will be $\pi a^2 I_i$. On the assumption that the sphere reflects this power uniformly in all directions, the intensity of the reflected wave at a distance r yd from the sphere will be the ratio of this power to the area of a sphere of radius r, or

$$I_r = \frac{\pi a^2 I_i}{4\pi r^2} = I_i \frac{a^2}{4r}$$

where I_r is the intensity of the reflection at range r. At the reference distance of 1 yd, the ratio of the reflected intensity I_r to the incident intensity is

$$\left.\frac{I_r}{I_i}\right|_{r=1} = \frac{a^2}{4}$$

and the target strength of the sphere becomes

$$\mathrm{TS} \equiv 10 \log \left.\frac{I_r}{I_i}\right|_{r=1} = 10 \log \frac{a^2}{4}$$

It is therefore evident that an ideal sphere of radius 2 yd ($a = 2$) has a target strength of zero db. In practical work, spheres make good reference targets for sonar when they can be used, because their target strengths are relatively independent of orientation.

9.1 The Echo as the Sum of Backscattered Contributions

For radar it was shown by Kerr (1) that the backscattering cross section of a radar target in integral form is

$$\sigma = \frac{4\pi}{\lambda^2} \left| \int_\alpha^\beta \frac{dA}{dz} e^{2ikz}\, dz \right|^2$$

where σ = ratio of scattered power to incident intensity
dA/dz = rate of change of cross-sectional area of body in direction of propagation z
k = wave number $2\pi/\lambda$
λ = incident wavelength
$z = \alpha$, $z = \beta$ are the range limits of the target

The return of an incident plane wave from the target can therefore be regarded as the sum of many wavelets, each originating at the changes in cross-sectional area of the target and added with respect to phase. The application of this expression to sonar is restricted to targets which are large, rigid, and immovable in the sound field, that is, to large targets that do not deform or move under the impact of the incident sound wave.

A thorough application of this approach to the backscattering of underwater sound targets has been made by Freedman (2). In his notation

$$\text{TS} = 10 \log |J|^2$$

where

$$J = \frac{1}{\lambda} \sum_{g=1} e^{-2ik(r_g - r_1)} \sum_{n=0}^{\infty} \frac{D_g^n(A)}{(2ik)^n}$$

and where the symbol $D_g^n(A)$ denotes the nth derivative of the cross-sectional area of the object at range r_g, and r_1 is the range of the closest point of the target. The target-strength factor J is thus proportional to the sum of all the derivatives of the cross-sectional area A of the object, measured at some range $r_g - r_1$ from the point nearest the source, weighted by the factor $1/(2ik)^n$, and then summed over all points of the object after allowance for phase by means of the factor $e^{ik}(r_g - r_1)$. Freedman also showed experimentally that echo envelopes can be predicted from the various contributions of the derivatives of the cross-sectional-area function occurring at times corresponding to the location of the cross sections along the object in the direction of propagation. A summation method for finding the reflection from irregular bodies has also been the subject of a paper by Neubauer (3).

9.2 Geometry of Specular Reflection

For objects of radii of curvature large compared to a wavelength, the echo originates principally by *specular reflection*, in which those portions of the target in the neighborhood of the point at which sound is normally incident give rise to a coherent reflected echo. One way to find the magnitude of the specular reflection is to construct Fresnel, or quarter-wave zones, on the surface of the body and to add their contributions, as has been done for underwater targets by Steinberger (4). A heuristic intuitive approach is to consider target strength as a measure of the spreading of an incident plane wave induced by specular reflection from a curved surface. If the power or energy-density contained within a small area A_i of the incident sound beam is spread, on reflection, over the area A_r at unit distance, then the target strength is $10 \log (A_r/A_i)$. These areas can be determined by drawing rays. This geometric view of specular reflection will be illustrated by computing the target strength of a sphere and a general convex surface, both satisfying the requirement of large radii of curvature compared to a wavelength.

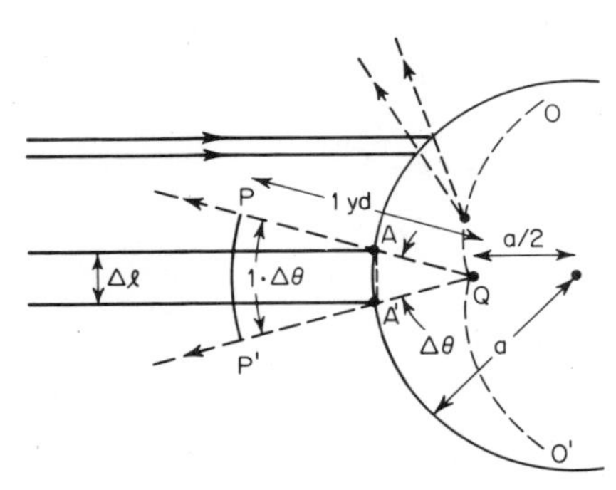

FIG. 9.3. ***Target strength of a sphere. The acoustic energy contained in the pencil of diameter*** Δl ***is spread on reflection over a portion*** PP' ***of a 1-yd sphere.***

Figure 9.3 shows a perfect, large, rigid sphere with a plane sound wave incident from the left. Adjacent incident rays intersect along the curve OQO', having a cusp at Q halfway from the surface to the center of the sphere. This *caustic* is the locus of the acoustic centers of the adjacent intersecting rays. Consider a small cylindrical bundle of parallel rays normally incident upon the sphere at AA' and of cross section Δl. If the intensity of the incident wave is I_i, the power contained in the bundle is $I_i\pi(\Delta l)^2/4$. This power will, in effect, be reradiated within an angle $\Delta\theta$ from the acoustic center Q for these rays. At unit distance from Q, this power is distributed over a portion of a sphere pp' of area $\pi(\Delta\theta \times 1)^2/4$ if $\Delta\theta$ is small, and the intensity there becomes

$$I_r = \frac{I_i\pi(\Delta l)^2/4}{\pi(\Delta\theta \times 1)^2/4} = I_i\left(\frac{\Delta l}{\Delta\theta}\right)^2$$

But, referring to the triangle AQA', it will be seen that

$$AA' = \Delta l = \frac{a}{2}\Delta\theta$$

for small $\Delta\theta$. Hence $\Delta l/\Delta\theta = a/2$ and

$$\text{TS} \equiv 10 \log \frac{I_r}{I_i}\bigg|_{r=1} = 10 \log \frac{a^2}{4}$$

This is the same as before, and indicates that a large sphere reflects the incident plane wave in the backward direction *as if* it were a uniform, or isotropic, reflector of sound. For this simple expression to hold strictly, distances must be reckoned from the acoustic center of the sphere located halfway from the surface to the center. For practical purposes in sonar, where ranges much greater than the radius of the sphere are involved, the exact location of the acoustic center is not usually significant. It should be observed that we have considered a sphere that is (1) *perfect* in shape, without irregularities, depressions, or protuberances, (2) *rigid*, or nondeformable by the impinging sound beam, (3) *immovable*, or does not partake of the acoustic motion of the field in which it is embedded, and (4) *large* compared to a wavelength $(2\pi a/\lambda \gg 1)$.

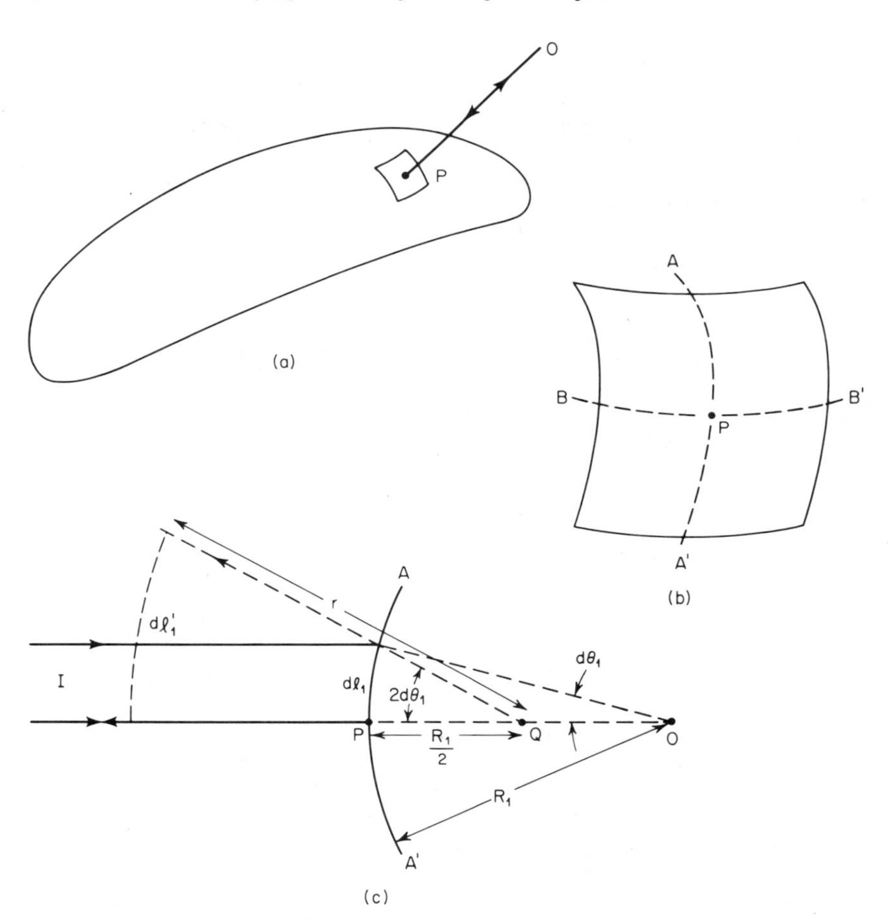

FIG. 9.4. Target strength of a convex target of large radius of curvature. (a) A target with sound normally incident along OP. (b) Plan view of the vicinity of P with AA' and BB' in the principal normal sections. (c) Cross section through AA' of radius of curvature R_1.

This same method can be extended to the reflection at normal incidence from any convex surface having all radii of curvature large compared to a wavelength. This requirement carries with it the absence of protuberances, corners, and angles, all of which involve small radii of curvature and which serve as scatterers rather than reflectors of sound. Such a large, smooth, convex object is shown in Fig. 9.4*a*, with sound normally incident at point P along the line OP. Imagine a series of planes through OP intersecting the object. Two of them lie at right angles to one another and intersect the object in the *principal normal* *sections* having a maximum and a minimum radius of curvature. Figure 9.4*b* is a plan view at P in which the principal normal sections are AA' and BB'; Fig. 9.4*c* is a cross section through AA' in which R_1 is the radius of curvature, O the center of curvature, and Q the acoustic center located halfway from O to P. Consider a small rectangular segment of the surface having one corner located at P, and of infinitesimal lengths dl_1 and dl_2 on each side. If a plane wave of intensity I_i is incident on the surface, the power dP intercepted by the segment will be

$$dP = I_i \, dl_1 \, dl_2$$

But from the cross section of Fig. 9.4*c*, it is apparent that

$$dl_1 = R_1 \, d\theta_1$$

and similarly in the perpendicular plane containing BB' we would have

$$dl_2 = R_2 \, d\theta_2$$

so that the power intercepted will be

$$dP = I_i R_1 R_2 \, d\theta_1 \, d\theta_2$$

On reflection from the sphere, this power is distributed, at range r from the acoustic center C, over an area

$$\begin{aligned} dA &= dl_1' \, dl_2' \\ &= 2r \, d\theta_1 \, 2r \, d\theta_2 \end{aligned}$$

Hence the intensity at r will be

$$I_r = \frac{dP}{dA} = \frac{I_i R_1 R_2}{4r^2}$$

and the target strength will be

$$\mathrm{TS} \equiv 10 \log \left. \frac{I_r}{I_i} \right|_{r=1} = 10 \log \frac{R_1 R_2}{4}$$

9.3 Target Strength of a Small Sphere

We turn now to the target strength of a small sphere, in which the return of sound back toward the source is a process of scattering rather than reflection.

The theory of sound scattering by a small, fixed, rigid sphere was first worked out by Rayleigh (5). By *small* is meant a sphere whose ratio of circumference to wavelength is much less than unity ($ka = 2\pi a/\lambda \ll 1$); by *fixed*, a sphere which does not partake of the acoustic motion of particles of the fluid in which the sphere is embedded; by *rigid*, a sphere that is nondeformable by the incident acoustic waves and into which the sound field does not penetrate. Under these conditions, Rayleigh showed that the ratio of the scattered intensity I_r at a large distance r to the intensity I_i of the incident phase wave is

$$\frac{I_r}{I_i} = \frac{\pi^2 T^2}{r^2 \lambda^4} \left(1 + \frac{3}{2} \mu\right)^2$$

where T = volume of sphere ($\frac{4}{3}\pi a^3$)
λ = wavelength
μ = cosine of angle between scattering direction and reverse direction of incident wave

For backscattering $\mu = +1$. On reducing to $r = 1$ and on taking 10 times the logarithm, we obtain

$$\text{TS} \equiv 10 \log \frac{I_r}{I_i}\bigg|_{r=1} = 10 \log \frac{\pi^2 T^2}{\lambda^4} \left(\frac{5}{2}\right)^2 = 10 \log \left[(1082) \frac{a^6}{\lambda^4}\right]$$

where a and λ are in units of yards. Thus, the target strength of a small sphere varies as the sixth power of the radius and inversely as the fourth power of the wavelength.

Defining the backscattering cross section of the sphere as

$$\sigma \equiv 4\pi \frac{I_r}{I_i}\bigg|_{r=1}$$

the ratio of backscattering cross section to geometric cross section becomes

$$\frac{\sigma}{\pi a^2} = 2.8(ka)^4$$

Figure 9.5 is a plot of this normalized ratio against the nondimensional quantity $ka = 2\pi a/\lambda$. It is seen that this quantity varies as the fourth power of ka, or as the fourth power of the frequency, for ka less than about 0.5, and is unity for ka greater than 5.0. Oscillations occur in the intermediate region ($ka \approx 1$).

Rayleigh also dealt with the scattering by small spheres, not fixed and rigid, but possessing a compressibility κ' and a density ρ' in a fluid of compressibility κ and density ρ. Fixed, rigid spheres are incompressible ($\kappa'/\kappa \ll 1$) and very dense ($\rho'/\rho \gg 1$) compared to the surrounding fluid; spheres having bulk moduli and densities comparable with that of the fluid oscillate to and fro in the incident sound field. They also pulsate, or change their volume, in the compressions and rarefactions of the incident wave. These motions of the sphere modify the scattered wave. Rayleigh showed (6) that the term $(1 + \frac{3}{2}\mu)^2$ in the expression for the ratio of intensities for a fixed, rigid sphere becomes

$$\left[1 - \frac{\kappa'}{\kappa} + \frac{3(\rho'/\rho - 1)}{1 + 2\rho'/\rho} \mu\right]^2$$

in terms of the ratios of compressibility κ'/κ and density ρ'/ρ. Figure **9.6** is a plot of the quantity

$$10 \log \left[1 - \frac{\kappa'}{\kappa} + \frac{3(\rho'/\rho - 1)}{1 + 2\rho'/\rho}\, \mu \right]^2 \Big/ \left(1 + \frac{3}{2}\, \mu\right)^2$$

This is the "correction" for compressibility and density to be applied to the target strength of a fixed, rigid sphere. This correction is seen

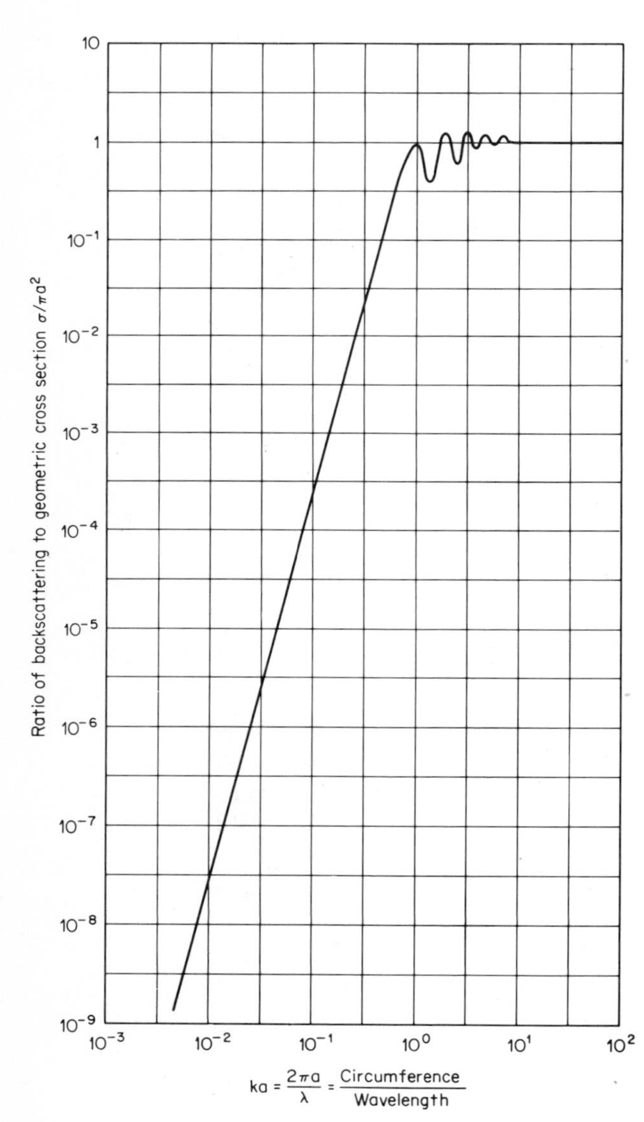

FIG. 9.5. Ratio of acoustic to geometric cross sections of a fixed, rigid sphere of radius a. The target strength equals $10 \log 4\pi\sigma$.

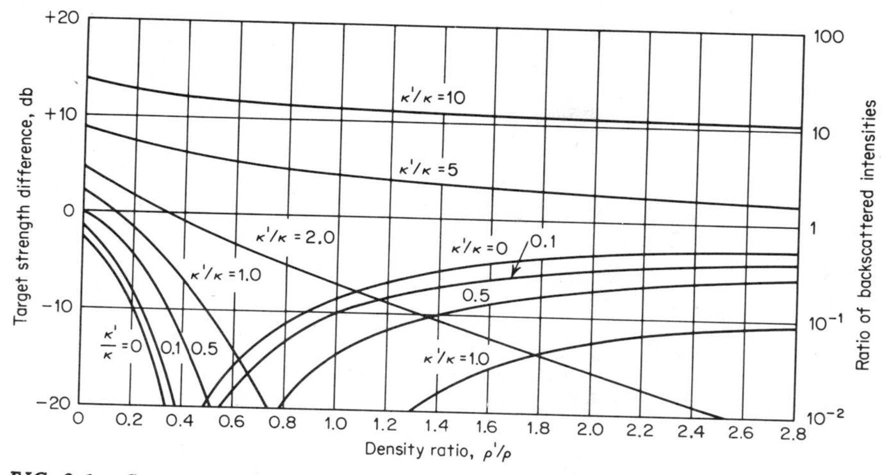

FIG. 9.6. Corrections to the target strength of a fixed rigid sphere for compressibility and density. The sphere is of compressibility ratio κ'/κ and density ratio ρ'/ρ relative to the surrounding fluid.

from Fig. 9.6 to be large in most instances. For a sand grain in water, for example, for which $\rho'/\rho = 2.6$ and $\kappa'/\kappa \approx 0.1$, the target strength is seen to be 4 db less than for the classic Rayleigh case of a fixed, rigid sphere. On the other hand, highly compressible spheres such as gas bubbles in water have enormously greater target strengths, even in the absence of resonance effects. For example, for a nonresonant air bubble in water at 1 atm, κ'/κ is approximately 19,000, and its backscattering is some 77 db greater than that for classic Rayleigh scattering by a sphere of the same size.

A thorough theoretical treatment of sound scattering from a fluid sphere is given in a paper by Anderson (7).

9.4 Target Strength of Simple Forms

The target strength of a number of geometric shapes and forms has been found theoretically, in most cases for applications to radar. Table 9.1 is a list of a number of mathematical forms for which the target strength has been determined, together with the appropriate literature references. These idealized expressions should be viewed as no more than crude approximations for targets of complex internal construction for which penetration and scattering are suspected to occur. Moreover, the mobility and nonrigidity of sonar targets, unlike radar targets, cause them to have target strengths different from what they would have if they were fixed and rigid as theory requires. Yet the expressions will

TABLE 9.1 Target Strength of Simple Forms

Form	t Target strength $= 10 \log t$	Symbols	Direction of incidence	Conditions	References
Any convex surface	$\frac{a_1 a_2}{4}$	$a_1 a_2$ = principal radii of curvature r = range $k = 2\pi/\text{wavelength}$	Normal to surface	$ka_1, ka_2 \gg 1$ $r > a$	1
Sphere Large	$\frac{a^2}{4}$	a = radius of sphere	Any	$ka \gg 1$ $r > a$	1
Small	$61.7 \frac{V^2}{\lambda^4}$	V = volume of sphere λ = wavelength	Any	$ka \ll 1$ $kr \gg 1$	2
Cylinder Infinitely long Thick	$\frac{ar}{2}$	a = radius of cylinder	Normal to axis of cylinder	$ka \gg 1$ $r > a$	1
Thin	$\frac{9\pi^4 a^4}{\lambda^2} r$	a = radius of cylinder	Normal to axis of cylinder	$ka \ll 1$	3
Finite	$aL^2/2\lambda$	L = length of cylinder a = radius of cylinder	Normal to axis of cylinder	$ka \gg 1$ $r > L^2/\lambda$	4
	$aL^2/2\lambda(\sin \beta/\beta)^2 \cos^2 \theta$	a = radius of cylinder $\beta = kL \sin \theta$	At angle θ with normal		

TABLE 9.1 (Continued)

Form	t Target strength $= 10 \log t$	Symbols	Direction of incidence	Conditions	References
Plate Infinite (plane surface)	$\frac{r^2}{4}$		Normal to plane		
Finite Any shape	$\left(\frac{A}{\lambda}\right)^2$	A = area of plate L = greatest linear dimension of plate l = smallest linear dimension of plate	Normal to plate	$r > \frac{L^2}{\lambda}$ $kl \gg 1$	5
Rectangular	$\left(\frac{ab}{\lambda}\right)^2 \left(\frac{\sin \beta}{\beta}\right)^2 \cos^2 \theta$	a, b = side of rectangle $\beta = ka \sin \theta$	At angle θ to normal in plane containing side a	$r > \frac{a^2}{\lambda}$ $kb \gg 1$ $a > b$	4
Circular	$\left(\frac{\pi a^2}{\lambda}\right)^2 \left(\frac{2J_1(\beta)}{\beta}\right)^2 \cos^2 \theta$	a = radius of plate $\beta = 2ka \sin \theta$	At angle θ to normal	$r > \frac{a^2}{\lambda}$ $ka \gg 1$	4
Ellipsoid	$\left(\frac{bc}{2a}\right)^2$	a, b, c = semimajor axes of ellipsoid	Parallel to axis of a	$ka, kb, kc \gg 1$ $r \gg a, b, c$	6
Conical tip	$\left(\frac{\lambda}{8\pi}\right)^2 \tan^4 \psi \left(1 - \frac{\sin^2 \theta}{\cos^2 \psi}\right)^{-3}$	ψ = half angle of cone	At angle θ with axis of cone	$\theta < \psi$	7

TABLE 9.1 (Continued)

Form	t Target strength $= 10 \log t$	Symbols	Direction of incidence	Conditions	References
Average over all aspects					
Circular disk	$\frac{a^2}{8}$	a = radius of disk	Average over all directions	$ka \gg 1$ $r > \frac{(2a)^2}{\lambda}$	5
Any smooth convex object	$\frac{S}{16\pi}$	S = total surface area of object	Average over all directions	All dimensions and radii of curvature large compared with λ	4 7
Triangular corner reflector	$\frac{L^4}{3\lambda^2}(1 - 0.00076\theta^2)$	L = length of edge of reflector	At angle θ to axis of symmetry	Dimensions large compared with λ	5

1. Physics of Sound in the Sea, part III, *Natl. Defense Res. Comm. Div. 6 Sum. Tech. Rept.* **8**:358–362 (1946). (Note: eqs. 49, 50, 53, and 56 in this reference are in error.)
2. Rayleigh, Lord: "Theory of Sound," vol. II, p. 277, Dover Publications, Inc., New York, 1945.
3. Rayleigh, Lord: "Theory of Sound," vol. II, p. 311, Dover Publications, Inc., New York, 1945.
4. Kerr, D. E. (ed.): "Propagation of Short Radio Waves," M.I.T. Radiation Laboratory Series, vol. 13, pp. 445–469, McGraw-Hill Book Company, New York, 1951.
5. Propagation of Radio Waves, Committee on Propagation, *Natl. Defense Res. Comm. Sum. Tech. Rept.*, **3**:182, (1946).
6. Willis, H. F.: Unpublished (British) report, 1941.
7. Spencer, R. C.: Backscattering from Conducting Surfaces, RDB Committee on Electronics, Symposium on Radar Reflection Studies, September, 1950.

often be found useful for the prediction of the target strength of new and unusual objects for which no measured data are available and which, it is felt, can be approximated well enough by an ideal geometric shape. This use of these expressions will be illustrated later on for mines and torpedoes.

9.5 Target-strength Measurement Methods

The obvious and direct method of measuring the target strength of an underwater object is to place a hydrophone at a distance of 1 yd from the object and to measure the ratio of the reflected (or scattered) intensity to the incident intensity. This method is impractical for a number of reasons. For many objects, it is difficult, if not impossible, to locate and place a hydrophone at this 1-yd point; even if it could be done, it would be difficult to separate the reflected sound from the incident sound at such a short distance. Moreover, the results would be invalid, in many cases, for use at longer ranges, since the target strength of objects like cylinders and submarines is different at short ranges than at long.

A more practical method is to use a reference target of known target strength, placed at the same range as the unknown, and to compare the levels of the echoes from the reference target and the target to be measured. The method is illustrated in Fig. 9.7. If the reference target is a sphere of diameter 4 yd, its target strength is zero db. Then the target strength of the target to be measured will be 10 times the logarithm of the ratio of the echo intensity of the unknown target to that of the zero-db reference target. Suitable corrections can easily be made for differences in range and size of the reference target. This straightforward comparison method is particularly suited to measurements on small objects at short ranges. For large objects, such as submarines at longer ranges, the handling and the size requirements of the reference sphere make the method impractical. In all cases, the construction and dimensions of the reference target must be carefully controlled to make sure that its target strength approximates that of the ideal shape. Because of its ease of handling and its higher

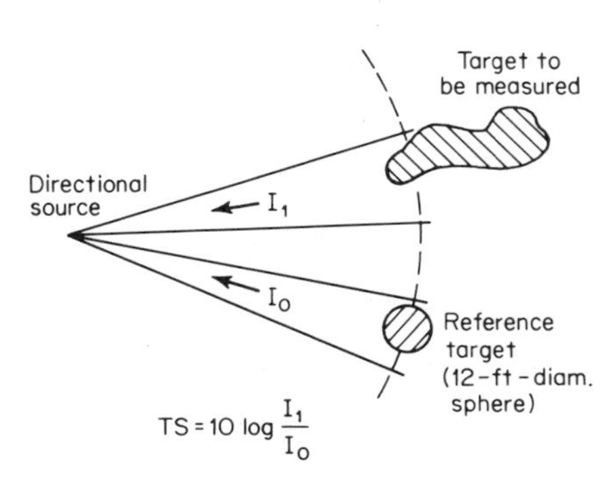

FIG. 9.7. Target-strength measurement using a reference sphere. A sphere of diameter 12 ft is placed at the same range as the target being measured. Using a directional active sonar, the intensities of the echoes are compared.

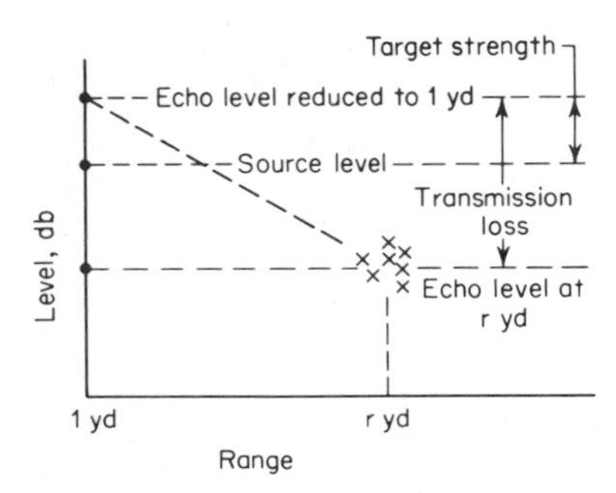

***FIG. 9.8.** Target-strength measurement using echo levels reduced to 1 yd. The range, source level, and transmission loss must be known.*

effective target strength, a calibrated transponder is a substitute for such a passive reference target.

Most target-strength measurements have been made by what may be called the conventional method, in which measurements of the peak or average intensity of the irregular echo envelope are made at some long range and then reduced to what they would be at 1 yd. This reduction requires, in effect, a knowledge of the transmission loss appropriate to the time and place the echoes were measured, together with the source level of the sound source producing the echoes. The method is illustrated in Fig. 9.8. A number of echo-level measurements of a sonar target are made at some range r. The average level of the echoes is reduced to 1 yd by adding the known or estimated transmission loss; the difference between the reduced 1-yd level and the known source level is the target strength desired. The method employs the active-sonar equation written in the form

$$\mathrm{EL} = \mathrm{SL} - 2\mathrm{TL} + \mathrm{TS}$$

where EL is the level of the echo. The equation is solved for the unknown, TS. This conventional method has the disadvantage of requiring an accurate knowledge of the transmission loss, which in turn requires either simple propagation conditions, or a special series of field measurements made for the purpose. Much of the scatter and diversity in existing target-strength data are doubtless attributable to erroneous transmission-loss assumptions and to indiscriminate use of peak and average levels of the echo. Nevertheless, the method is basically simple and straightforward and requires no special equipment or instrumentation.

A method requiring no knowledge of the transmission loss, but requiring special instrumentation, was used by Urick and Pieper (8). A measurement hydrophone and a transponder located about 1 yd from it were installed on the target submarine. On the measuring vessel (a surface ship) a hydrophone was suspended near the sound source producing the echoes to be measured. The relative levels of echo and transponder pulse were recorded aboard the surface ship; the relative levels of the incoming pulse and the transponder pulse were recorded aboard the submarine. As will be evident from Fig. 9.9, the target strength of the submarine is simply the difference in level between the two level differences recorded on the two vessels. The transponder serves, in effect, to

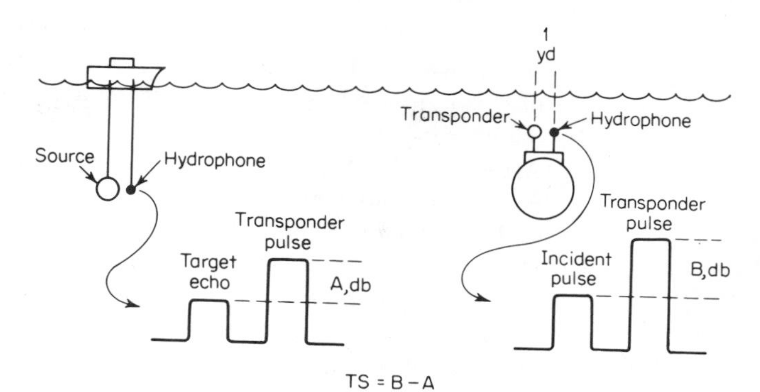

FIG. 9.9. A transponder method of target-strength measurement. No calibrations or knowledge of transmission loss are necessary.

"calibrate out" the underwater transmission path between the two vessels. No absolute calibration of the transducers used is needed, and the range separating the vessels need not be known.

9.6 Target Strength of Submarines

The target strength of submarines, of all underwater targets, has had the earliest historical attention and has been relatively well known from work done during World War II. This work has been well summarized in the NDRC Summary Technical Reports (9, 10). The present treatment of the subject will center about the variations of target strength with aspect, frequency, ping duration, depth, and range. The submarines considered will be limited to conventional fleet-type diesel-powered submarines.

It should be emphasized at the outset that submarine target strengths are perhaps most noteworthy for their variability. Not only do individual echoes vary greatly from echo to echo on a single submarine, but average values from submarine to submarine, as measured by different workers at different times and reduced to target strength, are vastly different. As a result, the discussion of the subject will be concerned with trends and broad effects, with the expectation that individual measurements will show great differences from the mean or average values.

Variation with Aspect Figure 9.10 shows two examples of the variation of target strength with aspect around a submarine. Example *A* is a typical World War II determination (11) at 24 kHz with each point representing an average of about forty individual echoes in a 15° sector of aspect angle. Example *B* is the result of a postwar measurement (8) in which about five echoes were averaged in each 5° sector. Plots of this kind are based on measurements at sea of echoes from a submarine which runs in a tight circle at a distance from the source-receiver ship. Alternatively, the latter may circle a submarine running on a straight course at a slow speed.

The two examples of Fig. 9.10 illustrate the variability of target-strength measurements mentioned above. There is first a point-to-point variability with aspect that is due to the variability in level of individual echoes. There is also an average difference of about 10 db between the two sets of measurements, with *A* being more typical of other determinations. A study of various determinations on different submarines and

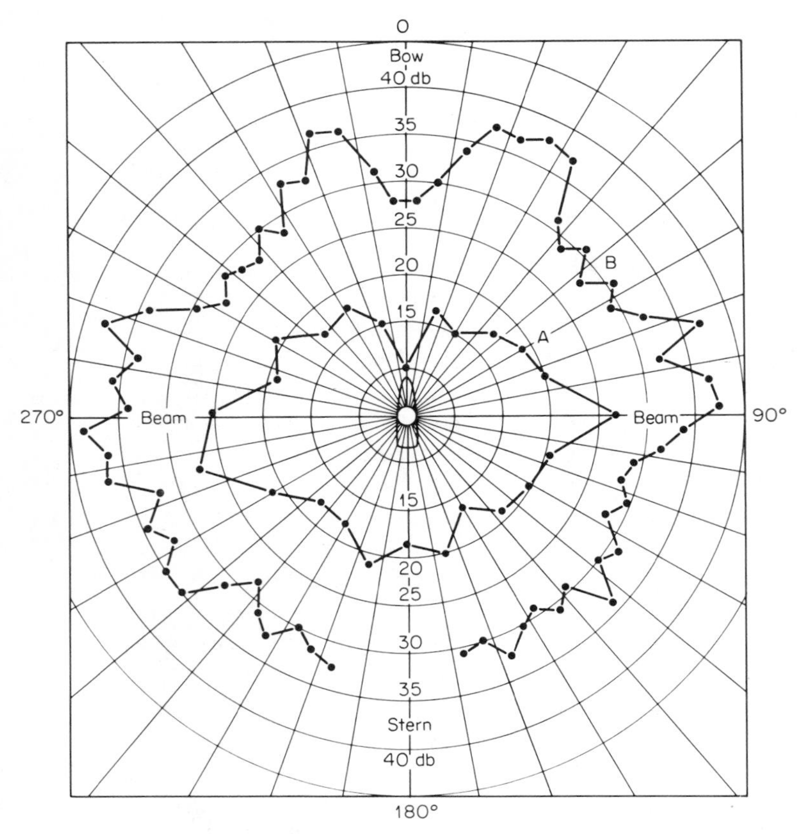

FIG. 9.10. Two determinations of the target strength of a submarine at various aspects. Example A (Ref. 10) is more typical of other determinations than B (Ref. 8).

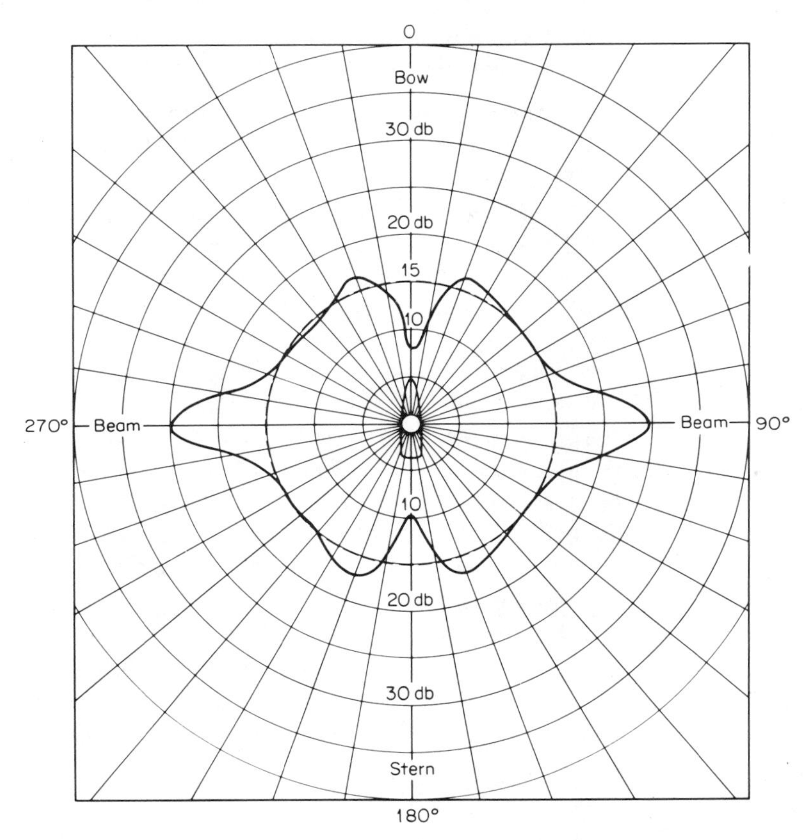

FIG. 9.11. The "butterfly" pattern of the aspect variation of submarine target strength.

at different frequencies indicates a spread in the measured values of 12 to 40 db at beam aspect, with a median value of 25 db.

Nevertheless, when viewed in a broad fashion, the aspect dependence of target strength may be considered to have the pattern shown in Fig. 9.11. This "butterfly" pattern has the following characteristics:

1. "Wings" at beam aspect, extending up to about 25 db and caused by specular reflection from the hull
2. Dips at bow and stern aspects caused by shadowing of the hull and wake
3. Lobes at about 20° from bow and stern, extending 1 or 2 db above the general level of the pattern, perhaps caused by internal reflections in the tank structure of the submarine

4. A circular shape at other aspects due to a multiplicity of scattered returns from the complex structure of the submarine and its appendages

All the characteristics of this idealized pattern are seldom seen in individual measurements. They may be absent entirely or lost in the scatter of the data.

Variation with Frequency Target-strength data were obtained during World War II (12) at 12, 24, and 60 kHz in an attempt to determine the effect of frequency on target strength. This attempt was unsuccessful, and it was concluded at that time that, if any frequency dependence existed, it was lost in the uncertainties of the data. The cause of this apparent frequency independence must be the many processes and sources that govern the return of sound from a submarine.

Variation with Depth Except for the contribution of sound returned from the submarine's wake, no effect of depth on target strength has been established, nor expected. Depth effects on the echo are more likely the result of changing sound propagation with changing depth in the sea, rather than a depth effect on the submarine itself.

Variation with Range For two reasons, the target strength of submarines is apt to be less at short ranges than at long ranges. One is the failure of a directional sonar, if one is used for the target-strength measurement, to insonify the entire target. The other is the fact that the echo of some geometrical forms does not fall off with range like the level from a

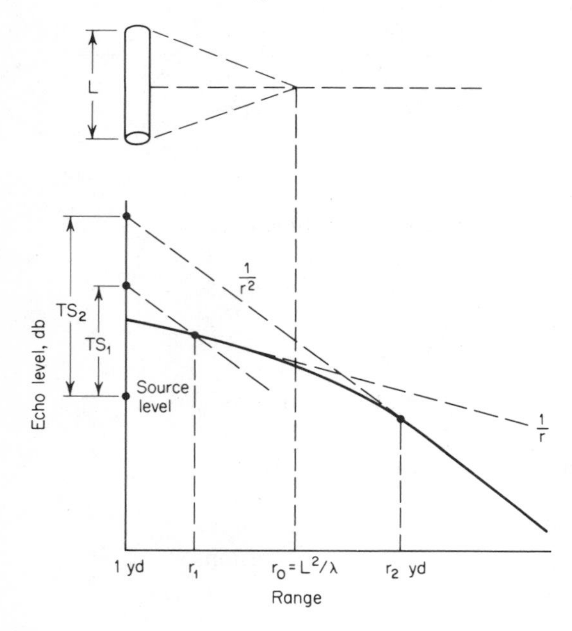

FIG. 9.12. Target strength versus range for a beam-aspect cylinder. At ranges less than r_0, the echo level varies with range like $1/r$ rather than $1/r^2$. This causes a lower target strength at short ranges than at long ranges ($TS_1 < TS_2$) if a $1/r^2$ transmission loss is used for the reduction to 1 yd.

point source. The solid curve in Fig. 9.12 shows the variation of echo intensity with range for a cylinder of length L at "beam" aspect. The intensity falls off like $1/r$ at short ranges ("cylindrical spreading") and like $1/r^2$ at long ranges ("spherical spreading") with a transition range approximately equal to L^2/λ. An echo at a short range r_1, when reduced to 1 yd by applying the transmission loss appropriate for a point source, would yield an apparent target strength TS_1 that would be less than the value TS_2 obtained with a long-range echo. Also, the target strength of an infinitely long cylinder, for which all points at a finite distance are in the near-field or Fresnel region of the cylinder, is seen in Table 9.1 to increase linearly with range. The former effect—that of failure to insonify all of the target—would apply to conditions of aspect and frequency where scattering by numerous scatterers on or in the target is the dominant echo-producing process; the latter effect applies to specular reflection. Both effects cause the short-range target strength of submarines to be less than the long-range target strength.

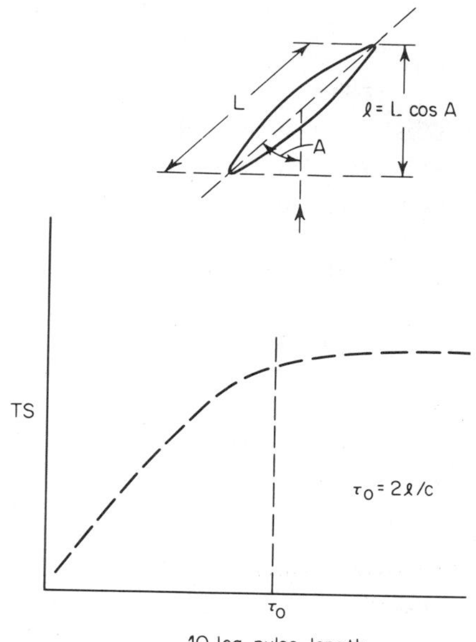

FIG. 9.13. Variation with pulse duration. The target strength increases until the pulse duration is great enough for the entire target to contribute to the echo at some one instant of time. This occurs at pulse duration $\tau_0 = 2l/c$.

Variation with Pulse Duration The target strength of an extended target, such as a submarine, may be expected to fall off with decreasing pulse duration, inasmuch as a short pulse must fail to insonify the entire target. Thus, Fig. 9.13 illustrates that the target strength of an extended target should increase with the pulse duration (in logarithmic units) until the pulse duration is just long enough for all points on the target to contribute to the echo simultaneously at some instant of time. This occurs when the ping duration τ_0 is such that

$$\tau_0 = \frac{2l}{c}$$

where l = extension in range of target
c = velocity of sound

If the target is of length L, then at aspect angle A,

$$l = L \cos A$$

This theoretical effect of reduced ping duration is not noticeable in target-strength measurements at beam aspect, where the extension in range is small and where specular reflection is the principal echo formation process, nor in measurements of peak target strength where individual target highlights are being measured.

9.7 Target Strength of Surface Ships

Because surface ships are not important targets for active sonars, comparatively little data are available on their target strengths. Three series of measurements at frequencies from 20 to 30 kHz were made during World War II on a total of 17 naval and merchant ships (13). They yielded beam-aspect values of 37, 21, and 16 db for the three different groups of measurements and values of 13, 17, and 14 db for corresponding off-beam target strengths. Standard deviations between 5 and 16 db are quoted for these values. The small amount of available data show similar effects of aspect and range, as for submarines.

9.8 Target Strength of Mines

Modern mines are quasi-cylindrical objects a few feet long and 1 to 2 ft in diameter, flat or rounded on one end, and containing protuberances, depressions, and fins superposed on their generally cylindrical shape. Such targets should be expected to have a high target strength at "beam" aspect, as well as at aspects where some flat portion of the shape is normal to the direction of incidence; at other aspects, a relatively low target strength should be had. Measured target strengths range from about +10 db within a few degrees of beam aspect, to much smaller values at intermediate aspects, with occasional lobes in the pattern attributable to reflections from flat facets of the mine shape. By Table 9.1, the target strength of a cylinder of length L, radius a, at wavelength λ is at long ranges

$$\text{TS} = 10 \log \frac{aL^2}{\lambda}$$

If a be taken as 0.2 yd, $L = 1.5$ yd, and $\lambda = 0.03$ yd, corresponding to a frequency of 56 kHz, then

$$\text{TS} = 10 \log \frac{0.2 \times (1.5)^2}{0.03} = 12 \text{ db}$$

in approximate agreement with measurements on mines at beam aspect. The principal effects of frequency, aspect, range, and pulse duration described for submarines apply for mines generally as well.

9.9 Target Strength of Torpedoes

A torpedo is, like most mines, basically cylindrical in shape with a flat or rounded nose. When specular reflection can be presumed to be the principal echo formation process, the target strength of a torpedo can be approximated by the theoretical formulas. An approaching torpedo of diameter 20 in. (radius = 0.28 yd) and having a hemispherical nose with a diameter equal to that of the torpedo, would have a target strength equal to $10 \log [(0.28)^2/4] = -17$ db. At beam aspect its target strength could be approximated by the cylindrical formula illustrated for mines.

9.10 Target Strength of Fish

Fish are the targets of fish-finding sonars and have been most carefully investigated by Japanese (14) and British (15) workers. Figure 9.14

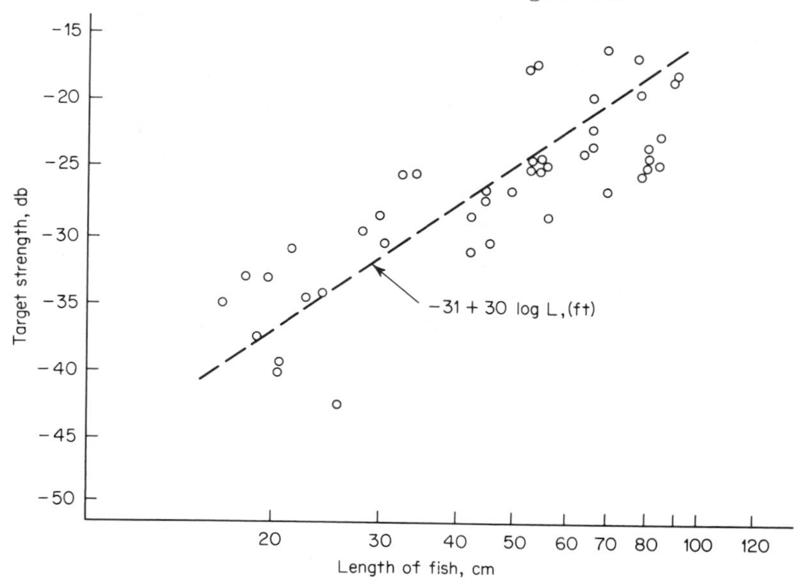

FIG. 9.14. Target strength measurements of various commercial fish (cod, plaice, perch, herring). Frequency 30 kHz, dorsal aspect, that is, "looking" down. The line drawn is a straight-line relationship between TS and 10 times the logarithm of the volume of the fish. (*Replotted from Ref. 16.*)

shows a series of measurements of various commercial fish caught in British waters, as reported by Cushing et al. (16). These measurements were made on dead fish, although some were fitted with artificial swim bladders of a thin plastic material, and refer to a dorsal "aspect" of the fish, in which an active sonar vertically overhead insonified a fish in normal swimming position. A school of n widely separated fish, each of target strength TS, would have a combined target strength equal to TS + 10 log n.

In the United States, the most extensive work on the target strength of fish has been done by Volberg (17). In this work, specimens of skipjack, yellowtail, and tuna were rotated about various axes and their target strengths measured at 40, 50, and 280 kHz. Although no clear evidence of a frequency or aspect dependence was found in this work, it should be noted that fish have been found by Coate (18) to produce strong attenuation of sound, and presumably, therefore, a high target strength, at lower frequencies where their swim bladders form resonant air cavities.

9.11 Echo Formation Processes

Complex underwater targets return sound back to the source by a number of processes. These processes will all occur, in general, for a complex target like a submarine, but only one or two will be dominant under any particular conditions of frequency and aspect angle.

Specular Reflection The most simple and best understood echo formation process is *specular reflection*. It is illustrated by the return of sound from large spheres or convex surfaces, as described earlier in this chapter. In this process, the target remains stationary and does not move in the sound field and thereby generates a reflected wave having a *particle velocity* just sufficient, in wave theory, to cancel that of the incident wave over the surface of the object. Alternatively, the reflecting surface may be *soft* instead of *hard* or stationary, so as to form a reflection in which the *pressures* of the two waves will cancel. A specular reflection has a waveform that is a duplicate of the incident waveform and can be perfectly correlated with it. For submarines and mines, specular reflection appears to be the dominant process at beam aspects, where a much-enhanced return is observed and where a short incident pulse has a faithful replica as an echo.

Scattering by Surface Irregularities Irregularities, such as protuberances, corners, and edges, that have radii of curvature small compared with a wavelength, return sound by scattering rather than reflection. Most real objects possess many such irregularities on their surface, and the scattered return is composed of contributions from a large number of such scattering centers. When only a few scatterers on the underwater target are dominant, they form *highlights*, which may be observable in the echo envelope. Submarines, especially the older types, possess numerous scattering irregularities on the hull, such as the bow and stern planes, railings, periscopes; of these protuberances, the conning tower is apt to be a strong reflector or scatterer of sound.

Penetration of Sound into the Target Underwater targets seldom remain rigid under the impact of an incident sound wave, but move or deform in a complex manner. This reaction may be thought of as penetration of sound into the target and a complex deformation of the target by the exciting sound wave.

In the case of solid metal spheres in water, Hampton and McKinney (19) found considerable experimental evidence that acoustic energy in the frequency range 50 to 150 kHz penetrated into solid spheres a few inches in diameter and thereby produced a complex echo envelope and a target strength which varied by as much as 30 db with frequency. A theoretical study by Hickling (20) of hollow metal spheres in water demonstrated that part of the echo originates from a kind of flexural wave moving around the shell. Such surface waves were also observed experimentally by Barnard and McKinney (21) in a study of solid and air-filled cylinders in water.

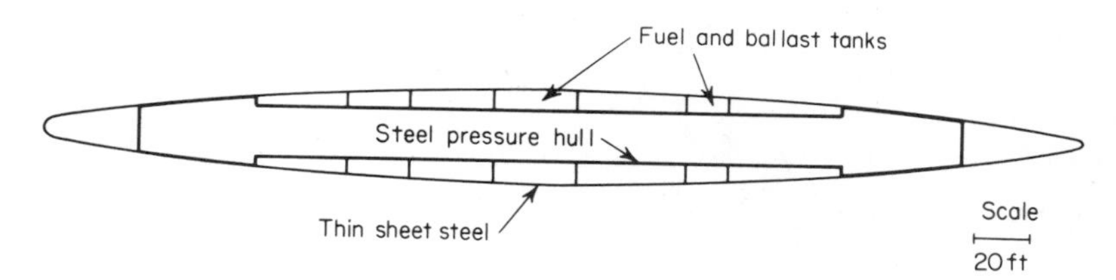

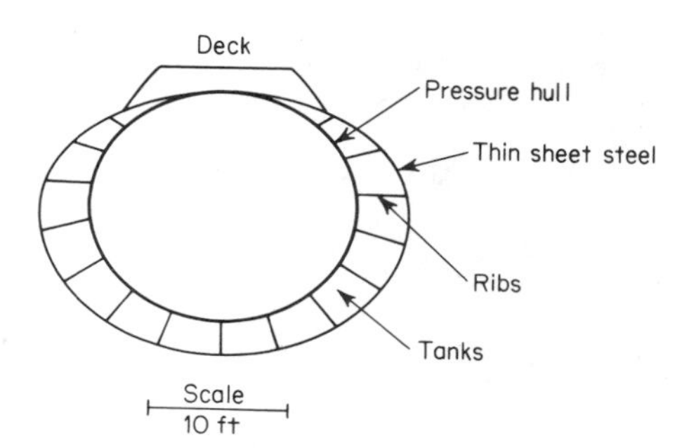

FIG. 9.15. Diagrammatic fore-and-aft and athwartship cross sections of a fleet submarine.

Submarines are structurally much more complicated than the simple forms just mentioned. Figure 9.15 shows cross-sectional views of a fleet-type submarine of basically World War II construction. The hull is essentially a quasi-cylindrical pressure hull surrounded by a tank structure of thin steel. Penetration of sound into the tank structure is relatively easy, and a return of sound by corner reflectors should be expected. These internal reflections may be the cause of the enhanced target strength at about 20° off bow and stern in the butterfly pattern of Fig. 9.11. In addition, scattering and resonance effects in the tank structure are likely to be sources of the echo at off-beam aspects. Finally, the pressure hull itself may be expected to contribute a scattered return at some aspect angles.

Resonant Effects Certain incident frequencies may correspond to various resonance frequencies of the underwater target. Such frequencies will excite different modes of oscillation or vibration of the target and give rise, in principle, to an enhanced target strength. Figure 9.16 is a pictorial description of a number of possible oscillatory modes of a compressible, flexible cigar-shaped object. When the sonar frequency is low enough, such modes may be excited by the incident sound wave and contribute to the target strength to an extent depending on the aspect

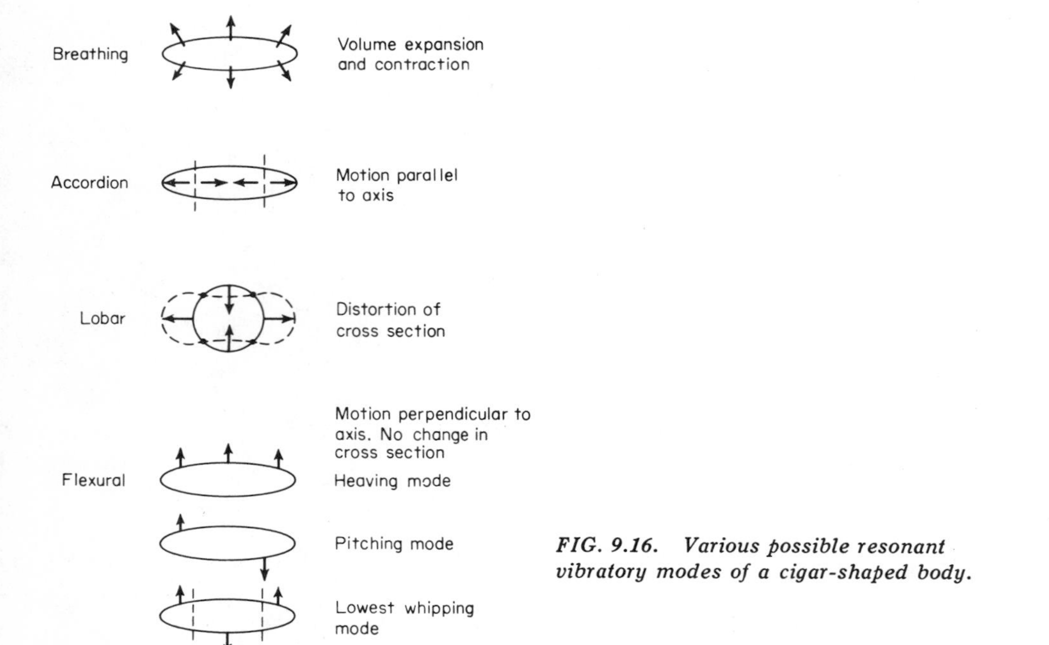

FIG. 9.16. Various possible resonant vibratory modes of a cigar-shaped body.

and the Q or damping constant of the oscillatory mode. However, it should be pointed out that those modes which are good radiators of sound have a high damping constant because of radiation loading of the surrounding fluid, and as a result they would not necessarily give rise to a much greater target strength. Similarly, slightly damped modes of high Q might be poor reradiators of sound and thereby be poor contributors to the target strength. Such slightly damped modes would require a long sonar pulse of closely matched frequency for their excitation. It is even possible that resonant vibratory modes may be accompanied by a *lower* target strength if high internal losses in the target cause absorption of the incident energy, or if the mode is such that sound is reradiated, at resonance, into directions other than into the specular direction.

A resonance of an entirely different type is the flexural resonance of the plates of which an underwater target like a submarine is composed. At certain angles a steel plate has been shown by the work of Finney (22) to be transparent to sound because of coupling between shear waves in the plate and sound waves in the surrounding water.

9.12 Echo Characteristics

Echoes from most underwater objects differ from the incident pulse in a number of ways other than intensity, as described by the parameter target strength. The reflecting object imparts its own characteristics to the echo; it interacts with the incident sound wave to produce an echo that is, in general, different in waveshape and other characteristics from the incident pulse. These differences are useful to the sonar engineer in two ways: they may be used as an aid in *detection*, as in filtering with narrow-band filters to enhance an echo buried in reverberation; they may be used to assist in target *classification* to distinguish one type of target from another, as in distinguishing a submarine from a school of fish.

Some distinguishing characteristics of echoes are:

Doppler Shift Echoes from a moving target are shifted in frequency by the familiar Doppler effect (23) by an amount equal to

$$\Delta f = \frac{2v}{c} f$$

where v = relative velocity or range rate of source and target
c = velocity of sound (in same units as v)
f = operating frequency

In practical terms, and for a sound velocity of 4,900 ft/sec,

$$\Delta f = \pm 0.69 \text{ Hz/(knot)(kHz)}$$

where the range rate is in units of knots and the sonar frequency is in kilohertz. An echo from an approaching target with a 10-knot relative velocity and a frequency of 10 kHz would thus be shifted higher in frequency by 69 Hz. The $\pm$ sign indicates that an approaching target produces an echo of higher frequency ("up-Doppler"), and a receding target one of lower frequency ("down-Doppler").

Extended Duration Echoes are lengthened by the extension in range of the target. Whenever an underwater target is such that sound is returned by scatterers and reflectors distributed all along the target, the entire area or volume of the target contributes to the echo. For a target of length L at an aspect angle θ, the incident pulse is lengthened in duration by the time interval

$$\frac{2L \cos \theta}{c}$$

for the monostatic case, and by

$$\frac{L}{c} (\cos \theta_i + \cos \theta_r)$$

for the bistatic case with incidence at aspect angle θ_i and an echo return at angle θ_r. This time elongation of the echo is most noticeable when short sonar pulses are employed. It occurs only for complex targets composed of numerous distributed scatterers and is negligible when specular reflection is the most important process of echo formation. Most sonar echoes are elongated, however, and so provide a clue as to the size, and therefore the nature, of the echoing object.

Irregular Envelope The echo envelope is irregular, especially where specular reflection is not important. This irregularity arises from acoustic interference between the scatterers of the target. For some targets, individual highlights in the echo may be identified as arising from individual strong echoing portions of the target; for example, the conning tower of a submarine may yield a recognizable strong return of its own as part of the echo. Most often, however, the sonar echo from a sinusoidal ping is an irregular blob, without distinguishing features, that varies in envelope shape from echo to echo as the changing phase relationships among the scatterers and the propagation paths to and from the target take effect.

Modulation Effects At stern aspects on propeller-driven targets, the propeller may amplitude-modulate the echo in the same way that the propeller of an aircraft is known (24) to modulate a radar echo. The propeller thus produces a cyclic variation in the scattering cross section of the target. Another possible form of modulation can arise from inter-

TABLE 9.2 Nominal Values of Target Strength

Target	Aspect	TS, db
Submarines	Beam	+25
	Bow-stern	+10
	Intermediate	+15
Surface ships	Beam	+25 (highly uncertain)
	Off-beam	+15 (highly uncertain)
Mines	Beam	+10
	Off-beam	+10 to −25
Torpedoes	Bow	−20
Fish of length L, ft	Dorsal view	$-31 + 30 \log L$

action between the echo from a moving vessel and the echo from its wake. The difference in frequency between the two may appear as beats, or amplitude variations, in the envelope of the combined echo from the hull and wake at certain aspect angles.

9.13 Summary of Numerical Values

Table 9.2 is a summary of target-strength values for the underwater targets described in this chapter. As previously mentioned, these are subject to considerable variation in individual measurements on targets of the same type, and the target-strength values given are to be regarded as nominal values useful for first-cut problem solving.

REFERENCES

1. Kerr, D. E. (ed.): "Propagation of Short Radio Waves," M.I.T. Radiation Laboratory Series, vol. 13, pp. 445–481, McGraw-Hill Book Company, New York, 1951.
2. Freedman, A.: Recent Approaches to Echo-structure Theory, *J. Acoust. Soc. Am.*, **36**:2000(A) (1964). Also, A Mechanism of Acoustic Echo Formation, *Acoustica*, **12**:10 (1962).
3. Neubauer, W. G.: A Summation Formula for Use in Determining the Reflection of Irregular Bodies, *J. Acoust. Soc. Am.*, **35**:279 (1963).
4. Steinberger, R. L.: Theoretical Analysis of Echo Formation in the Fluctuation Environment of the Sea, *U.S. Naval Res. Lab. Rept.* 5449, 1960.
5. Rayleigh, Lord: "Theory of Sound," vol. 2, p. 277, eqs. 26 and 27, Dover Publications, Inc., New York, 1945.
6. Rayleigh, Lord: "Theory of Sound," vol. 2, p. 24, eq. 13, Dover Publications, Inc., New York.
7. Anderson, V. C.: Sound Scattering from a Fluid Sphere, *J. Acoust. Soc. Am.*, **22**:426–431 (1950).
8. Urick, R. J., and A. G. Pieper: "Determination of the Target Strength of a Submarine by a New Method," Office of Naval Research Informal Report, April, 1952.
9. Principles of Underwater Sound, *Natl. Defense Res. Comm. Div. 6 Sum. Tech. Rept.* 7, chap. 8, 1946.

10. Physics of Sound in the Sea: Reflection of Sound from Submarines and Surface Vessels, *Natl. Defense Res. Comm. Div. 6 Sum. Tech. Rept.* 8, 1946.
11. Physics of Sound in the Sea: Reflection of Sound from Submarines and Surface Vessels, *Natl. Defense Res. Comm. Div. 6 Sum. Tech. Rept.* 8, fig. 3, p. 391, 1946.
12. Physics of Sound in the Sea: Reflection of Sound from Submarines and Surface Vessels, *Natl. Defense Res. Comm. Div. 6 Sum. Tech. Rept.* 8, pp. 408–410, 1946.
13. Physics of Sound in the Sea: Reflection of Sound from Submarines and Surface Vessels, *Natl. Defense Res. Comm. Div. 6 Sum. Tech. Rept.* 8, chap. 24, 1946.
14. Hashimoto, T.: Characteristics of Ultrasonic Waves Transmitted Vertically in the Water, *Japan. Fishing Boat Lab. Rept.* 1, December, 1953.
15. Cushing, D. H., and I. B. Richardson: Echo Sounding Experiments on Fish, *Brit. Min. Agr. Fisheries Invest.*, (2)**18**(4) (1955).
16. Cushing, D. H., et al.: Measurements of the Target Strength of Fish, *J. Brit. Inst. Radio Engrs.*, **25**:299 (1963).
17. Volberg, H. W.: "Acoustic Properties of Fish," paper given at 14th Pacific Tuna Conference, September, 1963.
18. Coate, M. M.: Effect of a Single Fish on Low Frequency Sound Propagation, *Naval Ordnance Lab. Rept.* NAVORD 4514, 1957.
19. Hampton, L. D., and C. M. McKinney: Experimental Study of the Scattering of Acoustic Energy from Solid Metal Spheres in Water, *J. Acoust. Soc. Am.*, **33**:664 (1961).
20. Hickling, R.: Analysis of Echoes from a Hollow Metallic Sphere in Water, *J. Acoust. Soc. Am.*, **36**:1124 (1964).
21. Barnard, G. R., and C. M. McKinney: Scattering of Acoustic Energy by Solid and Air-filled Cylinders in Water, *J. Acoust. Soc. Am.*, **33**:226 (1961).
22. Finney, W. J.: Reflection of Sound from Submerged Plates, *J. Acoust. Soc. Am.*, **20**:626 (1948).
23. Kinsler, L. E., and A. R. Frey: "Fundamentals of Acoustics," p. 453, John Wiley & Sons, Inc., New York, 1950.
24. Lawson, J. L., and G. E. Uhlenbeck: "Threshold Signals," M.I.T. Radiation Laboratory Series, vol. 24, p. 288, McGraw-Hill Book Company, New York, 1950.

TEN

Radiated Noise of Ships, Submarines, and Torpedoes: Radiated-noise Levels

Ships, submarines, and torpedoes are excellent sources of underwater sound. Being themselves machines of great complexity, they require numerous rotational and reciprocating machinery components for their propulsion, control, and habitability. This machinery generates vibration that appears as underwater sound at a distant hydrophone after transmission through the hull and through the sea. Of particular importance is the machinery component called the propeller, which serves to keep the vehicle in motion, and in doing so, generates sound through processes of its own.

Radiated noise is of particular importance for passive sonars, which are designed to exploit the peculiarities of this form of noise and to distinguish it from the background of self-noise or ambient noise in which it is normally observed.

In this chapter the principal characteristics of radiated noise will be discussed, and some data on the level of this noise for various vehicles will be presented. The discussion will be restricted to conventional designs and conventional propulsion systems, omitting any discussion of the noise of nuclear-propelled vessels or unusual propulsion methods. The word "vessel" will be used to refer to surface ships, submarines, or torpedoes indiscriminately when it is unnecessary to distinguish between them.

10.1 Source Level and Noise Spectra

The parameter *source level* for radiated noise in the sonar equations is the intensity, in decibel units, of the noise

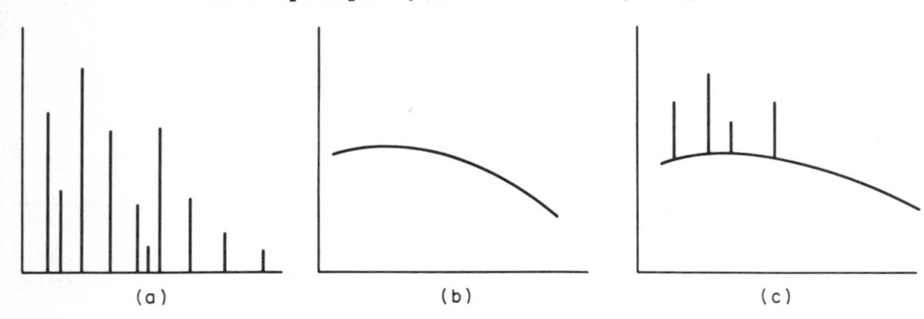

FIG. 10.1. Diagrams of (a) a line-component spectrum, (b) a continuous spectrum, and (c) the composite spectrum obtained by superposing (a) and (b).

radiated to a distance by an underwater source, when measured at an arbitrary distance and reduced to a distance of 1 yd from the *acoustic center* of the source. By acoustic center is meant the point inside or outside of the vessel from which the sound, as measured at a distance, *appears* to be radiated. Practically speaking, radiated-noise measurements must of necessity be made at a distance from the radiating vessel, typically 50 to 200 yd, and must be reduced to 1 yd by applying an appropriate spreading or distance correction. Source levels are specified in a 1-Hz band, and will be referred to a reference level of 1 dyne/cm^2. The source levels for radiated noise are accordingly *spectrum* levels relative to the 1 dyne/cm^2 reference level.

Noise spectra are of two basically different types. One type is *broadband noise* having a continuous spectrum. By continuous is meant the fact that the level is a continuous function of frequency. The other basic type of noise is *tonal noise* having a discontinuous spectrum. This form of noise consists of tones or sinusoidal components having a spectrum containing *line components* occurring at discrete frequencies. These two spectral types are illustrated diagrammatically in Fig. 10.1. The radiated noise of vessels consists of a mixture of these two types of noise over much of the frequency range and may be characterized as having a continuous spectrum containing superposed line components.

10.2 Methods of Measurement

The noise radiated by vessels is nearly always measured by running the vessel past a stationary distant measurement hydrophone. Various types of hydrophones and hydrophone arrays have been employed for this purpose. The simplest arrangement uses a single hydrophone hung from a small measurement vessel. More elaborate configurations involve an array of hydrophones, either strung in a line along the bottom in shallow water, or hung vertically in deep water, as illustrated in Fig. 10.2. The former arrangement is suitable for measuring the noise of surface ships, and the latter is useful for measuring the noise of submarines or torpedoes running at deep depths. In both cases the vessel under test is arranged to run at a constant speed and course so as to pass the measurement hydrophones at a known distance. Suitable techniques are used to determine the range of the vessel while its radiated-noise output is being measured. During the run broadband tape recordings are made, and later subjected to analysis in different frequency bands.

Although radiated noise is commonly expressed in *spectrum levels*, that is, in 1-Hz bands, frequency analyses are more conveniently made in wider bands. The results are reduced to a band of 1 Hz by applying a bandwidth reduction factor equal to 10 times the logarithm of the bandwidth used. That is to say, if BL is the noise level measured in a band w Hz wide, the spectrum level in a 1-Hz band is BL $-$ 10 log w (Sec. 1.3). This reduction process is valid for continuous "white" noise having a flat spectrum; it can be shown to be also valid for noise having a continuous spectrum falling off at the rate of -6 db/octave, if the center frequency of the band is taken to be the geometric mean of the two ends of the frequency band. But for line-component noise containing one or several strong lines within the measurement bandwidth, this reduction process is not valid and yields spectrum levels lower than the level of

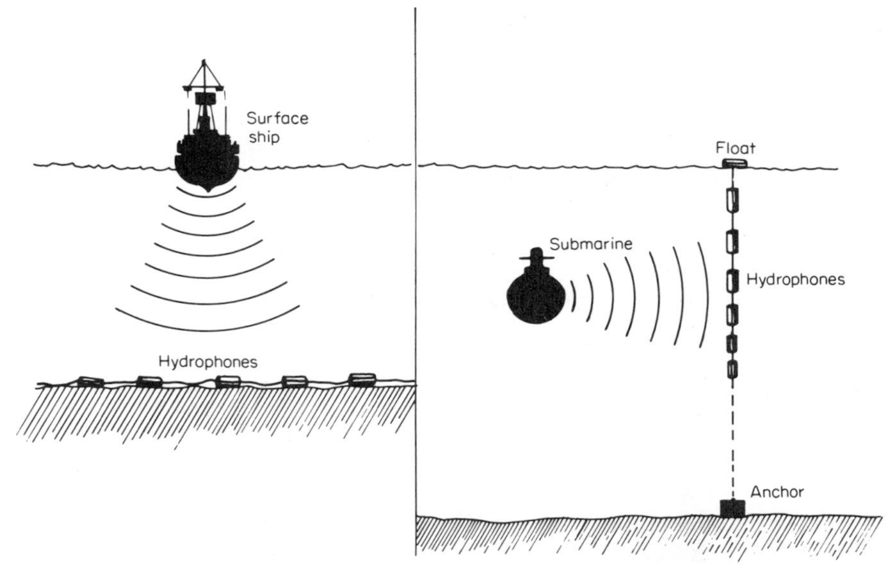

***FIG. 10.2.** Arrangements of hydrophones used at sound ranges for measuring radiated noise.*

the line component dominating the spectrum. In short, the nature of the spectrum must in principle be known before the reduction is made. Much old reported data are almost useless at low frequencies because an excessively broad frequency band had been employed in the original analysis.

Similarly, a correction for distance is required to reduce the measurements to the 1-yd reference distance. The ubiquitous spherical-spreading law is generally applied for this purpose. Several investigations have indicated that spherical or inverse-square spreading is a good rule of thumb for expressing the variation of ship noise with range at close distances, even for low frequencies in shallow water (1). In any case, in spite of the apparent artificiality of the reduction processes in many instances, the original levels can be recovered from the published reduced values if the reduction process, bandwidth, and measurement distance are stated.

Studies of the sources of noise under various operating conditions, as distinct from measurements of noise level alone, have employed various clever schemes to pinpoint the dominant sources of noise. For example, when ships and torpedoes are run very close to the measuring hydrophone, the principal source of noise—whether machinery noise originating amidships or propeller noise originating near the stern—can be identified by the correspondence between the peak of the noise and the closest part of the ship at that instant. Overside surveys (2) of surface ships and submarines, in which a hydrophone is lowered over the side of the moored vessel, are sometimes made to measure the noise radiated into the water by different pieces of ship's machinery. Various modifications to a running vessel have been made to study their effect on the noise output; such modifications include towing a surface ship without its propeller and operating a torpedo in a "captive" condition alongside a measurement platform. The effect of a bubble screen has been simulated by turning a ship so as to cross its own wake and noting the effect on the self-noise and radiated noise. All of these methods have contributed to our present knowledge of the sources of radiated noise.

10.3 Sources of Radiated Noise

The sources of noise on ships, submarines, and torpedoes can be grouped into the three major classes listed in Table 10.1. *Machinery noise* comprises that part of the total noise of the vessel caused by the ship's machinery. *Propeller noise* is a hybrid form of noise having features and an origin common to both machinery and hydrodynamic noise. It is convenient to consider propeller noise separately because of its importance. *Hydrodynamic noise* is radiated noise originating in the irregular

TABLE 10.1 Sources of Radiated Noise (Diesel-electric Propulsion)

Machinery noise:
- Propulsion machinery (diesel engines, main motors, reduction gears)
- Auxiliary machinery (generators, pumps, air-conditioning equipment)

Propeller noise:
- Cavitation at or near the propeller
- Propeller-induced resonant hull excitation

Hydrodynamic noise:
- Radiated flow noise
- Resonant excitation of cavities, plates, and appendages
- Cavitation at struts and appendages

flow of water past the vessel moving through it and causing noise by a variety of hydrodynamic processes.

Machinery Noise Machinery noise originates as mechanical vibration of the many and diverse parts of a moving vessel. This vibration is coupled to the sea via the hull of the vessel. Various paths, such as the mounting of the machine, connect the vibrating member to the hull. Machine vibration can originate in the following ways:

1. Rotating unbalanced parts, such as out-of-round shafts or motor armatures
2. Repetitive discontinuities, such as gear teeth, armature slots, turbine blades
3. Reciprocating parts, such as the explosions in cylinders of reciprocating engines
4. Cavitation and turbulence in the fluid flow in pumps, pipes, valves, and condenser discharges
5. Mechanical friction, as in bearings and journals

The first three of these sources produce a *line-component spectrum* in which the noise is dominated by tonal components at the fundamental frequency and harmonics of the vibration-producing process; the other two give rise to noise having a *continuous spectrum* containing superposed line components when structural members are excited into resonant vibration. The machinery noise of a vessel may therefore be visualized as possessing a low-level continuous spectrum containing strong line components that originate in one or more of the repetitive vibration-producing processes listed above.

A diagrammatic view of the sources of machinery noise aboard a diesel-electric vessel is shown in Fig. 10.3. Each piece of machinery

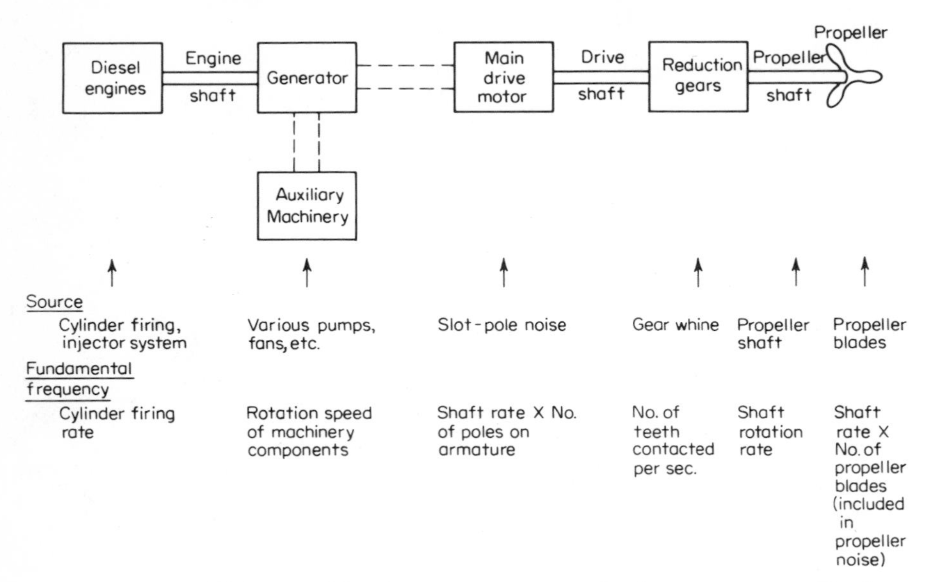

FIG. 10.3. Machinery components and noise sources on a diesel-electric vessel.

produces periodic vibrational forces at the indicated fundamental frequency and thereby generates a series of line components at this frequency and at its harmonics. However, at a distance in the sea, the sound produced by these vibrational forces depends not only on their magnitude, but also on how such forces are transmitted to the hull and coupled to the water. A notable example is the resonant excitation of large sections of the hull by machinery vibration—called "hull drone"—such as that produced by the rotation of the massive propeller shaft, wherein certain frequencies of the excitation spectrum are reinforced by a sort of sounding-board effect. The manner of mounting of the machine and the resulting vibration of the hull are determining factors in the radiation of sound. Another source of variability is in the propagation of different frequencies to a distant point in the sea. Because of these various effects, the harmonic structure of radiated noise is complex, and the line-component series generated by even a single source of noise is irregular and variable. When many noise sources are present, as in a vessel under way, the machinery-noise spectrum contains line components of greatly different level and origin, and is, consequently, subject to variations of level and frequency with changing conditions of the vessel.

Propeller Noise Even though the propeller is a part of the propulsion machinery of a vessel, the noise it generates has both a different origin and a different frequency spectrum from machinery noise. As just described, machinery noise originates *inside* the vessel and reaches the water by various processes of transmission and conduction through the hull. Propeller noise, on the other hand, originates *outside* the hull as a consequence of the propeller action and by virtue of the vessel's movement through the water. The location of the sources of noise along the hull is different as well. When a vessel passes close to a nearby hydrophone it is observed that noises ascribable to the vessel's machinery reach a peak level *before* those originating at the propellers, in keeping with the place of origin aboard the noise-producing vessel.

The source of propeller noise is principally the noise of cavitation induced by the rotating propellers. When a propeller rotates in water, regions of low or negative pressure are created at the tips and on the surfaces of the propeller blades. If these negative (tensile) pressures become high enough, physical rupture of the water takes place and cavities in the form of minute bubbles begin to appear. These cavitation-produced bubbles collapse a short time later—either in the turbulent stream or up against the propeller itself—and in so doing emit a sharp pulse of sound. The noise produced by a great many such collapsing bubbles is a loud "hiss" that usually dominates the high-frequency end of the spectrum of ship noise when it occurs. The production and collapse of cavities formed by the action of the propeller is called propeller cavitation.

Propeller cavitation may be subdivided into *tip-vortex cavitation*, in which the cavities are formed at the tips of the propeller blades and are intimately associated with the vortex stream left behind the rotating propeller, and *blade-surface cavitation*, where the generating area lies at front or back sides of the propeller blades. Of these two types of cavitation, the former has been found by laboratory measurements with model propellers and by analyses of field data (3) to be the more important noise source with propellers of conventional design.

Because cavitation noise consists of a large number of random small bursts caused by bubble collapse, cavitation noise has a continuous spectrum. At high frequencies, its spectrum level *decreases* with frequency at the rate of about 6 db/octave, or about 20 db/decade. At low frequencies, the spectrum level of cavitation noise *increases* with frequency, although this reverse slope tends to be obscured in measured data by other sources of noise. There is, therefore, a peak in the spectrum of cavitation noise which, for ships and submarines, usually occurs within the frequency decade 100 to 1,000 Hz. The location of the peak in the spectrum shifts to lower frequencies at higher speeds and (in the case of submarines) at smaller depths. Figure 10.4 shows diagrammatic cavitation-noise spectra for three combinations of speed and depth for a hypothetical submarine. The behavior of the spectral peak is asso-

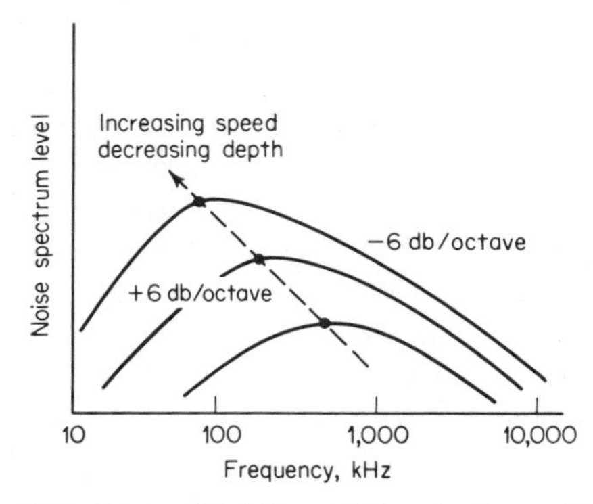

FIG. 10.4. Variation of the spectrum of cavitation noise with speed and depth.

ciated with the generation of larger cavitation bubbles at the greater speeds and the lesser depths and with the resulting production of a greater amount of low-frequency sound. On an actual submarine, the variation of propeller noise with speed at a constant depth and with depth at a constant speed is illustrated by the measured curves of Fig. 10.5, which were obtained with a hydrophone placed 4 ft from the propellers on the two submarines "Hake" and "Hoe."

It has long been known that as the speed of the ship increases, there is a speed at which propeller cavitation begins and the high-frequency

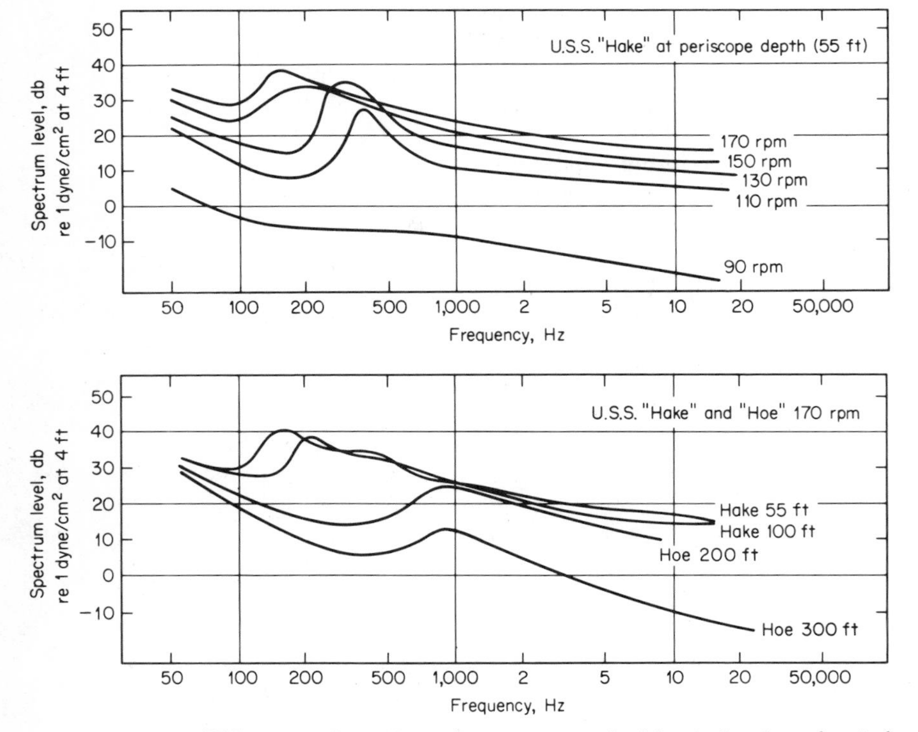

FIG. 10.5. Propeller noise as measured with a hydrophone located 4 ft from the tips of the propeller blades. World War II data on the submarines "Hake" and "Hoe." (Ref. 4.)

radiated noise of the vessel suddenly and dramatically increases. This speed has been called the *critical speed* of the vessel. The submarines measured during World War II were observed to have critical speeds between 3 and 5 knots when operating at periscope depth and to have well-developed propeller cavitation at 6 knots and beyond. At speeds well beyond the critical speed, the noise of cavitation increases more slowly with speed. The noise-speed curve of a cavitating vessel accordingly has a shape like the letter S, with an increase of 20 to 50 db at high frequencies when the speed is a few knots beyond the critical speed and with a slower rise, at a rate of 1.5 to 2.0 db/knot, beyond. Measured data illustrating the S-shape characteristic of the noise-speed curve for submarines are shown in Fig. 10.6. The flattening of the curve at high speeds may be surmised to be due to a self-quenching, or internal absorption, effect of the cloud of cavitation bubbles. Surface ships do not exhibit the S-shape feature in their noise-speed curves, but show, instead, a more gradual, and nondescript, increase of level with speed.

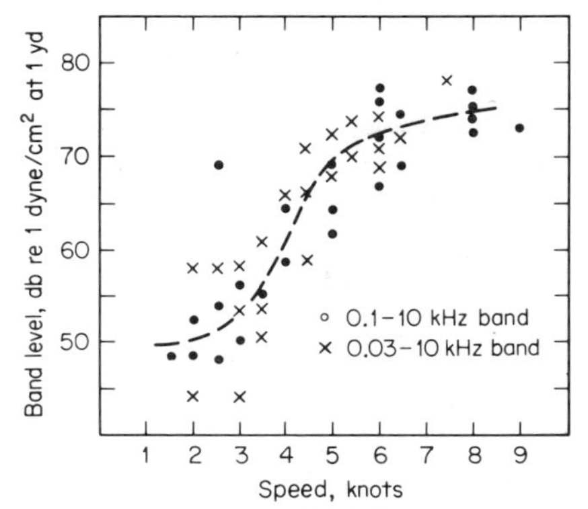

FIG. 10.6. ***Broadband measurements of the radiated noise of a number of World War II submarines operating at periscope depth. 200-yd data reduced to 1 yd.*** (*Ref. 4.*)

Cavitation noise is suppressed, and the critical speed is increased, by submerging to a greater depth. This effect has long been well known to submariners as a means to avoid detection. By simple hydrodynamic theory, it is known that the intensity of tip-vortex cavitation is governed by the *cavitation index*, defined by

$$K_T = \frac{p_0 - p_v}{(\frac{1}{2})\rho v_T^2}$$

where p_0 = static pressure at propellers
p_v = vapor pressure of water
ρ = density of water
v_T = tip velocity of propeller blades

When K_T lies between 0.6 and 2.0, tip cavitation begins; when K_T is less than 0.2, the occurrence of cavitation is certain; when K_T is greater than 6.0, cavitation is unlikely. If it is assumed that cavitation begins at some fixed value of K_T and if p_v is neglected, we conclude that the critical speed must vary as the square root of the static pressure at the

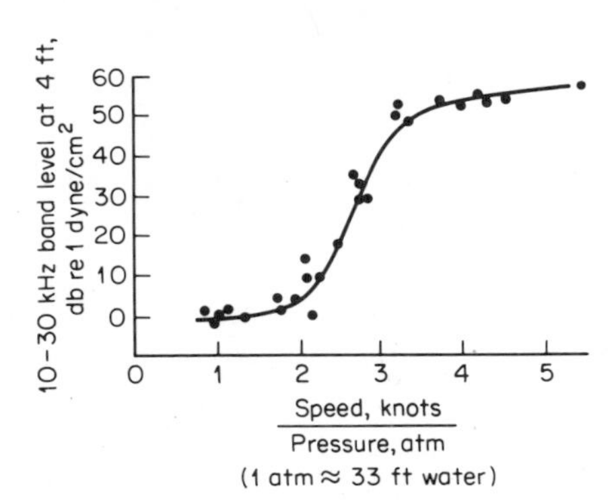

FIG. 10.7. Normalized curves of cavitation noise. (Ref. 4.)

depth of the propellers. This is illustrated by the measurements on "Hake" and "Hoe" previously referred to. For these submarines, Fig. 10.7 shows noise levels in the 10- to 30-kHz band observed with a hydrophone mounted 4 ft from the propeller tips and normalized by dividing by the square root of the hydrostatic pressure at the operating depth.

Although the cavitation noise of submarines is suppressed by depth, the reduction of noise with depth does not occur uniformly. When strong cavitation at high speeds occurs, the radiated noise of submarines is observed to first *increase* as the submarine dives, before the onset of the normal suppression of noise with depth. Figure 10.8 illustrates the effect of depth on the cavitation noise of a German type XXI submarine. At 8 knots, for example, this particular submarine had to submerge to 150 ft before experiencing the beginning of lower noise levels. This has been called the *anomalous depth effect*, although the term is now a misnomer, since the effect can be accounted for by theoretical analysis of cavitation-noise formation.

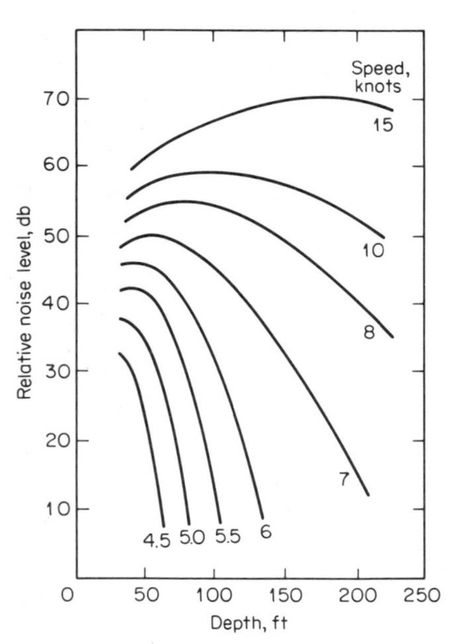

FIG. 10.8. The anomalous depth effect, illustrated by measured wartime data at audio frequencies on type XXI German submarine. Relative levels only.

Many factors other than speed and depth affect propeller noise. A damaged propeller makes more noise than an undamaged one. More noise is made during turns and accelerations in speed than during uniform cruising. Truly anomalous conditions are sometimes found—most notably, the occurrence of "singing" propellers, in which the propeller blades are excited by the flow into vibrational resonance and "sing" with vigor. The singing of propellers can be

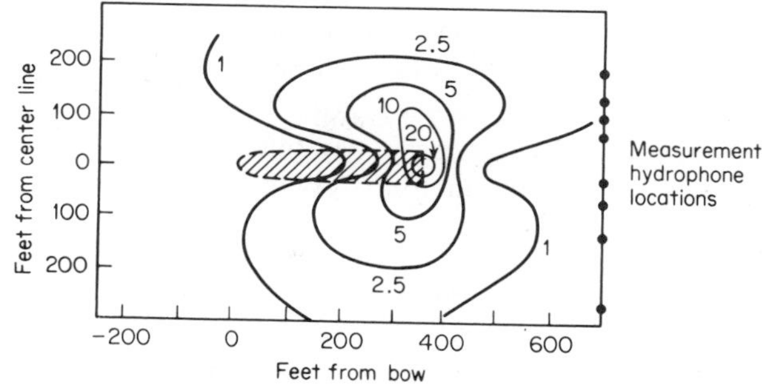

FIG. 10.9. Equal pressure contours on the bottom in 40 ft of water of a freighter at a speed of 8 knots. Contour values are pressures, in dynes per square centimeter in a 1-Hz band, at a point on the bottom, measured in the octave band 2,500 to 5,000 Hz. (*Ref. 6.*)

corrected by proper tip design or alleviated by filing down the propeller blade tips to a chisellike point.

Propeller noise is not radiated uniformly in all directions, but has a characteristic directional pattern in the horizontal plane around the radiating vessel. Less noise is radiated in the fore-and-aft directions than abeam, probably because of screening by the hull in the forward direction and by the wake at the rear. The dips in the pattern generally occur within 30° of the fore-and-aft direction, with the bow dip a few decibels deeper than the dip at the stern. A directivity pattern in the 2.5- to 5-kHz band for a freighter traveling at 8 knots is given in Fig. 10.9, where the contours show the locations, relative to the ship, of equal sound intensity on the bottom in 40 ft of water.

Propeller noise has been known for many years to be amplitude-modulated and to contain "propeller beats," or periodic increases of amplitude, occurring at the rotation speed of the propeller shaft, or at the propeller blade frequency equal to the shaft frequency multiplied by the number of blades. Propeller beats have long been used by listening observers for target identification and for estimating target speed. They have been observed (7) to be present in the noise of torpedoes as well as in the noise of ships and submarines. Propeller beats are most pronounced at speeds just beyond the onset of cavitation and diminish in the great roar of steady cavitation noise at high speeds.

Propeller noise, with its origin in the flow of water about the propeller, creates tonal components in addition to the continuous spectrum of cavitation noise. One tonal component is the "singing" tone of a vibrating singing propeller just mentioned. More normally, at the low-frequency end of the spectrum, propeller noise contains discrete spectral "blade-rate" components occurring at multiples of the rate at which any irregularity in the flow pattern into or about the propeller is intercepted by the propeller blades. The frequency of the blade-rate series of line components is given by the formula

$$f_m = mns$$

where f_m is the frequency, in hertz, of the mth harmonic of the blade-rate series of lines, n is the number of blades on the propeller, and s is the propeller rotation speed in number of turns per second. In one Russian experiment (8), these "blade" lines were observed with and without the emission of air bubbles around the propeller. At speeds when cavitation was well developed, they were found to be intimately associated with the cavitation process; in addition, line components not falling in the expected series were observed as well. Blade-rate line components were long ago observed (9) to be the dominant source of the noise of submarines in the 1- to 100-Hz region of the spectrum. It should be noted in passing that line components generated by the rotating propeller shaft—a form of machinery noise—fall in the same harmonic series as the blade-rate lines. Propeller noise, like machinery noise, can excite, and be reinforced by, the vibrational response of mechanical structures in the vicinity of the propeller.

Hydrodynamic Noise Hydrodynamic noise originates in the irregular and fluctuating flow of fluid past the moving vessel. The pressure fluctuations associated with the irregular flow may be radiated directly as sound to a distance, or, more importantly, may excite portions of the vessel into vibration. The noise created by the turbulent boundary layer is sometimes called "flow noise."

The excitation and reradiation of sound by various structures of the vessel are an important source of hydrodynamic noise. One kind of such noise is propeller singing, mentioned above. In addition, the flow of fluid may have an "aeolian harp" effect on other structures of the vessel, such as struts, and excite them into a vibrational resonance. Like Helmholtz resonators, cavities may be excited by the fluid flow across their openings, in the manner that a bottle can be made to "sing" by blowing over its opening. These resonant occurrences can sometimes be easily diagnosed and curative measures applied.

The form of hydrodynamic noise called flow noise is a normal characteristic of flow of a viscous fluid and occurs in connection with smooth bodies without protuberances or cavities. Flow noise can be radiated directly or indirectly as flow-induced vibrations of plates or portions of the body. The former—direct radiation to a distance—is an inefficient process not likely to be important at the low Mach numbers (ratio of speed of the vessel to the speed of sound in water) reached by vessels moving through water, since it is of quadrupole origin and is therefore not efficiently radiated to a distance. The latter process—flow excitation of a nonrigid body—depends on (1) the properties of the pressure fluctuation in the turbulent boundary layer, (2) the local response of the structure to these fluctuations, and (3) the radiation of sound by the vibrating portion. The response of plates to a random pressure field has been studied theoretically by Dyer (10). Flow noise is a more important contribution to self-noise than to radiated noise, and will be discussed more fully as a part of self-noise.

Other kinds of hydrodynamic noise are the roar of the breaking bow and stern waves of a moving vessel and the noise originating at the intake and exhaust of the main circulating water system.

Under normal circumstances, hydrodynamic noise is likely to be only a minor contributor to radiated noise, and is apt to be masked by machinery and propeller noises. However, under exceptional conditions, such as when a structural member or cavity is excited into a resonant source of line-component noise, hydrodynamic noise becomes a dominant noise source in the region of the spectrum in which it occurs.

10.4 Summary of the Sources of Radiated Noise

Of the three major classes of noise just described, machinery noise and propeller noise dominate the spectra of radiated noise under most conditions. The relative importance of the two depend upon frequency, speed, and depth. This is illustrated by Fig. 10.10 which shows the characteristics of the spectrum of submarine noise at two speeds. Figure 10.10*a* is a diagrammatic spectrum at a speed when propeller cavitation has just begun to appear. The low-frequency end of the spectrum is

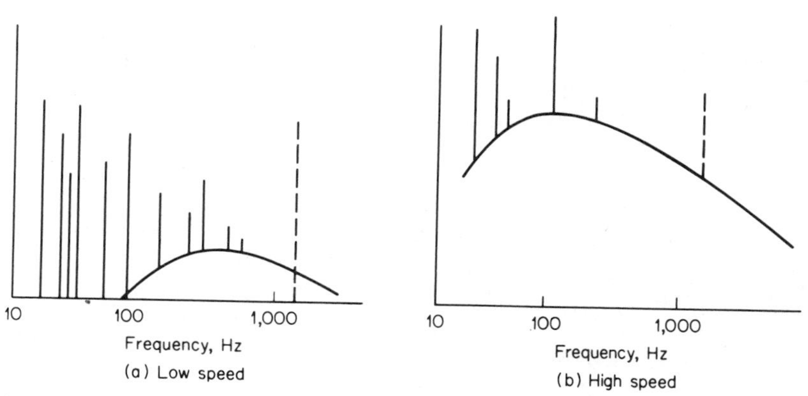

FIG. 10.10. Diagrammatic spectra of submarine noise at two speeds.

dominated by machinery lines, together with the blade-rate lines of the propeller. These lines die away irregularly with increasing frequency and become submerged in the continuous spectrum of propeller noise. Sometimes, as indicated by the dotted line, an isolated high-frequency line or group of lines appear amid the continuous background of propeller noise. These high-frequency lines result from a singing propeller or from particularly noisy reduction gears, if the vessel is so equipped.

At a higher speed (Fig. 10.10*b*), the spectrum of propeller noise increases and shifts to lower frequencies. At the same time, some of the line components increase in both level and frequency, whereas others, notably those due to auxiliary machinery running at constant speed, remain unaffected by an increase in ship speed. Thus, at the higher speeds, the continuous spectrum of propeller cavitation overwhelms many of the line components and increases its dominance over the spectrum. A *decrease* in depth at a constant speed has, as indicated above, the same general effect on the propeller-noise spectrum as an *increase* in speed at constant depth.

For a given speed and depth, therefore, a "crossover" frequency may be said to exist, below which the spectrum is dominated by the line components of the ship's machinery plus its propeller, and above which the spectrum is in large part the continuous noise of the cavitating propeller. For ships and submarines, this frequency lies roughly between 100 and 1,000 Hz, depending on the individual ship and its speed and depth; for torpedoes, the crossover frequency is higher and the line components extend to higher frequencies because of the generally higher speeds of operation of torpedo machinery.

For illustrating the nature of ship spectra, a frequency-time analyzer, such as that used for speech analysis, is particularly convenient. This kind of analyzer, called a sound spectrograph, was first described by

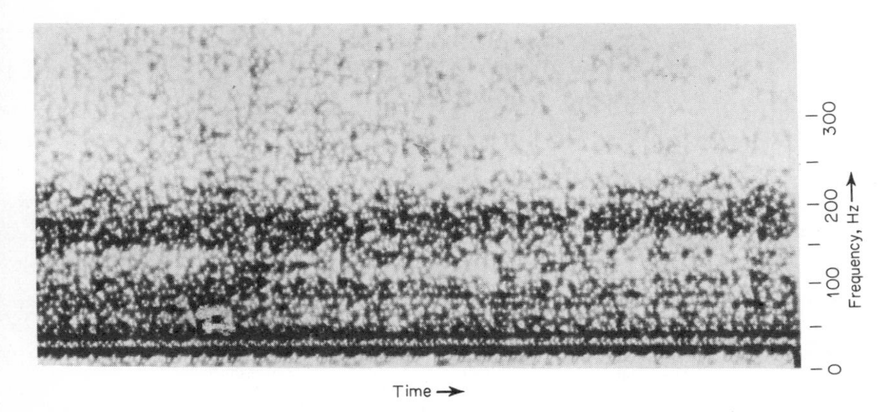

FIG. 10.11. Sound spectrogram of a freighter.

Koenig, Dunn, and Lacy (11) and is widely used for the analysis of speech (12). It gives a plot of frequency against time and shows the intensity of the sound in the analysis bandwidth by a darkening of the record. Figure 10.11 is an example of a sound spectrogram extending over a total time interval of 15 sec for a freighter passing at a speed of 14 knots near a hydrophone on the bottom in 50 ft of water. Of particular note is the occurrence of line components in the spectrum at 200 Hz and below and the absence of such lines in the continuous spectrum at higher frequencies.

10.5 Total Radiated Acoustic Power

It is of interest to determine how much total acoustic power is radiated by a moving vessel and how it compares with the power used by the vessel for propulsion through the water. This can easily be done by integration of the spectrum. When the spectrum is continuous and has a slope of −6 db/octave, the intensity dI in a small frequency band df centered at frequency f can be written

$$dI = \frac{A}{f^2}\,df$$

where $1/f^2$ represents the −6 db/octave slope of the spectrum and A is a constant.

Integrating between any two frequencies f_1 and f_2, the total intensity between f_1 and f_2 becomes

$$I_{(f_1,f_2)} = \int_{f_1}^{f_2} \frac{A}{f^2}\,df = A\left(\frac{1}{f_1} - \frac{1}{f_2}\right)$$

Letting $f_2 \to \infty$, we find that the total intensity for all frequencies above f_1 is

$$I_{\text{tot}} = \frac{A}{f_1}$$

Recognizing that A/f_1^2 is the intensity in a 1-Hz band centered at frequency f_1, we note that

$$I_{\text{tot}} = (I_{f_1})(f_1)$$

where $10 \log I_{f_1}$ is the spectrum level at f_1.

As an example (Fig. 10.12), we take a spectrum having a spectrum level of 60 db at 200 Hz with slope of $1/f^2$ above this frequency. The total level of the radiated noise above 200 Hz becomes

$$\begin{aligned} 10 \log I_{\text{tot}} &= 10 \log I_{f_1} + 10 \log f_1 \\ &= 60 + 23 = 83 \text{ db} \end{aligned}$$

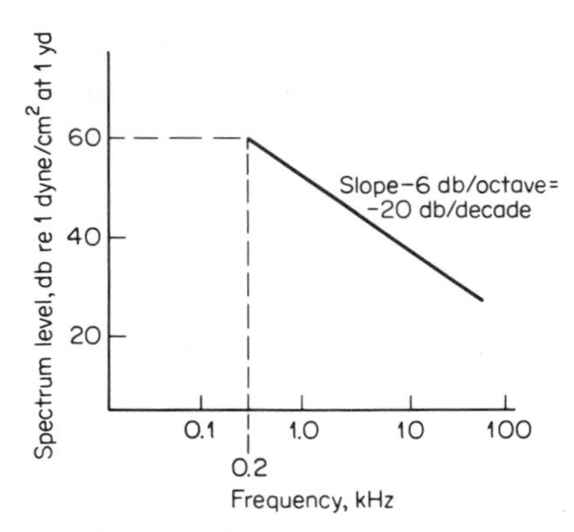

FIG. 10.12. Idealized spectrum integrated to obtain the total acoustic power radiated by a vessel.

The total acoustic power represented by this source level, if assumed to be radiated nondirectionally, amounts to 12 watts. It may be observed that if the continuous spectrum extended below 200 Hz to zero Hz at the constant level of 60 db, as shown by the dashed line in Fig. 10.12, the total radiated power at frequencies *below* 200 Hz would also be 12 watts, and the total radiation over the entire spectrum would become 24 watts.

If the vessel is regarded as a mechanical source of sound, this radiated power may be compared with the shaft horsepower developed by the radiating vessel. Let us assume that the selected spectrum corresponds to that of a destroyer at a speed of 20 knots. At this speed, an average destroyer is known to generate a shaft horsepower of approximately 14,000, or about 10^7 watts. Comparing this power with that radiated as sound in the continuous part of the spectrum, we observe that the efficiency of the vessel as a sound producer is only of the order of 10^{-6}. Surface ships are therefore extremely inefficient radiators of sound in terms of the developed shaft power of their propulsion system, even when allowance is made for the power represented by the tonal components of the radiation.

10.6 Radiated Noise Levels

During World War II, the United States and Great Britain, motivated by the design needs of acoustic mines, made a great many measurements of the radiated noise of surface ships at a number of acoustic ranges. At locations such as Wolf Trap, Virginia; Treasure Island, California; Thames River, New London, Connecticut; Puget Sound, Washington; and Waipio Point, Honolulu, Hawaii; literally thousands of "runs" on hundreds of ships of all types were made. Far fewer wartime measurements were made on submarines and torpedoes. Although much of this old data is obsolete because many of the vessels measured are no longer in existence, the general run of sound levels and their variation with speed, frequency, and depth will still be pertinent to many presently existing vessels, if only as a guide for approximation.

In the following sections, a few extracts have been selected from this

TABLE 10.2 Typical Average Source Levels for Several Classes of Ships in db vs. 1 dyne/cm² in a 1-Hz band at 1 yd.* (Ref. 5)

Frequency	Freighter 10 knots	Passenger 15 knots	Battleship 20 knots	Cruiser 20 knots	Destroyer 20 knots	Corvette 15 knots
100 Hz	52	62	76	69	63	57
300 Hz	42	52	66	59	53	47
1 kHz	31	41	55	48	42	36
3 kHz	21	31	45	38	32	26
5 kHz	17	27	41	34	28	22
10 kHz	11	21	35	28	22	16
25 kHz	3	13	27	20	14	8

* Originally reported at 20 yd.

vast wartime literature to illustrate the main quantitative aspects of the subject. Particular attention is called to two excellent summaries on the radiated-noise levels of submarines (4) and surface ships (5) and to one of the NDRC Summary Technical Reports (13) on the radiated noise of torpedoes.

Surface Ships Table 10.2 shows typical radiated source levels for various classes of ships current during World War II, as reduced to 1 yd from an original reference distance of 20 yd by the conventional assumption of spherical spreading.

In graphical form, Fig. 10.13 gives in the upper set of curves average spectrum levels at 5 kHz as a function of speed for a number of classes of surface ships. The lower curve is a relative spectrum for use in obtaining values at other frequencies. The standard deviation (in decibels) of individual measurements from the line drawn is indicated for each class of ships.

Empirical expressions also were devised to fit the mass of data. In terms of the propeller-tip speed V, in feet per second, the displacement tonnage T of the ship, the frequency f, in kilohertz, and distance D, in yards, the source level for the average radiated noise of large ships was found to be given by

$$\mathrm{SL} = 51 \log V + 15 \log T - 20 \log F - 20 \log D - 113.5$$

This formula, based on 157 runs of 77 ships of 11 different classes (mostly freighters, tankers, and large warships), was found to fit individual measurements to a standard deviation of 5.4 db. It is applicable only at frequencies above 1 kHz where propeller cavitation is the principal source of noise. A more convenient formula, in terms of the speed of the ship, for use when information concerning the propeller-tip speed

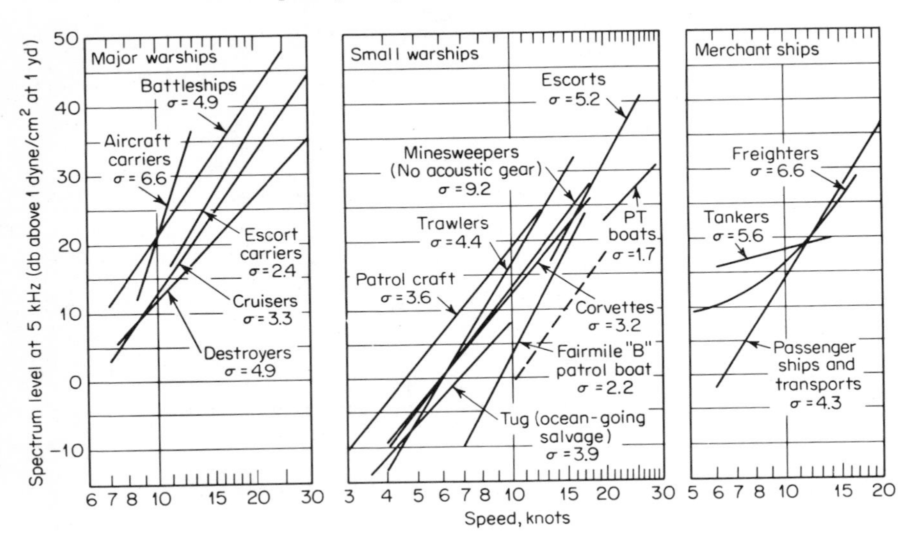

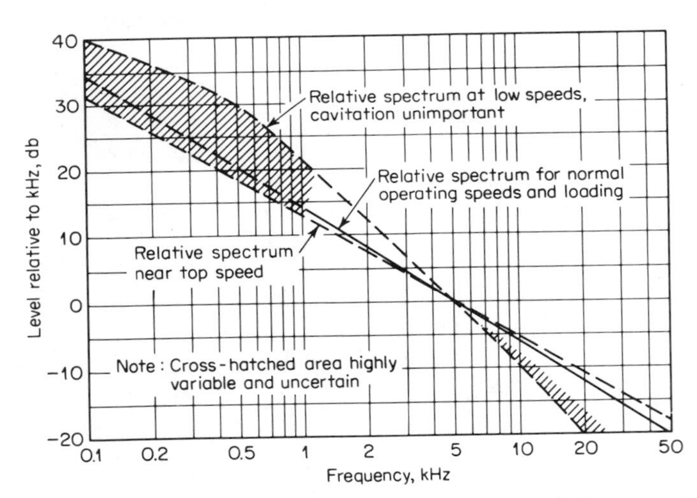

FIG. 10.13. Average radiated spectrum levels for several classes of surface ships. (*Ref. 5.*)

is lacking, was found to be

$$\mathrm{SL} = 60 \log K + 9 \log T - 20 \log F - 20 \log D - 65$$

In this formula K is the forward speed of the ship, in knots. This expression was found to fit the measured levels of passenger ships, transports, and warships at a frequency of 5 kHz to a standard deviation of 5.5 db, but to be unreliable for freighters and tankers.

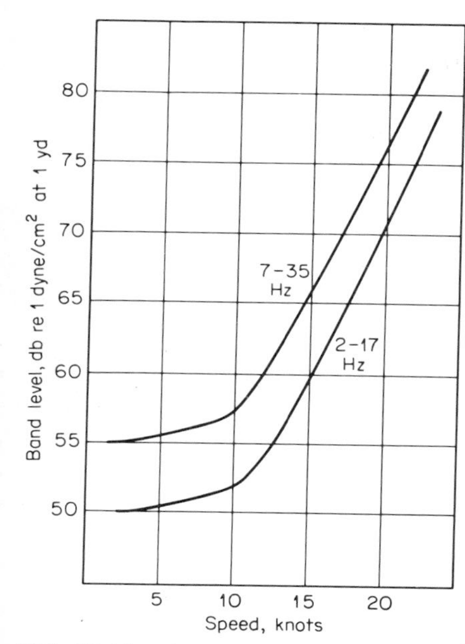

***FIG. 10.14.** Average levels on four destroyers in two frequency bands as measured on the bottom in 105 ft of water at the Puget Sound acoustic range. Values reduced from 105 ft to 1 yd. (Ref. 1.)*

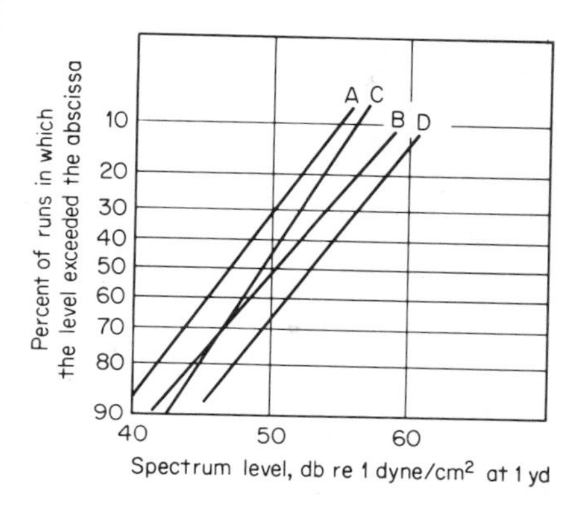

		Ships	Speed
A	(88 runs)	< 5,000 tons	< 10 knots
B	(132 runs)	< 5,000 tons	≥ 10 knots
C	(72 runs)	> 5,000 tons	< 10 knots
D	(129 runs)	> 5,000 tons	≥ 10 knots

***FIG. 10.15.** Cumulative distribution curves of the radiated noise of surface ships in the octave 75 to 150 Hz. (Ref. 14.) Horizontal scale is spectrum level obtained from band levels by subtracting 10 log bandwidth and reduced from 100 ft to 1 yd by assuming spherical spreading.*

At the low-frequency end of the spectrum, Fig. 10.14 illustrates other measured data (1) on four destroyers measured in the frequency bands 2 to 17 Hz and 7 to 35 Hz. The levels shown are band levels giving the total intensity measured in each band. The levels of any individual line components in the spectrum would be an indefinite number of decibels higher than the level indicated, depending on the number and relative strengths of the lines occurring in each band.

The distribution of radiated level in the octave 75 to 150 Hz, based on a large number of ship runs at a number of wartime acoustic ranges (14), is shown in Fig. 10.15. In terms of the spectrum level at 1 yd, the ordinate in this figure gives the percentage of runs in which the measured level exceeded that shown on the horizontal scale. The levels have been reduced to spectrum levels at 1 yd from octave-band data given at a distance of 100 ft beneath the ship.

Regarding such averaged, reduced data, it should be borne in mind that individual ships occasionally deviate greatly from the average

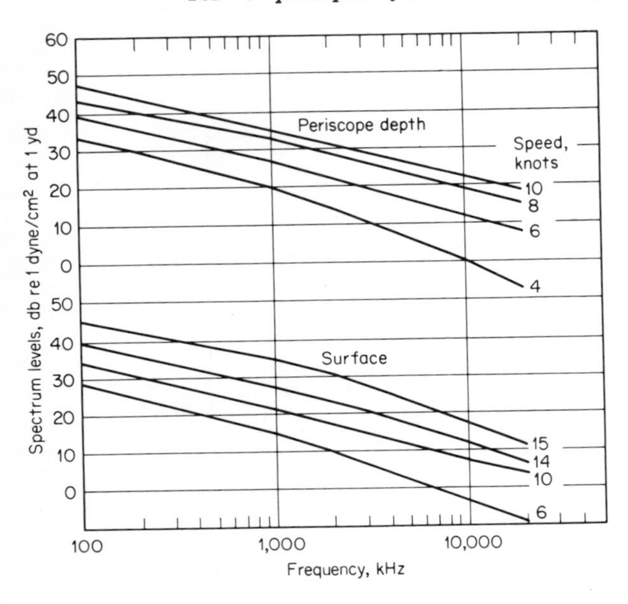

FIG. 10.16. Smoothed spectra of three submarines (USS S-48, "Hake," and "Runner") on electric drive. Levels, originally reported at 200 yd, reduced to 1 yd. (Ref. 4.)

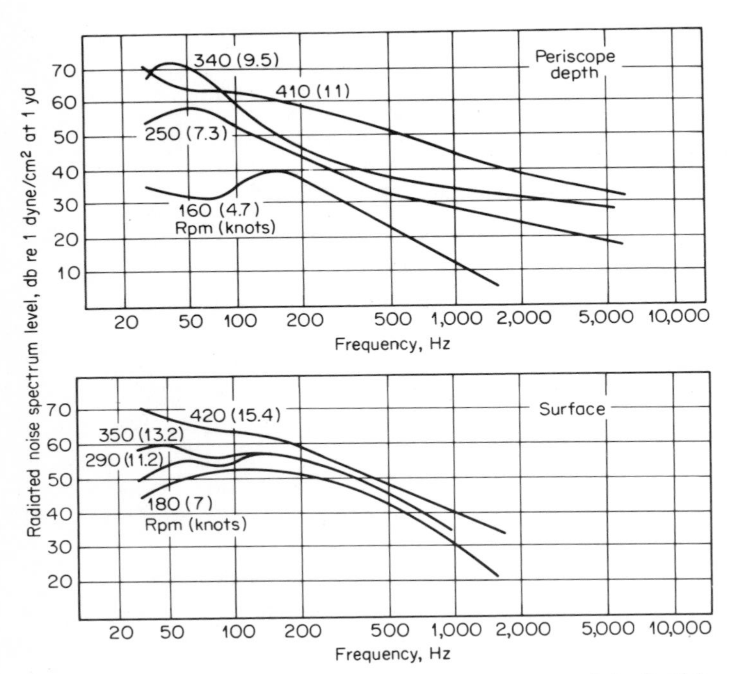

FIG. 10.17. Radiated-noise spectra of the British submarine HMS "Graph" at periscope depth and on the surface. 200-yd reported data reduced to 1 yd. (Ref. 4.)

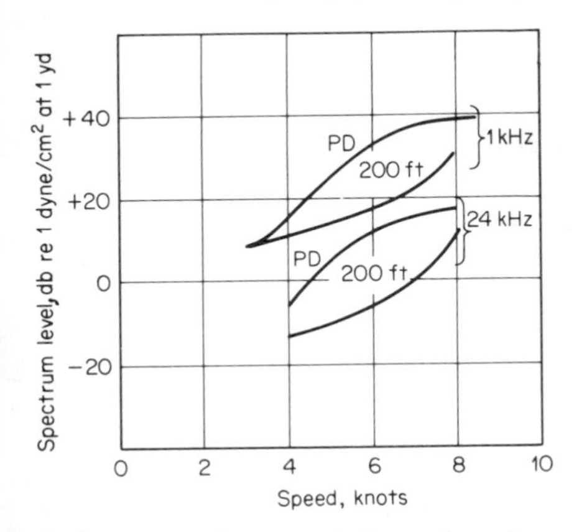

FIG. 10.18. Representative submarine-radiated-noise spectrum levels at two frequencies at periscope depth (PD) and at a depth of 200 ft on electric drive. (Ref. 15.)

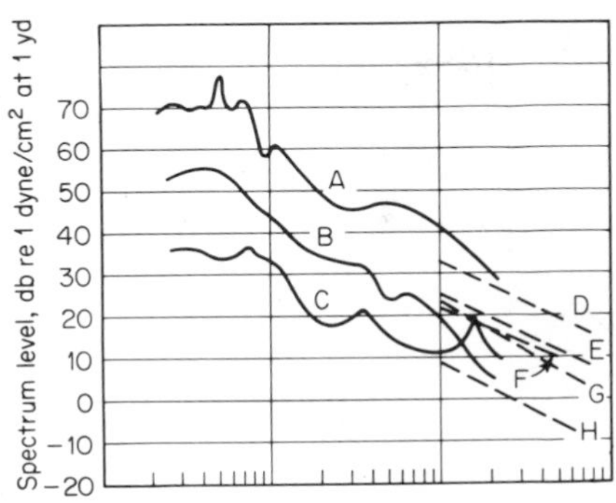

FIG. 10.19. Noise spectra of various World War II torpedoes running at shallow depths. [(A–C), Ref. 16; (D–H), Ref. 13.] (A) Highest measured values for several U.S. torpedoes. (B) Japanese Mark 91, 30 knots. (C) U.S. Mark 13, 30 knots. (D) U.S. Mark 14, 45 knots. (E) British Mark VIII, 37 knots. (F) U.S. Mark 18, 30 knots. (G) U.S. Mark 13, 33 knots. (H) British Mark VIII, 20 knots.

levels. Moreover, at low frequencies, the use of spectrum levels is questionable because of the likely tonal content of the radiated sound. Finally, the data are entirely based on measurements on the bottom in shallow water averaging 20 yd in depth, and does not necessarily indicate the levels that would be observed in deep water.

Submarines The available data on submarines are much less extensive and involve only a few submarines under limited operating conditions. Figure 10.16 shows average spectra for three World War II submarines at periscope depth (about 55 ft to the keel) and at the surface. Spectra appreciably higher, though similar in shape, were obtained for the single British submarine, HMS "Graph," as illustrated in Fig. 10.17.

The effect of depth on the radiated levels of submarines is illustrated in Fig. 10.18, which shows the reduction in noise involved in submerging from periscope depth to 200 ft. The effect of depth may also be determined from Fig. 10.7, given previously, in which the spectrum level was plotted as a function of speed divided by the square root of the hydrostatic pressure. These depth effects occur only for the noise produced by the cavitating propeller; when machinery noise predominates, as at very low speeds or very low frequencies, little or no quieting on submergence to deep depths is expectable.

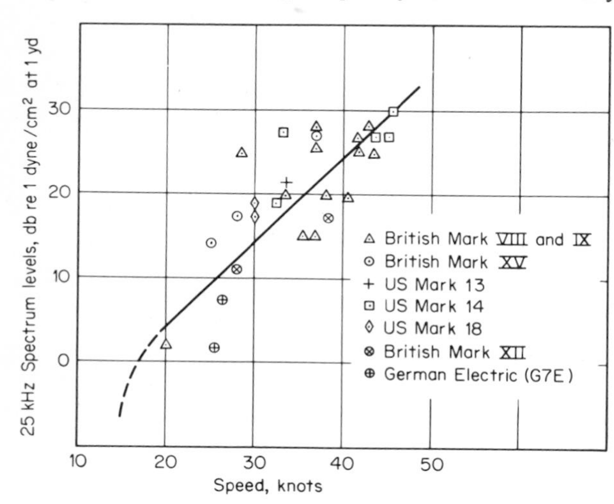

FIG. 10.20. 25-kHz spectrum levels of various torpedoes as a function of speed. The straight line has a slope of 1 db/knot increase of speed. (*Ref. 13.*)

Torpedoes Figure 10.19 illustrates measured radiated-noise spectra over a wide frequency range for a variety of torpedoes running at different speeds, and Fig. 10.20 is a compilation of torpedo noise levels at 25 kHz as a function of speed. Although the torpedoes listed had a variety of propulsion systems, the noise radiated at kilohertz frequencies must be presumed to be dominated by propeller cavitation.

REFERENCES*

1. Pomerantz, J., and G. F. Swanson: Underwater Pressure Fields of Small Naval Vessels in the 2–17 and 7–35 cps Bands at the Puget Sound Acoustic Ranges, *Naval Ordnance Lab. Rept.* 1022, 1945.*
2. Ship Acoustical Surveys, *U.S. Navy Bur. Ships Rept.* NAVSHIPS 250–371, no date.
3. Strasberg, M., and W. J. Sette: Measurements of Propeller Noise on Three Submarines of the SS212 Class, *David Taylor Model Basin Rept.* R-205, 1944.*
4. Knudsen, V. C., R. S. Alford, and J. W. Emling: Survey of Underwater Sound No. 2: Sounds from Submarines, *Natl. Defense Res. Comm. Div. 6 sec.* 6.1–NDRC-1306, 1943.*
5. Dow, M. T., J. W. Emling, and V. O. Knudsen: Survey of Underwater Sound No. 4: Sounds from Surface Ships, *Natl. Defense Res. Comm. Div. 6 sec.* 6.1–NDRC-2124, 1945.*
6. Pomerantz, J.: An Analysis of Data in the 2500–10,000 cps Region Obtained at the Wolf Trap Range, *Naval Ordnance Lab. Rept.* 733, 1943.*
7. Irish, G.: "Low Frequency Modulation of the Supersonic Pressure Field of Torpedoes," Naval Ordnance Laboratory unpublished memorandum, 1944.*
8. Aleksandrov, I. A.: Physical Nature of the Rotation Noise of Ship Propellers in the Presence of Cavitation, *Soviet Phys. Acoust.*, **8**(1):23 (July–September, 1962).

* References indicated by an asterisk were originally issued during World War II as classified reports. These reports have been declassified and are available through the U.S. Department of Commerce, Federal Clearinghouse for Scientific and Technical Information, Springfield, Va., 22151.

9. Jaques, A. T.: Subsonic Pressure Variations Produced by Submarines, *Naval Ordnance Lab. Rept.* 744, 1943.*
10. Dyer, I.: Response of Plates to a Decaying and Convecting Random Pressure Field, *J. Acoust. Soc. Am.*, **31**:922 (1959).
11. Koenig, W., H. K. Dunn, and L. Y. Lacy: The Sound Spectrograph, *J. Acoust. Soc. Am.*, **18**:19 (1946).
12. Potter, R. K., G. A. Kopp, and H. C. Green: "Visible Speech," D. Van Nostrand Company, Inc., New York, 1947.
13. Acoustic Torpedoes, *Natl. Defense Res. Comm. Div. 6 Sum. Tech. Rept.*, vol. 22, 1946.*
14. Pomerantz, J.: "Cumulative Percentage Distribution of Ship's Sound Pressure in the 75–150 cps Octave," Naval Ordnance Laboratory unpublished memorandum, 1944.
15. Prediction of Sonic and Supersonic Listening Ranges, *Natl. Defense Res. Comm. Div. 6 sec.* 6.1–ser. 1131–1884, 1944.*
16. Listening Systems, *Natl. Defense Res. Comm. Div. 6 Sum. Tech. Rept.*, vol. 14, 1946.*

ELEVEN

Self-noise of Ships, Submarines, and Torpedoes: Self-noise Levels

Self-noise differs from radiated noise in that the measurement hydrophone is located on board the noise-making vessel and travels with it, rather than being fixed in the sea at a location some distance away. Although the fundamental causes of noise are the same, the relative importance of the various noise sources is different. Moreover, in self-noise, the paths by which the noise reaches the hydrophone are many and varied and play a dominant role in affecting the magnitude and kind of noise received by the hydrophone on the moving vessel.

In the sonar equations, radiated noise occurs as the parameter *source level* SL where it is the level of the source of sound used by passive sonar systems. By contrast, self-noise is a particular kind of background noise occurring in sonars installed on a noisy vehicle; in the sonar equations, self-noise occurs quantitatively as the *noise level* NL. Self-noise exists in fixed hydrophones as well, whenever the manner of mounting or suspension creates noise of its own.

Self-noise is one of many different kinds of undesired sound in sonar and originates in a variety of ways. Figure 11.1 illustrates the interrelationships of the various kinds of sonar backgrounds. Self-noise refers to those noise sources between the dashed lines.

Self-noise depends greatly upon the directivity of the hydrophone, its mounting, and its location on the vehicle. On surface ships, the sonar transducer is located in a streamlined dome projecting below the keel of the ship. On older submarines, the principal passive sonar was the JT and JP sonars, having horizontal line transducers

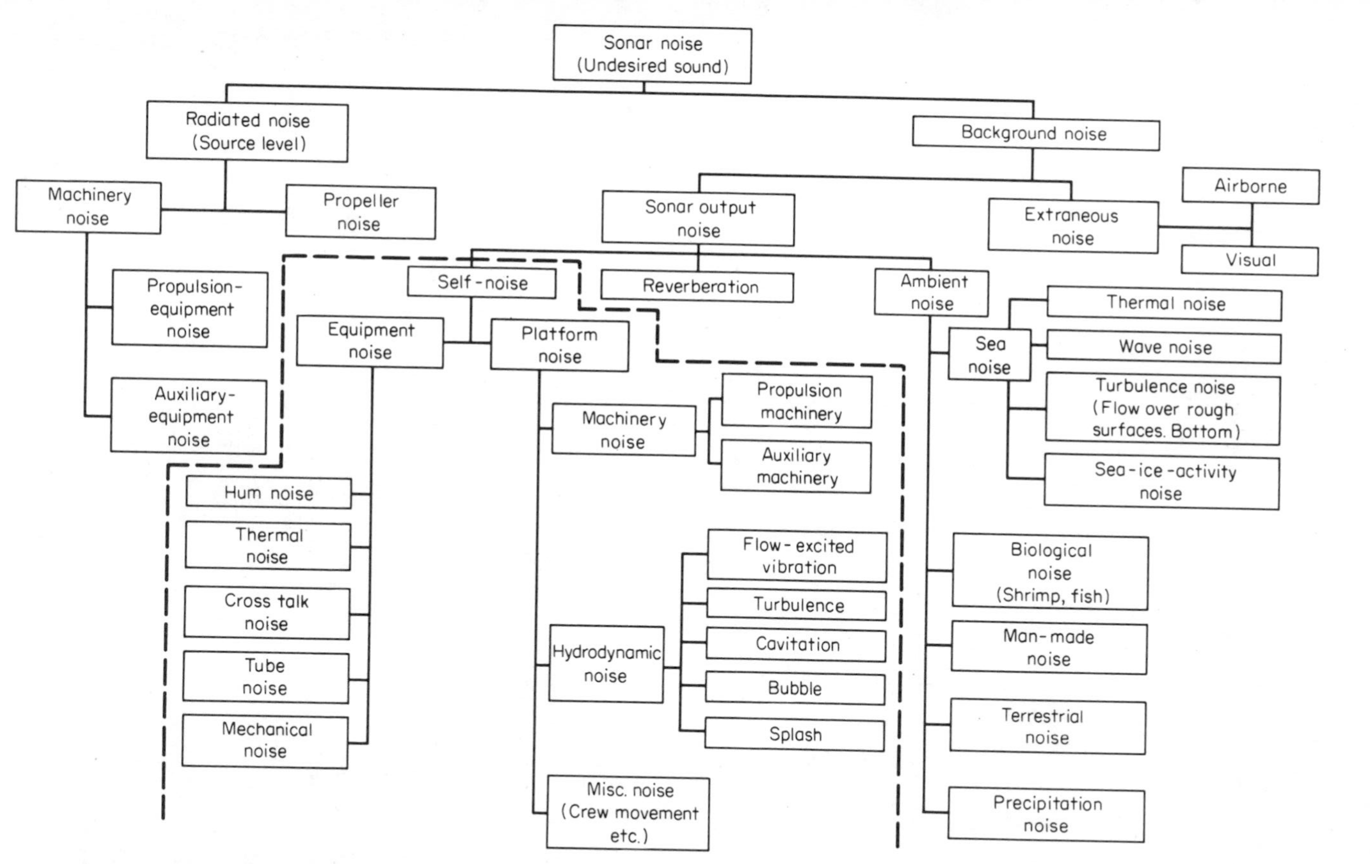

FIG. 11.1. Interrelationships of various sonar noise sources. (Ref. 1.)

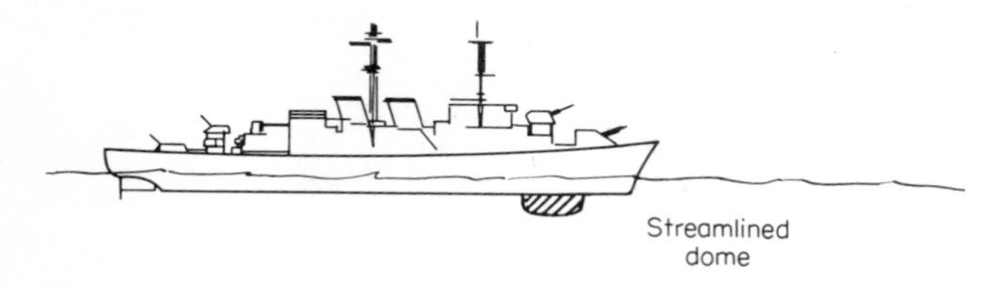

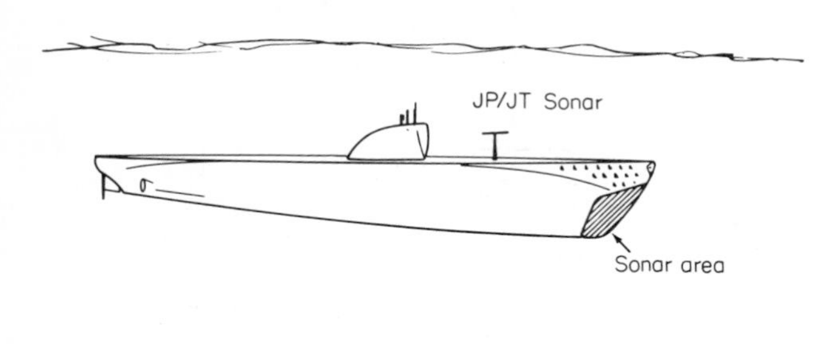

FIG. 11.2. Locations of the sonar transducer aboard surface ships, submarines, and torpedoes.

placed topside toward the bow of the vessel; on modern submarines, a large portion of the bow of the vessel is taken over by the sonar transducer. Acoustic homing torpedoes often have their transducers in the nose of the torpedo and face forward in the direction of travel. On all these vessels, the sonar transducer is placed as far forward on the vessel as is practicable, so as to be removed as much as possible from the propulsion machinery and the propeller noises of the vessel. Figure 11.2 is a pictorial view of the location of the sonar transducer on these three kinds of vessels.

11.1 Self-noise Measurements and Reduction

Self-noise measurements on these vehicles have been made in the past with sonar hydrophones of varying sizes and shapes and hence of differing directivity. In order to make these various measurements compatible with each other, and at the same time to make them useful for other directional sonars, it is convenient to express self-noise levels as *equivalent isotropic levels*. The equivalent isotropic self-noise level is the level that would be indicated by a nondirectional hydrophone of sensitivity equal to that of the directional transducer with which the self-noise measurements were made. If the noise level measured with a directional hydrophone is NL′, the equivalent isotropic level is

$$\mathrm{NL} = \mathrm{NL}' + \mathrm{DI}$$

In converting to equivalent isotropic levels, therefore, the measured levels NL′ are corrected for directivity by applying the directivity index of the measurement hydrophone. The isotropic level NL is the level required for use in the sonar equations previously written.

Although this convention serves to bring self-noise measurements made on the same kind of vessel with different transducers into some degree of agreement, some hesitancy must be felt when applying self-noise data obtained with one sonar to a sonar of different design, directivity, mounting, and location on the same vehicle. Self-noise, like ambient noise, is seldom isotropic in its directional and coherence characteristics, and the DI is not always a satisfactory measure of the discrimination of transducers against it. Often noise coming from a single direction will predominate, and the DI will be almost meaningless as a measure of the discrimination against noise. In torpedoes, for example, it has been found that the front-to-back ratio of the transducer beam pattern—defined as the difference in response between the forward and backward directions—is a more useful measure of the discrimination against self-noise than is the directivity index. In general, it is necessary to understand the sources and paths of the prevailing kind of noise before transferring self-noise data from one sonar to another. When this knowledge is lacking, the equivalent isotropic self-noise level must be used in the sonar equations, with recognition of its crude nature and of the likelihood that large errors may occur in some circumstances.

11.2 Sources and Paths of Self-noise

The three major classes of noise—*machinery noise*, *propeller noise*, and *hydrodynamic noise*—apply for self-noise as well as radiated noise. The sound and vibration generated by each kind of noise reach the sonar hydrophone through a variety of different acoustic paths.

Figure 11.3 shows a surface ship and the paths in the ship and through the sea by which sound generated at the propeller and in the machinery spaces of the ship can reach the sonar hydrophone. Path *A* is an all-hull path by which the vibration produced by the machinery, the propeller shaft, and the propeller itself reach the vicinity of the sonar array at a forward location. Here it may be reradiated by the hull or, more importantly, cause vibration of the wall of the streamlined dome and the mounting of the hydrophone array. Path *B* is an all-water path leading directly from the ship's propellers to the hydrophone. Path *C* shows propeller noise backscattered by volume scatterers located in the volume of the sea. In general, these scatterers are the same as those causing volume reverberation. Path *D* is the bottom-reflected or scattered path by which propeller noise can reach the vicinity of the hydrophone. This path is likely to be a major contributor to self-noise on surface ships operating in shallow water.

For submarines and torpedoes, the upward analog of Path *D*—reflection and scattering from the sea surface—is, similarly, an important acoustic path when the vehicle is running at a shallow depth. An example of this kind of path is shown in Fig. 11.4. This is an artist's conception of a torpedo with three areas of forward and side scattering on the surface by which sound from the propellers can reach hydrophones located near the nose. Of all the vessels of interest, the self-noise of torpedoes long ago received a considerable amount of analytical attention, probably because of the relative ease of performing experiments upon them. For example, torpedoes were "run" with and without propellers, in water and in air, and with various modifications, all during the World War II years (2).

Both machinery noise and propeller noise are prominent contributors to self-noise. The self-noise contribution of the vessel's machinery occurs principally at low frequencies as tonal components in the overall noise.

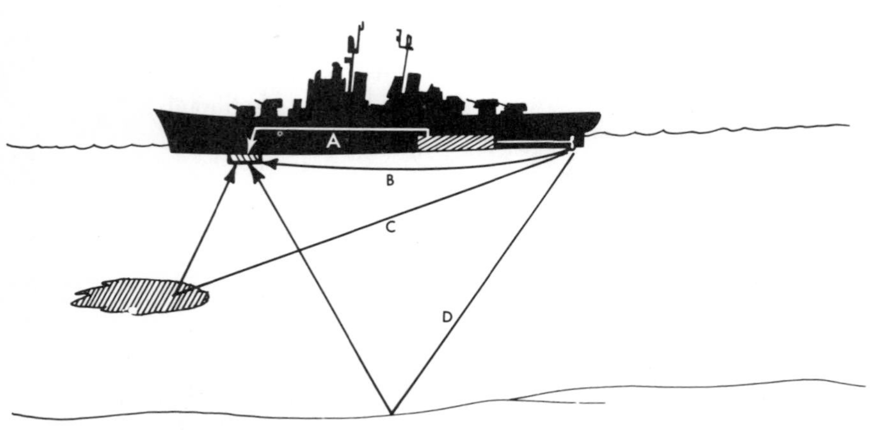

FIG. 11.3. Paths of self-noise on a surface ship.

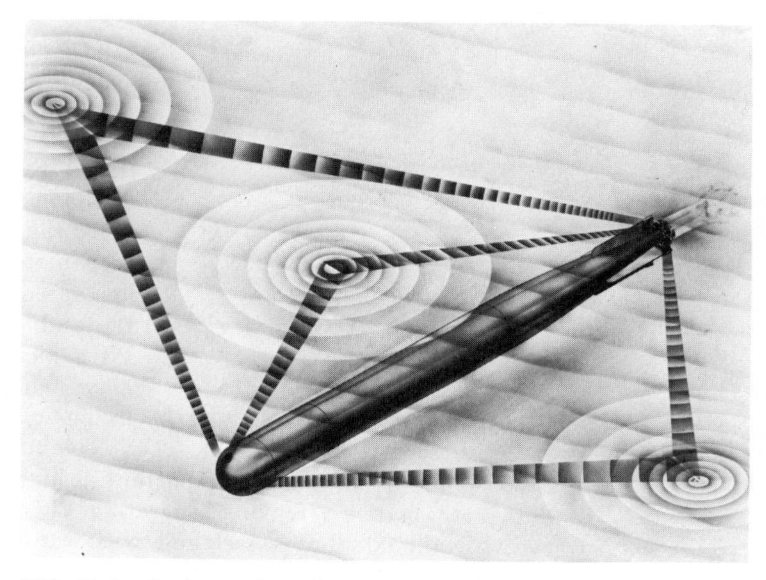

FIG. 11.4. Surface-reflected paths in torpedo self-noise.

Unlike other kinds of noise, machinery noise tends to be relatively independent of speed, since much of it originates in the constant-speed auxiliary machinery of the vessel. Hence, at slow speeds, where other kinds of noise are of low level, the noise of the vessel's auxiliary machinery is often a troublesome source of self-noise. In the wartime JP equipment on submarines operating below 5 knots, it was observed that listening was affected by power operation of the bow and stern planes, the steering machinery, and certain rotating equipment aboard the submarine. At higher speeds, propeller noise becomes the dominant contributor to self-noise under conditions of high frequencies, shallow water depths, and stern bearings. At high speeds also, the many and diverse forms of hydrodynamic noise become important.

Hydrodynamic noise includes all those sources of noise resulting from the flow of water past the hydrophone and its support and the outer hull structure of the vessel. It includes the turbulent pressures produced upon the hydrophone face in the turbulent boundary layer of the flow (flow noise), rattles and vibration induced by the flow in the hull plating, cavitation around appendages, and the noise radiated to a distance by distant vortices in the flow. Hydrodynamic noise increases strongly with speed, and because the origin of this noise lies close to the hydrophone, it is the principal source of noise at high speeds whenever the

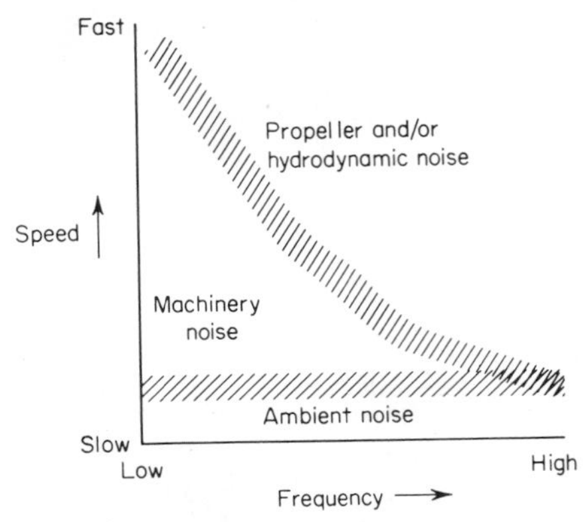

FIG. 11.5. Regions of dominance of the sources of self-noise.

noise of propeller cavitation—itself a form of hydrodynamic noise—is insignificant.

The relative importance of the sources of self-noise can be illustrated by showing their areas of dominance on a plot having vessel speed and frequency as the two coordinates (Fig. 11.5). At very low speeds, the hydrophone "sees" the ambient noise of the sea itself. With increasing speed, machinery noise tends to dominate the low-frequency end of the spectrum, and a combination of propeller and hydrodynamic noise becomes important at high frequencies. The scales and the crosshatched regions separating the noise sources in Fig. 11.5 vary with the kind of vessel and with the directivity, location, and mounting arrangement of the measuring hydrophone.

Hydrodynamic noise occurs in stationary hydrophones as well as in those on moving vessels. Around the case of an acoustic mine on a rocky bottom, for example, a swift tidal current was found (3) to produce pressures of 1,000 dynes/cm^2 (+60 db) in the 1- to 4-Hz band and 50 dynes/cm^2 (+34 db), in the 5- to 32-Hz band on the crystal hydrophone of the mine. Another kind of hydrodynamic noise is the *flutter noise* observed in cable-suspended hydrophones (4), in which cable vibration caused by eddies shed by the cable in a water current appears as self-noise in a hydrophone at the end of the cable. This kind of noise occurs at the frequency f, at which eddies are shed by the cable, given by

$$f = \frac{sv}{d}$$

where v = water current speed, fps
d = diameter of cylindrical cable, ft
s = Strouhal number (dimensionless)

The Strouhal number is equal to 0.18 over a wide range of v and d. Flutter noise can be alleviated by some simple remedies, such as providing a cable fairing or isolating the hydrophone from the cable-transmitted vibrations.

Electrical noise is occasionally bothersome in sonar sets as a form of self-noise. However, its existence indicates a pathological condition whose cure is normally obvious. Except under exceptionally quiet conditions, as at times in the Arctic under a uniform ice cover, electrical noise is not a serious problem in well-designed sonars.

11.3 Flow Noise

A particular kind of hydrodynamic noise has been called *flow noise.* Because it is amenable to theoretical and experimental study and has an urgent application to the reduction of cabin noise inside aircraft, it has received relatively abundant attention in the literature. Flow noise, in actual practice, may be said to be what is "left over" after all other sources of hydrodynamic noise aboard the vessel have been accounted for or removed.

Flow noise consists of the pressures impinging upon the hydrophone face created by turbulent flow in the turbulent boundary layer about the hydrophone. Figure 11.6 illustrates a rigid, flat boundary containing a flush-mounted pressure hydrophone, above which a viscous fluid is flowing. Between the free stream and the boundary lies a turbulent boundary layer within which fluctuating pressures are created and are transmitted to the hydrophone in the boundary. While these turbulent pressures are not true sound, in that they are not propagated to a distance, they form what has been termed "pseudosound" (5) and give rise to a fluctuating noise voltage at the output of the pressure hydrophone. In the following sections, some of the salient characteristics of flow noise will be briefly summarized. The interested reader is referred to the quoted literature for more detailed information.

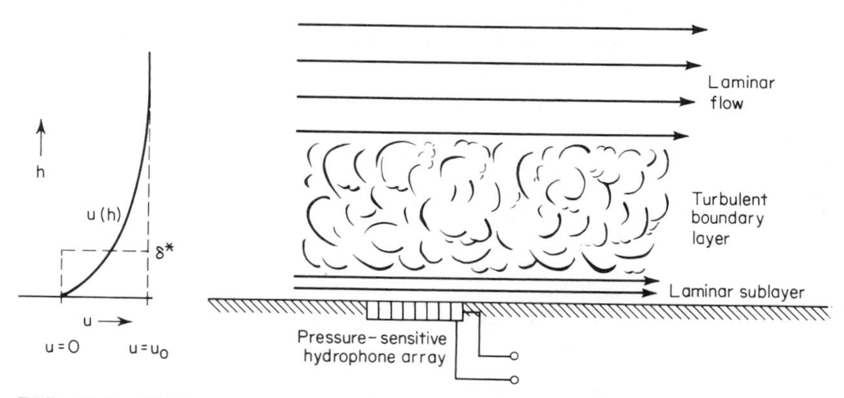

FIG. 11.6. Flow structure in a fluid moving over a stationary surface. The diagram at the left shows the variation of flow velocity with height above the boundary.

Total Fluctuating Pressure The rms pressure p_{rms} on the boundary due to the turbulent flow is related to the free-stream dynamic pressure by

$$\frac{p_{\mathrm{rms}}}{\frac{1}{2}\rho u_0^2} = 3 \times 10^{-3}\alpha$$

where ρ = fluid density
u_0 = free-stream flow velocity

α is a constant, called the Kraichman constant after a pioneering investigator of turbulent flow, ranging from 0.6 to 4 in different measurements using different data (6), but centering roughly about unity. In any one series of measurements, α is found to be a constant over a wide range of flow velocities.

Spectrum of Flow Noise The power spectrum of flow noise, or the distribution in frequency of the mean-squared pressure, is found to be flat at low frequencies and to slope strongly downward at high frequencies at the rate of f^{-3}, or as -9 db/octave. Examples of flow-noise spectra observed with a flush-mounted hydrophone ½ in. in diameter inside a rotating cylinder are shown in Fig. 11.7. The transition frequency f_0 between the flat and the sloping portions of the spectrum is given by

$$f_0 = \frac{u_0}{\delta}$$

where δ is the thickness of the boundary layer. Since the flow velocity is a continuous function of distance away from the wall (Fig. 11.6), the boundary-layer thickness is more exactly expressed by the "displacement

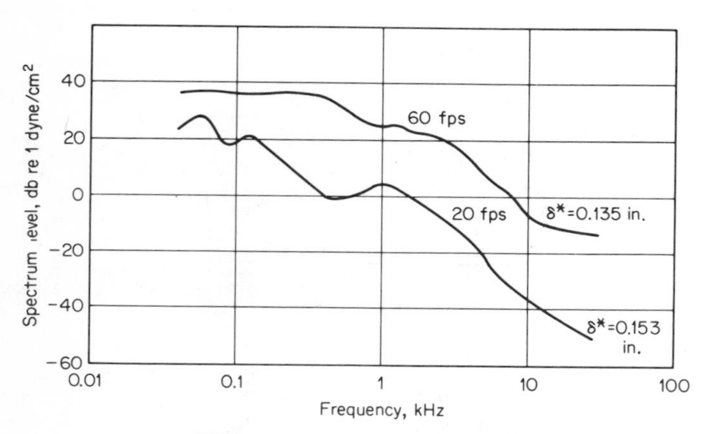

FIG. 11.7. Spectra of flow noise at two speeds inside a rotating cylinder. (From Ref. 6.)

thickness" δ^* such that

$$\delta^* u_0 = \int_0^\infty u(h)\, dh$$

where $u(h)$ is the flow velocity at a distance h normal to the wall. The actual boundary layer is thus replaced by an equivalent layer (in the above sense) of zero velocity (that is, the layer is attached to the wall) and of thickness δ^*. It is found that $\delta = 5\delta^*$, approximately. In experiments of Skudryzk and Haddle (6), δ^* was found to be 0.153 in. at a speed (u_0) of 20 ft/sec and 0.135 in. at 60 fps.

Variation with Speed At frequencies less than f_0, the spectrum level of flow noise varies as the cube of the speed u_0; at frequencies appreciably greater than f_0, it varies as the sixth power of the speed. The latter rate of increase amounts to 18 db per speed doubled. This is equivalent to a straight-line rate-of-rise of 1.8 db/knot in the speed range 10 to 20 knots, in approximate agreement with observations of self-noise on large naval surface ships over this range of speed.

Effect of Surface Roughness Flow noise across a rough surface has been studied by Skudryzk and Haddle (6) using a rotating cylinder having different degrees of surface roughness obtained by cementing grit of various sizes to the outside It was found that at 24 kHz the noise produced by the roughnesses became equal to the flow noise across the perfectly smooth surface when the height of the roughness was such that

$$h = \frac{0.06}{u'}$$

where u' = flow velocity, knots
h = roughness height, in.

Surfaces need not be optically smooth to be considered "smooth" for flow noise, but must be free of roughnesses high enough to extend above the laminar boundary layer (Fig. 11.6) and affect the turbulent flow.

Coherence of Turbulent Pressures Measurements of the longitudinal and transverse correlation of the pressures due to the flow have been made with small hydrophones in the walls of tubes and pipes (7, 8). In a frequency band w Hz wide centered at frequency f, the crosscorrelation coefficient of the pressure measured at two points along the walls a distance d apart has been found (7) to be

$$\rho(d,w) = \rho(s) \frac{\sin(\pi w d/u_c)}{\pi w d/u_c} \cos 2\pi s$$

where s is the (nondimensional) Strouhal number defined as

$$s = \frac{fd}{u_c}$$

in which u_c is the "convection velocity" equal to the velocity at which turbulent patches are carried past the hydrophone by the flow. u_c is somewhat smaller than the free-stream velocity u_0 and varies from 0.6 u_0 to 1.0 u_0, depending on the frequency f. The correlation function $\rho(s)$ has been found experimentally to be

$$\rho_L(s) = e^{-0.7|s|}$$

for longitudinal separations d parallel to the flow (7), and

$$\rho_T(s) = e^{-5|s|}$$

for transverse separations at right angles to the flow (8).

Discrimination against Flow Noise A pressure hydrophone of finite size, that is, an array, will discriminate against flow noise to an extent determined by the spatial correlation coefficients ρ_L and ρ_s described above. The magnitude of this discrimination β is defined as

$$\beta = \frac{R'}{R}$$

where R' is the mean-square voltage output of an array placed in a flow noise field, and R is the mean-square voltage output of a very small pressure hydrophone placed in the same noise field and having a sensitivity equal to the plane-wave axial sensitivity of the array. The quantity 10 log $(1/\beta)$ is equivalent to the *array gain for flow noise;* the ratio β measures the reduction of flow noise experienced by an array of pressure-sensitive elements relative to the noise pickup of a single small element alone. The magnitude of the discrimination factor β has been worked out by Corcos (8) for circular and square arrays, on the assumption that the correlation function in oblique directions to the flow is given by the product of the two principal components ρ_L and ρ_T. White (9) has extended this work by means of a unified theory and has carried out computations for rectangular arrays with long side parallel to and perpendicular to the direction of flow. Figure 11.8 shows the quantity β as defined above for a rectangular array (after White) and for a circular array (after Corcos). For a large square hydrophone of side equal to L, Corcos shows that

$$\beta = \frac{0.659}{\gamma^2}$$

where $\gamma = 2\pi fL/u_c$, and for a circular hydrophone of radius r

$$\beta = \frac{0.207}{\gamma^2}$$

where $\gamma = 2\pi fr/u_c$. In both cases, γ must be much greater than unity.

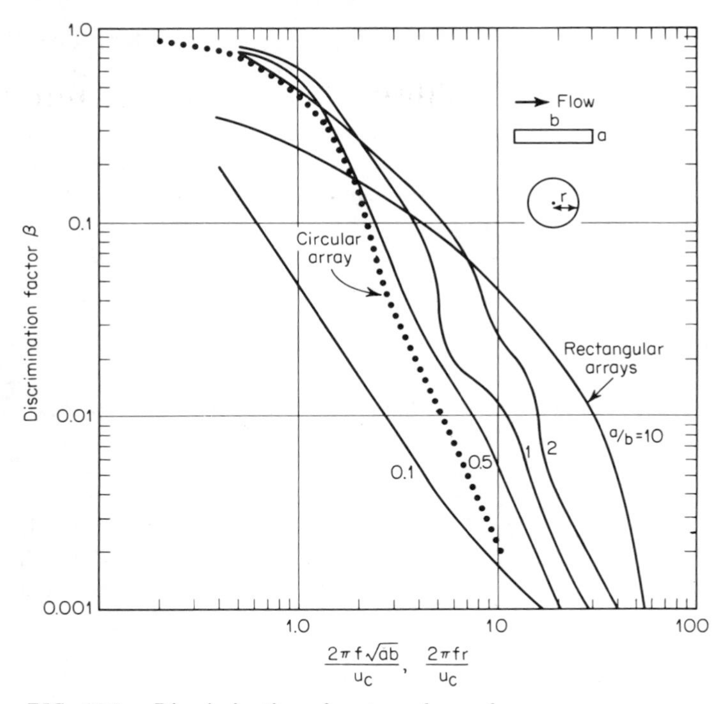

FIG. 11.8. Discrimination of rectangular and circular arrays against flow noise. The sides a and b of the rectangular array are oriented perpendicular and parallel to the direction of flow, respectively. [After White (9) and Corcos (8).]

Comparison with an Isotropic Sound Field If now we define a *convection wavelength* λ_c such that

$$\lambda_c = \frac{u_c}{f}$$

in analogy with the acoustic wavelength

$$\lambda_s = \frac{c}{f}$$

where c is the velocity of sound, the expression for the discrimination factor of a circular hydrophone becomes

$$\beta = 0.207 \left(\frac{\lambda_c}{2\pi r}\right)^2$$

For isotropic noise, the corresponding expression is (Table 3.2)

$$\beta_{iso} = \left(\frac{\lambda_s}{2\pi r}\right)^2$$

where 10 log $(1/\beta_{iso})$ is the ordinary directivity index of the circular array. Comparing the two expressions, we observe that $\beta = 0.207\beta_{iso}$ *when the appropriate wavelength is used for the two types of noise.* But λ_c is far smaller than λ_s for vehicles traveling in the sea; the ratio λ_c/λ_s is approximately equal to the Mach number of the vessel, or the ratio of its speed to the speed of sound in the sea. Since M is a small quantity, it follows that the discrimination against flow noise for a large array of a given size is much greater than it is for isotropic noise. In terms of the Mach number M,

$$\frac{\beta}{\beta_{iso}} = 0.207M^2$$

As an example, at a speed of 20 knots, $M = 7 \times 10^{-4}$, and the quantity 10 log (β/β_{iso}) becomes -70 db. In actual practice on a moving vessel, however, such great benefits of large arrays for reducing noise are not likely to be observed because of the existence of sources of noise other than the turbulent-flow pressures on the rigid wall. If the wall is not rigid, resonant wall vibration as well as the radiated noise of distant turbulence are likely to overwhelm the ideal flow-noise pressures picked up by a large array.

11.4 Domes

It was observed many years ago that large reductions of what is now known as hydrodynamic noise could be had on surface ships by surrounding the sonar transducer by a streamlined housing. These housings are called *sonar domes.* They reduce self-noise by minimizing turbulent flow, by delaying the onset of cavitation, and by transferring the source of flow noise to a distance from the transducer. Sonar domes were originally spherical in shape but were soon streamlined into a teardrop shape to prevent the occurrence of cavitation at high speeds. Some examples of domes used during World War II are shown in Fig. 11.9. The domes pictured in this figure are of all-metal construction and have a thin stainless steel window to permit the ready exit and entrance of sound out of and into the transducer inside. Modern domes are constructed of rubber reinforced with thin steel ribs. Many have baffles, such as those seen in Fig. 11.9, to reduce machinery and propeller noises coming from the rear.

The acoustic and mechanical requirements of dome design are severe. The dome must be acoustically transparent, so as to introduce only a small transmission loss and produce no large side lobes in the directivity pattern of the enclosed transducer. The latter requirement means the absence of internal specular reflection from the dome walls. At the same

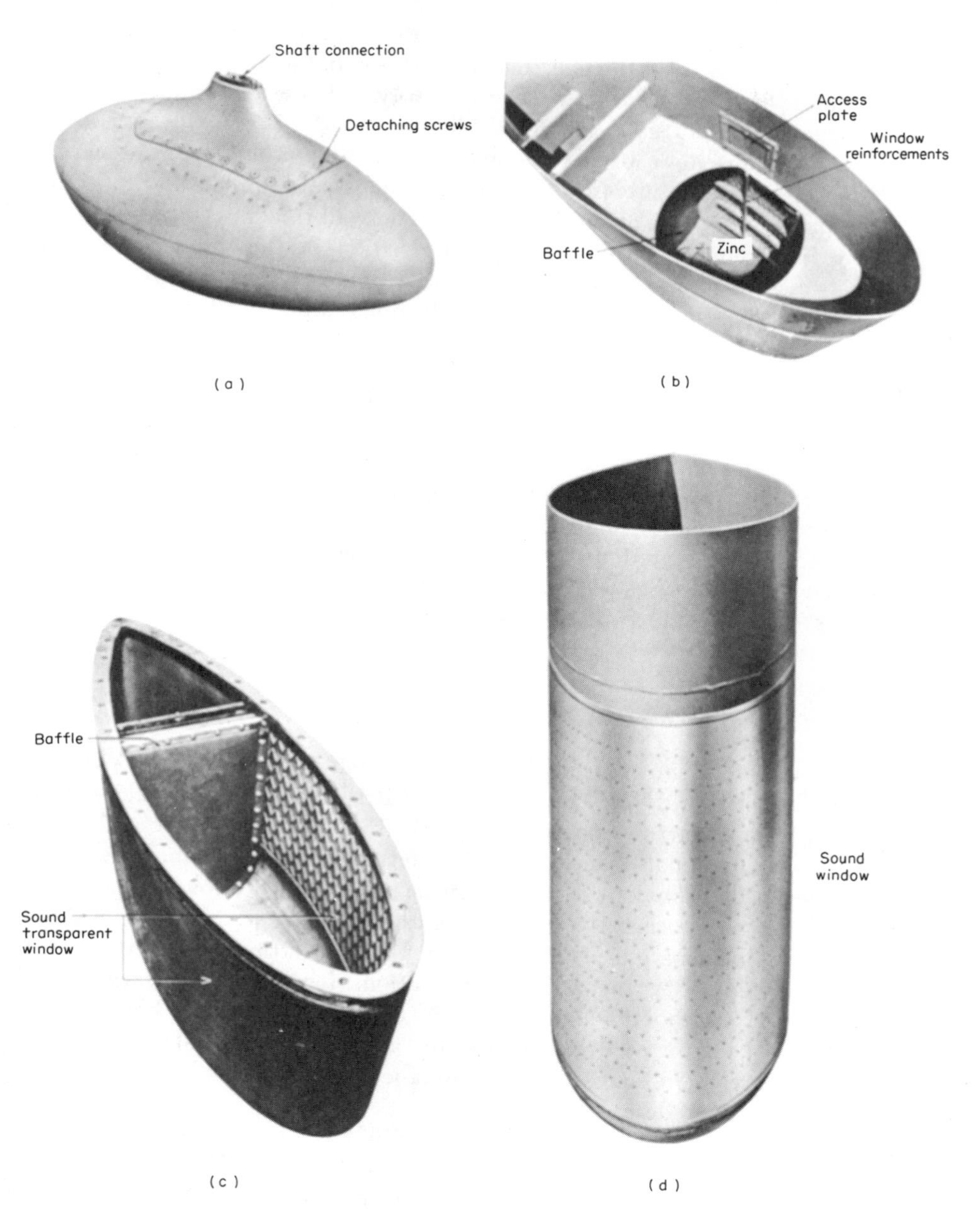

FIG. 11.9. Photographs of streamlined domes. (From Ref. 12.) (a) QCU dome. (b) 100-in. QGA dome. (c) QBF dome. (d) 100-in. QGA dome.

time the dome should be sufficiently streamlined to delay the onset of cavitation on its surface beyond the highest speed reached by the vessel and should be of sufficient mechanical strength to resist the hydrodynamic stresses upon it when under way. These requirements are to a large extent mutually incompatible.

Expressions have been obtained theoretically (10, 11) and generally verified experimentally for the transmission loss and the specular reflection produced by a dome of a given material and wall thickness on a transducer of given frequency, directivity, and position in the dome. Both the transmission loss and reflectivity of the dome increase with frequency and with the thickness and density of the dome walls. Internal reflections in domes can be greatly reduced by increasing the horizontal, and particularly the vertical, curvature of the dome walls. Hence, both for acoustic and hydrodynamic reasons, sonar domes employ materials as thin and light as possible formed into curved streamlined shapes. The acoustic windows of sonar domes used during World War II, like those of Fig. 11.9, ranged from 0.020 to 0.060 in. in thickness (12).

For low self-noise, sonar domes must be kept undamaged and free of marine fouling. A dome with a rough exterior surface will produce a higher flow noise, as well as noise caused by local cavitation at "hot spots" on its surface at the higher speeds of the vessel on which it is mounted.

11.5 Self-noise Levels

Figure 11.10 shows equivalent isotropic self-noise levels at 25 kHz on a number of World War II American and British destroyers. These data, taken from a wartime study by Primakoff and Klein (13), were obtained with various sonar transducers of differing directivity, and have been corrected to equivalent isotropic levels, as discussed above, by allowing for (adding) the directivity index of the transducer. The curved line shows an increase of self-noise intensity as the sixth power of the speed. in agreement with theoretical expectations for the speed variation of flow noise. Although higher noise levels are observed at high speeds. probably because of dome cavitation and other sources of noise, the general agreement suggests the dominance of some form of flow noise in these data at moderate speeds.

Figure 11.11 is a similar plot for small warships of the PC and SC classes and for British DE-type ships and frigates. Here the increase of noise with speed is much more rapid and suggests the dominance of propeller cavitation noise in these smaller ships, on which the sonar dome, its distance from the propellers, and the amount of screening by the hull are all much smaller than on destroyers.

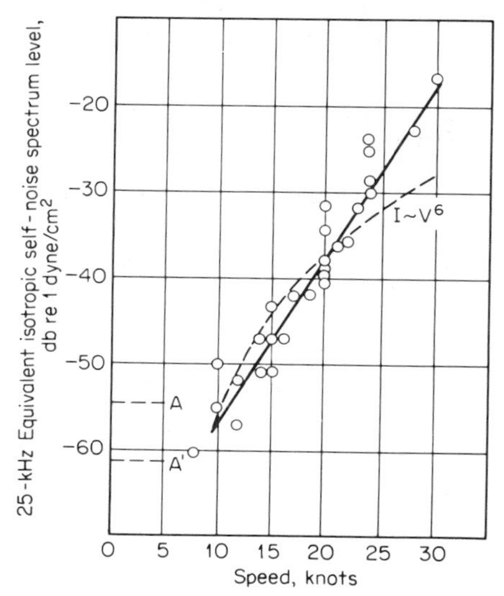

FIG. 11.10. Equivalent isotropic spectrum levels at 25 kHz on destroyers (Ref. 13). The levels A and A′ are deep-sea ambient levels at sea states 3 and 6. The dashed curve shows the theoretical variation with speed of flow noise.

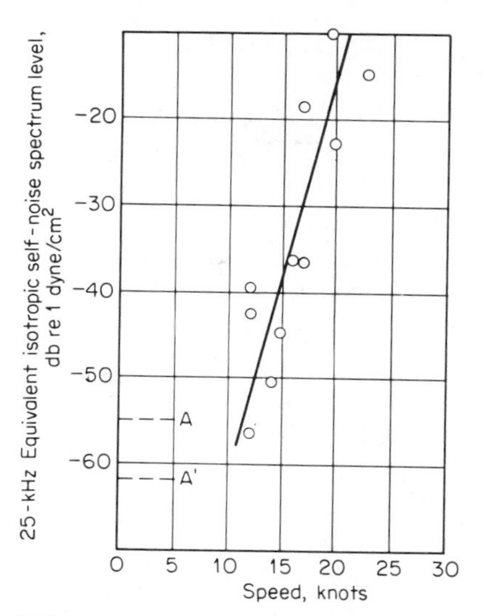

FIG. 11.11. Equivalent isotropic spectrum levels at 25 kHz on American PC and SC class ships and on British DE types and frigates. A and A′ are ambient levels at sea states 3 and 6. (Ref. 13.)

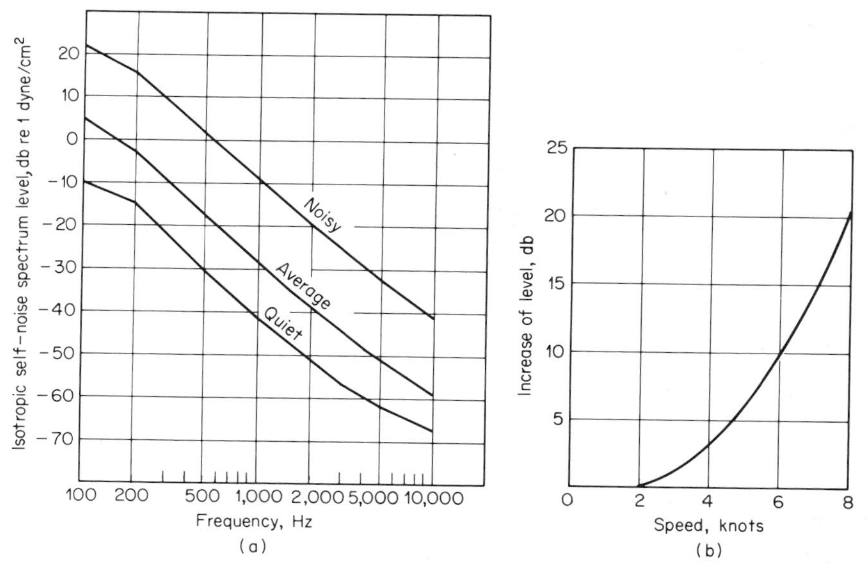

FIG. 11.12. Average self-noise spectra on submarines. (a) JP-1 data, forward bearings, speed 2 knots, periscope depth. (b) Increase of noise level with speed relative to 2 knots. (Ref. 15, Figs. 9 and 10.)

The levels of Figs. 11.10 and 11.11 apply for forward bearings, when the directional transducer in its dome is trained in a forward direction. When trained toward the stern of the ship, higher levels are observed, particularly on small ships, due to the noise of the cavitating propeller.

Self-noise levels on submarines are illustrated by Figs. 11.12*a* and *b*. These show average spectra and the increase of noise with speed measured during World War II with the JP-1 listening equipment. The JP sonar had a 3-ft horizontal line hydrophone mounted on the deck of the submarine forward of the conning tower (Fig. 11.2) and could be trained manually so as to listen in different directions. The spectra of Fig. 11.12*a* are for noisy, average, and quiet installations at a speed of 2 knots. They approximate the levels of deep-water ambient noise at the high-frequency end, but rise more rapidly with decreasing frequency, probably as a result of machinery-noise contributions. The extremely rapid rise with speed suggests the influence of propeller cavitation as the speed increases, as does the fact that the self-noise of the JP-1 sonar decreased with increasing depth of submergence (14). In other submarine sonars, less exposed to cavitation noise, the effect of speed is less marked and

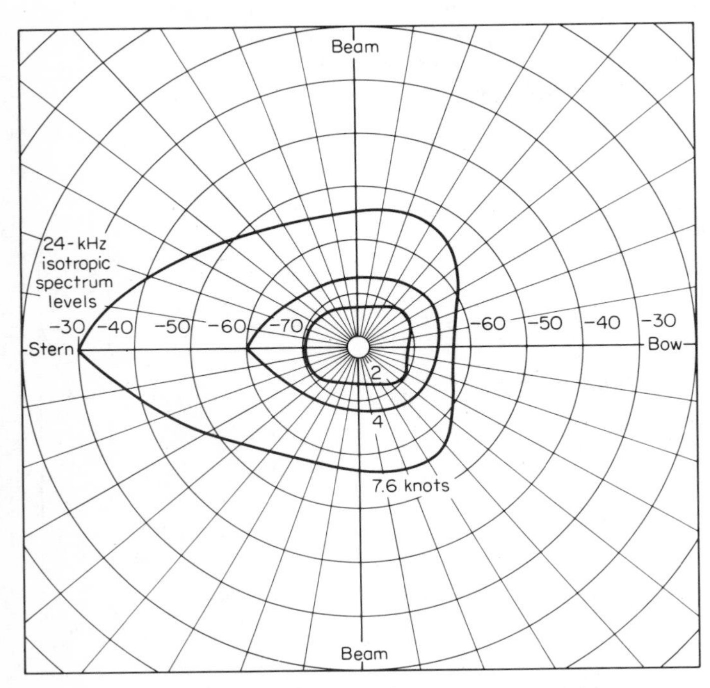

FIG. 11.13. Directionality of high-frequency self-noise at various speeds. Hydrophone directivity index assumed to be 20 db. (Ref. 14.)

approximates the rate of rise of about 2 db/knot found on destroyer-like surface ships (Fig. 11.10).

Just as for destroyers, the self-noise of submarines observed with the JP-1 sonar was found to be independent of bearing on forward bearings, but to increase sharply as the hydrophone was trained toward the stern. Examples of the "directivity pattern" of self-noise at several speeds are given in Fig. 11.13.

REFERENCES*

1. Halley, R.: Physics of Passive Sonar, *U.S. Dept. Navy Electronics Lab. Rept.* NAVPERS 92934, 1962.
2. Hunt, F. V.: Completion Report on Machinery Noise of the Electric Torpedo, *Natl. Defense Res. Comm. Div. 6 sec.* 6.1–ser. 287–2053, 1945.*
3. Childs, W. B., and A. T. Jaques: Water Noise around Mine Cases in Strong Tidal Currents, *Naval Ordnance Lab. Rept.* 889, 1944.*
4. Urick, R. J.: Flutter Noise in Suspended Hydrophones, *J. Acoust. Soc. Am.*, **32**:1498(A) (1960).
5. Lighthill, M. J.: Sound Generated Aerodynamically—the 1961 Bakerian Lecture, *Proc. Roy. Soc. London,* **A267**:147 (1962).
6. Skudryzk, E. J., and G. P. Haddle: Noise Production in a Turbulent Boundary Layer by Smooth and Rough Surfaces, *J. Acoust. Soc. Am.*, **32**:19 (1960).
7. Bakewell, H. P.: Longitudinal Space Time Correlation Function in Turbulent Airflow, *J. Acoust. Soc. Am.*, **35**:936 (1963).
8. Corcos, G. M.: Resolution of Pressure in Turbulence, *J. Acoust. Soc. Am.*, **35**:192 (1963).
9. White, F. M.: A Unified Theory of Turbulent Wall Pressure Fluctuations, *U.S. Navy Underwater Sound Lab. Rept.* 629, 1964.
10. Primakoff, H.: The Acoustic Properties of Domes, Parts I and II, Office of Scientific Research and Development, *Natl. Defense Res. Comm. Div. 6 sec.* 6.1–ser. 1130–1197, 1944.*
11. Basic Methods for the Calibration of Sonar Equipment, *Natl. Defense Res. Comm. Div. 6 Sum. Tech. Rept.*, vol. 10, chap. 9, 1946.
12. A Manual of Calibration Measurements of Sonar Equipment, *Natl. Defense Res. Comm. Div. 6 Sum. Tech. Rept.*, vol. 11, chap. 3, 1946.*
13. Primakoff, H., and M. J. Klein: The Dependence of the Operational Efficacy of Echo-ranging Gear on Its Physical Characteristics, *Natl. Defense Res. Comm. Div. 6 sec.* 6.1–ser. 1130–2141, 1945.*
14. Principles of Underwater Sound, *Natl. Defense Res. Comm. Div. 6 Sum. Tech. Rept.*, vol. 7, fig. 5, p. 248, 1946.
15. Prediction of Sonic and Supersonic Listening Ranges, *Natl. Defense Res. Comm. Div. 6 sec.* 6.1–ser. 1131–1884, 1944.*

* References indicated by an asterisk were originally issued during World War II as classified reports. These reports have been declassified and are available through the U.S. Department of Commerce, Federal Clearinghouse for Scientific and Technical Information, Springfield, Va., 22151.

TWELVE

Detection of Signals in Noise and Reverberation: Detection Threshold

Sonar signals, whether the echoes of active sonars or the target sounds of passive sonars, must almost always be observed amid a background of noise or reverberation. Before any other system function is performed, the sonar system must first *detect* the presence of the signal in this background, that is, the system must determine, or help a human observer to determine, whether or not the signal is occurring within some given interval of time. When a human observer is used, the process of detection calls for a decision on the part of the observer using certain criteria as to whether, in his judgment, the signal really occurred, or not, during the observation interval.

In this chapter, some simple expressions will be obtained for the input signal-to-noise ratio required for making this decision at some preassigned level of correctness of the decision as to "target present" or "target absent." At the preassigned level, the signal-to-noise ratio is called the *detection threshold*, and refers to the input terminals of the receiver-display-observer combination. It is the term in the sonar equations which satisfies the equality in the equation when the signal is just being detected. Although other functions besides detection will sometimes be of prime concern to the sonar, such as "classifying" or determining the identity of a target, detection requires the least signal-to-noise ratio of all and places the greatest demands on the design engineer. Moreover, the detection process is the one that has received greatest theoretical attention in the literature. The words "*detection threshold*" imply two of the most important aspects involved in extracting a signal from the background in which it is embedded: (1) the function of detection itself and (2) the existence of a threshold somewhere near the output of the receiving system.

In seeking expressions for the detection threshold, the method of approach is to view the decision as the end product that must be related to the signal-to-noise ratio at the input of the processing system (1). Mathematical detail will be avoided in what follows. The reader interested in the statistical theory of signals in noise is referred particularly to a book by Helstrom (2), as well as to an excellent review paper by Bennett (3), and to numerous other books on the subject (4–11). At the same time, only peripheral attention will be given to sonar processing methods, since a number of papers have appeared on the theory of correlators and matched filters for application to active sonar (12–16).

12.1 Definition of Detection Threshold

Figure 12.1 illustrates in diagrammatic form the components of a sonar system that lie between the hydrophone array and the decision "target present" or "target absent." These components are (1) a *receiver*, designed to process the signal appearing across its terminals *A-A′* in the most advantageous manner, (2) a visual or an aural *display* of present and past signals and backgrounds, and (3) a human *observer* who, based on the display, makes the required *decision*.

Referring to Fig. 12.1, *detection threshold* is defined as the ratio, in decibel units, of the signal power (or mean-squared voltage) in the *receiver bandwidth* to the noise power (or mean-squared voltage), *in a*

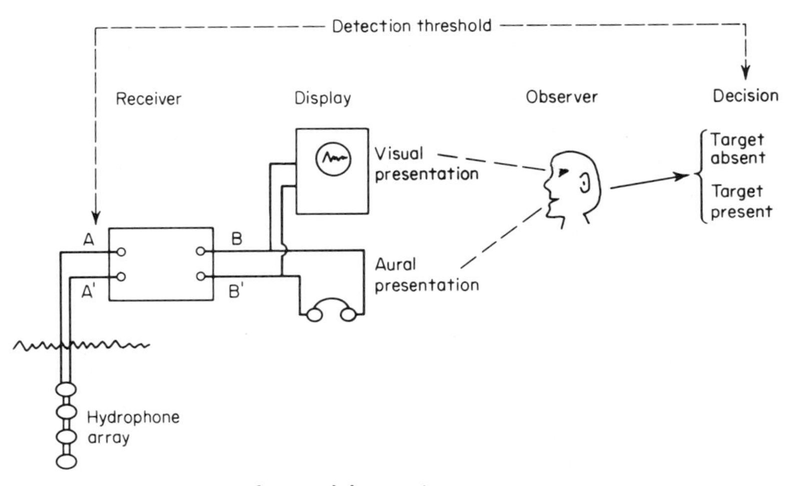

FIG. 12.1. Elements of a receiving system.

		Decision:	
		Signal present	Signal absent
At input:	Signal present	Correct detection p(D)	Miss 1-p(D)
	Signal absent	False alarm p(FA)	Null decision 1-p(FA)

FIG. 12.2. Binary decision matrix.

1-*Hz band*,* measured at the receiver terminals, required for detection at some preassigned level of correctness of the detection decision. If S is the signal power in the receiver bandwidth at A-A' of Fig. 12.1 and N is the noise power in a 1-Hz band at A-A', then

$$\mathrm{DT} = 10 \log \frac{S}{N}$$

when the decision is made under certain probability criteria of correct decisions and errors.

When a signal is in fact present at the receiver input terminals, two decisions, absent or present, are possible; when a signal is in fact absent, the same two decisions can be made. In making this binary, forced-choice decision (the observer *has* to make a "yes" or "no" choice), four possibilities occur, as shown by the decision matrix of Fig. 12.2. Two decisions are correct, and two incorrect; they appear as the diagonal elements of the matrix. The probability that if a signal is present, the *correct* decision, "signal present," is made is called the *detection probability* $p(D)$; the probability that if a signal is absent, the *incorrect* decision, "signal present," is made is called the *false-alarm probability*, $p(\mathrm{FA})$. The detection threshold depends upon these two independent probabilities in a way that will be described below.

An alternate, though less satisfactory, way of describing the performance of a processing system is by means of the term *processing gain*. The processing gain of a receiver is defined as the difference between the signal-to-noise ratio in decibels at the input and the signal-to-noise ratio in decibels at the output, when *both* are referred to or measured in the receiver bandwidth. In Fig. 12.1, the input and output terminals are A-A' and B-B'. Although useful for comparing receivers, processing gain describes the performance of only a part of the input to the decision chain and has only limited value as a working concept for the overall detection system.

12.2 The Threshold Concept

The decision process requires the setting of a threshold such that when it is exceeded, the decision "target present" will be made. An example of such a decision is the closing of a relay and the sounding of an alarm

* Alternatively, signal and noise may *both* be taken in the receiver bandwidth. As defined above, however, DT is appropriate for use in the sonar equations when noise backgrounds are expressed in 1-Hz bands (spectrum levels).

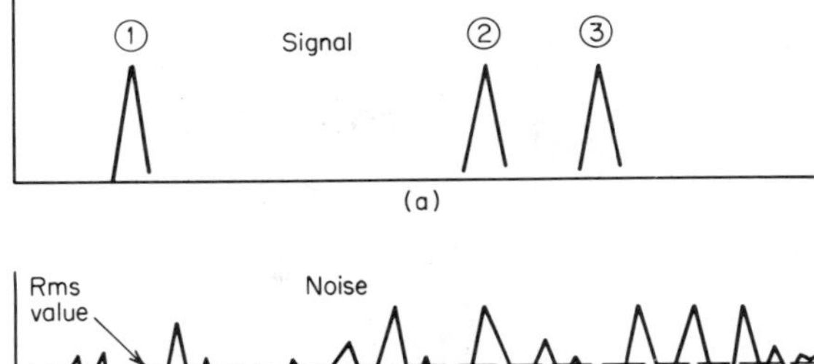

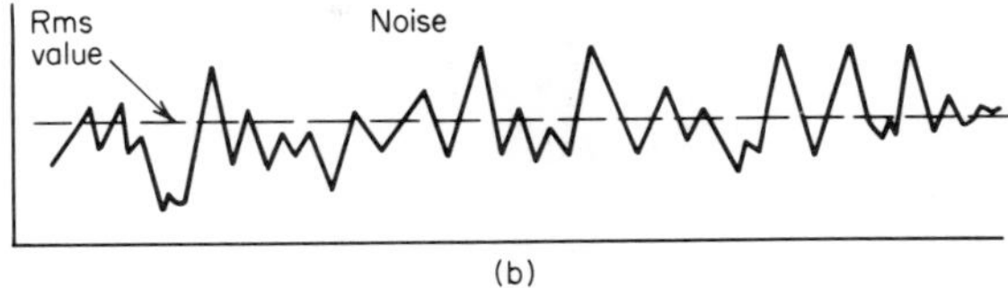

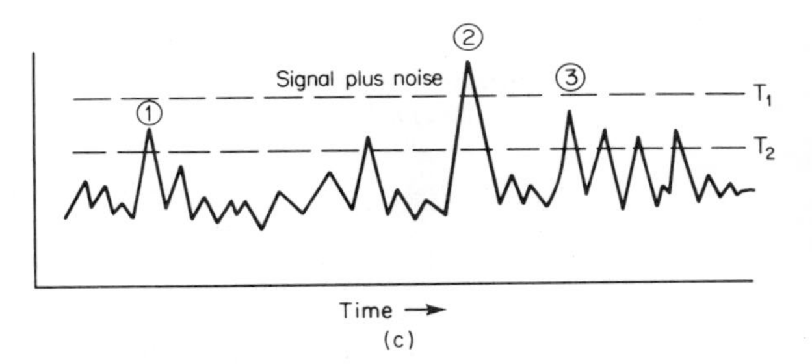

***FIG. 12.3.** Signal and noise at two threshold settings T_1 and T_2. (a) shows three target signals, which, when superposed on noise (b), appear as signal plus noise in (c).*

when the threshold is exceeded. If the threshold is set too high, however, only strong targets will be detected; if set too low, too many "false alarms" will be sounded. At a high threshold setting, both the probability of target detection and the probability of a false alarm are low; at a low threshold, both probabilities become high. This effect of threshold setting is illustrated in Fig. 12.3. The three target signals having the envelope shown in (*a*), when added to the noise envelope (*b*), appear as signal plus noise in (*c*). At threshold T_1, only the second target is detected, and there are no false alarms within the time interval shown in the figure. At a lower threshold T_2, all three targets are detected, but a number of false alarms will occur as well. For a fixed output signal-to-noise ratio, various threshold settings correspond to different pairs of values of the two probabilities. On a plot of detection probability (defined as the probability that a signal, when present, will be detected) and false-alarm probability (defined as the probability that a threshold crossing is caused by noise), a curve will be traced out as the threshold setting is varied. This curve is one of a family of curves called *Receiver-operating-characteristic* (ROC) curves. An example of the ROC curve that would result from Fig. 12.3 as the threshold is changed from T_1 to T_2 is shown in Fig. 12.4. Figure 12.5 shows the family of ROC curves on a linear scale as originally drawn by Peterson and Birdsall (17); Fig.

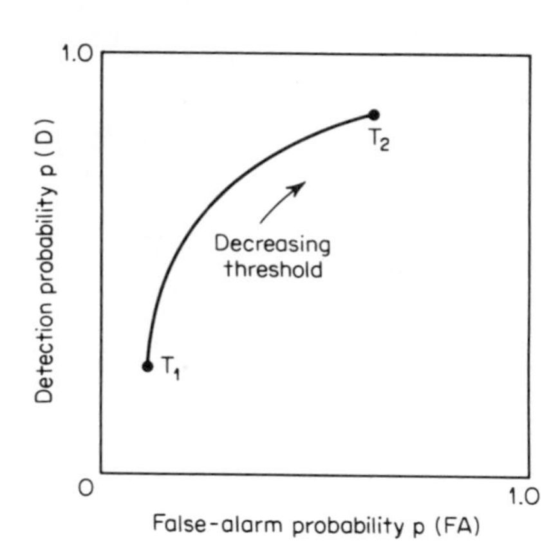

FIG. 12.4. A sample ROC curve that might be constructed from Fig. 12.3.

12.6 is the equivalent family of straight lines on a probability scale extending down to lower values of $p(D)$ and $p(\mathrm{FA})$.

ROC curves are strictly determined by the probability-density functions of signal and noise at the receiver output terminals where the threshold setting is made. Consider Fig. 12.7, which shows curves of probability density $P(a)$ plotted against amplitude a for noise alone and signal plus noise. $P(a)$ is the probability that the amplitude of the envelope of the output lies within a small unit interval of amplitude centered at a. The mean noise amplitude is $M(N)$ and the mean signal-plus-noise amplitude is $M(S + N)$. Both noise and signal plus noise are assumed* to be Gaussian with equal variance σ^2.

* By the central-limit theorem (18), the distribution of the sum of a large number of samples is Gaussian, regardless of the distribution of the population from which the samples were drawn. In the present context, the sum refers to the envelopes of noise and signal plus noise and implies time integration in the processing such that large sample sizes (large bandwidth-time products) exist.

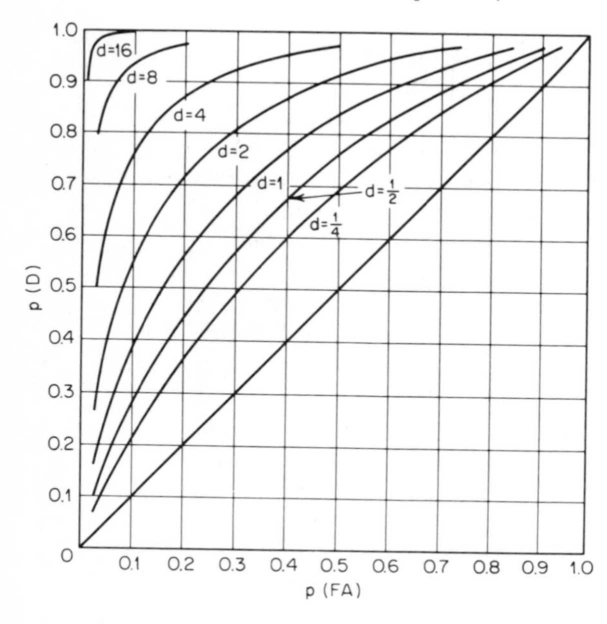

FIG. 12.5. Receiver-operating-characteristic (ROC) curves of detection probability $p(D)$ against false-alarm probability $p(\mathrm{FA})$, with detection index d as a parameter. (Ref. 17.)

When repeated signals are used for detection, as when a number of successive echoes in different samples of background are incoherently added before the detection decision is made, the effect is to lengthen the signal duration t. Hence, the effect of using n incoherently added signals is to decrease the detection threshold by 5 log n, or 1.5 db, for each doubling of the number of signals. When a human observer is used in the detection chain, however, this relationship is found to hold only approximately. For example, for visual detection of radar pips on an A scan oscilloscope screen, it was found during World War II (21) that the detection index increased by 5 log n for small n, but was less than this for large values of n. This implies a deterioration in the integration ability of the eye for large n, as might be expected. However, it must be pointed out that no improvement appears to occur when the same signal *in the same background* is presented repeatedly to an alert human observer.

12.4 Estimating Detection Threshold

In estimating a detection threshold for a particular sonar, it is first necessary to decide upon an acceptable detection probability and an acceptable false-alarm probability for the system *under the conditions in which it is to be used.* These two probabilities are determined by the values of the two possible correct responses (Fig. 12.2) and the costs of the two incorrect responses. Decision theory (2, 22) provides a method for finding the likelihood ratio, and hence the threshold setting, that is optimum for the values and costs involved in the decision process. This optimization process will not be gone into here. In an actual sonar design problem, these quantities are exceedingly difficult to estimate, and will vary greatly with the tactics, strategy, and environmental factors applying for the conditions likely to be encountered. Most often, $p(D)$ and $p(\text{FA})$ are selected by a combination of experience and intuition and a clear understanding of the use to be made of the system under design. Moreover, in operational exercises with a sonar set where a human observer is involved in the decision process, the threshold criteria under which the human observer says "target present" or "target absent" are often ill-defined, and the detection threshold is, correspondingly, uncertain.

In selecting a value for the false-alarm probability, it should be remembered that $p(\text{FA})$ occurring as a ROC-curve coordinate is the probability of occurrence of a false alarm in an *interval of time equal to the signal duration.* Practically speaking, for many sonars this probability is extremely small. Of greater interest is the probability of occurrence of a false alarm within some longer period. Let this longer time interval be denoted by t' and let the signal duration be t. Then the probability $p'(\text{FA})$ of a false alarm in the interval t' is related to that in the signal duration $p(\text{FA})$ by

$$1 - p'(\text{FA}) = [1 - p(\text{FA})]^{t'/t}$$

since the signal may occur in any one of the t'/t intervals with equal probability. When $p(\text{FA})$ is small ($p(\text{FA}) \ll 1$)

$$p'(\text{FA}) \approx \frac{t'}{t}\, p(\text{FA})$$

so that

$$p(\text{FA}) \approx \frac{t}{t'}\, p'(\text{FA})$$

As an example, let an echo-ranging sonar producing an echo 0.1 sec in duration be required to operate with no more than a 50 percent chance of a false alarm during an operating period of 1 hr. Then $p'(\text{FA}) = 0.50$, and

$$p(\text{FA}) = \frac{0.1}{3{,}600}(0.50) = 1.4 \times 10^{-5}$$

If $p(D)$ be taken at 0.50, the required value of d from Fig. 12.6 is 20. Fortunately, the variation of d with $p(\text{FA})$ is extremely slight when $p(\text{FA})$ is small.

12.5 Effect of Duration and Bandwidth

In the design of receivers and displays, the objective is to achieve the *minimum* detection threshold consistent with system requirements, among which are the detection probability and false-alarm probability compatible with the practical use of the sonar under design. The expressions for DT given above show that, for minimum DT, *it is desirable to use the longest signal duration permitted by system considerations.* For active sonars, this means the longest pulse duration; for passive sonars, it means the longest observation time possible before the detection decision must be made. In active sonars, the maximum allowable pulse duration is usually determined by the emergence of reverberation as the background that masks the echo; thereafter no increase in echo-to-reverberation ratio is had by an increased pulse duration. The use of long pulses without a corresponding buildup of the coherent reverberation background is achieved in sonars using pseudo-random noise for transmission and clipped time-compressed correlators for reception (13); such sonars are useful against reverberation backgrounds or against targets of low Doppler shift. In passive sonars, long observation times are limited by the memory of the processing system and the display, by the changing target sounds over a long period of time, and by the necessity of making an immediate detection decision.

Concerning bandwidth, it will be observed that *the detection threshold for the case of the completely known signal* (*Case I*) *is independent of bandwidth.* Although the processing gain, as defined above in terms of signal-to-noise ratios *in the receiver bandwidth*, is 10 log $2Wt$ and increases with bandwidth, the detection threshold does not. In practice, however, a filter ahead of the correlator is used to remove noise contributions outside of the frequency band occupied by the signal. *For Case II, it may be seen that minimum DT is obtained when the minimum possible bandwidth is used*, provided that the signal power in the receiver bandwidth remains constant. The latter condition obtains for sinusoidal signals for bandwidths wider than $1/\tau$. For broadband signals, even though the signal power in the receiver bandwidth tends to increase as the first power of the bandwidth, depending on the signal-power spectrum, the improvement in detection threshold is only as the square root of the bandwidth; hence for broadband signals, it is desirable to use as broad a frequency band as possible whenever both signal and noise have about the same spectral distribution. This would occur, for example, in detecting the broadband sound of a cavitating vessel against a background of ambient noise in the region of the spectrum where both have the same spectral shape.

Detection threshold, defined as a ratio of powers or intensities, is not entirely appropriate as a measure of performance of all kinds of processors. It is sometimes incomplete, since it ignores properties of signal and noise other than their power over a period of time. An example is the degradation of performance of correlation sonars caused by multiple propagation paths in the sea. Such multiple paths tend to destroy the correlation between received and transmitted signals, even though they may leave unaltered, or indeed may enhance, the signal intensity. In general, detection threshold must be defined in terms of the particular processing system employed, as well as the appropriate properties of signal and background. Although, as a power ratio, detection threshold is adequate for processors using signal and noise power or energy, such as a square-law detector, it is only an approximation—if that—for sophisticated processing systems using other characteristics of signal and background.

12.6 Example of a Computation

As examples of a computation of DT in echo ranging for the two cases considered above, we consider first the idealized case of a fixed-point target at a known range in an ideal medium free from distortion or multiple-path propagation effects. In this idealized case, the echo from a sonar pulse of known wave shape will be a perfect replica of the transmitted pulse and will occur at a known instant of time. We take as parameters the following:

$p(D) = 0.5$ (50% probability of detection)
$p(\text{FA}) = 0.0001$ (1 chance in 10^4 of a false alarm occurring in the 0.1-sec signal duration)
$t = 0.1$ sec (signal duration)
$w = 500$ Hz (receiver bandwidth)

Referring to the ROC curves of Fig. 12.6, we find $d = 15$ for the specified detection and false-alarm probabilities. Since the wave shape of the echo is known, the optimum processor is a correlator in which signal plus noise is correlated against a noise-free replica of the echo. The detection threshold is then

$$\text{(Case I)} \quad \text{DT} = 10 \log \frac{d}{2t} = 10 \log 75 = +19 \text{ db}$$

At the other end of the scale of knowledge is the completely unknown echo, in which neither the wave shape (frequency), time of occurrence, or duration can be specified. Such a situation occurs in echo ranging with a short sinusoidal pulse against a target of unknown range and range rate, where a single frequency band 500 Hz wide is required for reasons of simplicity to accommodate the Doppler shift of the target. Now

$$\text{(Case II)} \quad \text{DT} = 5 \log \frac{dw}{t} = 5 \log \frac{15 \times 500}{0.1} = +24\tfrac{1}{2} \text{ db}$$

If the output averager were not "matched" to the signal duration t, but had an averaging time T of, say, 0.05 sec, the mismatch would amount to $|5 \log (T/t)| = |5 \log {}^{0.05}\!/_{0.1}| = 1\tfrac{1}{2}$ db, and DT would rise to 26 db.

It should be noted that these cases represent the extremes of knowledge concerning the signal. When this knowledge is complete, the processing can be tailored to the signal, and the minimum DT for the probability levels and signal duration demanded by system requirements can be achieved. In the example, this minimum threshold amounts to 19 db. The additional 7 db needed for detecting the echo whose frequency and duration are only approximately known is the price that must be paid for an incomplete knowledge about the signal to be detected.

More elaborate processing can be done to approach the minimum DT. Examples are a bank of 50 adjacent narrow-band filters each of bandwidth $1/t = 10$ Hz, instead of the single broadband filter; another example is a deltic correlator (13) to provide continuous time correlation for use when the range is unknown. In such processing schemes, the maximum gain in DT can be found from the two extremes presented above, and the decibel improvement can be weighed against the increased cost and complexity required. Actual processing systems usually involve some degradation in DT relative to the theoretical values. Examples that may be mentioned are the *clipping loss* experienced when hard clipping is done before correlating, as in a polarity coincidence correlator (23), and the *correlation loss* caused by multipath propagation effects in the underwater medium (24).

12.7 Tabular Summary

Table 12.1 gives expressions for the detection threshold for one input and for two inputs from one and two hydrophones, respectively, together with the type of detector and the signal-to-noise ratios before and after the output averager, and for m hydrophones combined in various ways, as adapted from a previously referenced report by Faran and Hills (20). In this table the expressions given for DT, for the case of hydrophone outputs added together before additional processing, include the term $5 \log 1/m^2$, equivalent to the directivity index of an m-element array in an incoherent noise field. Hence the expressions, in effect, refer to the two terms DT and DI in the sonar equations. For multiplicative processing (Sec. 3.9), it is not possible to separate the effects of the array and of the processing system on the overall system performance.

12.8 Auditory Detection

The ear historically has been the earliest detection device used in sonar. Before the use of amplifiers and the perfection of visual displays, the hearing of a human observer provided the only means of detecting sonar signals. In recent years, the human ear has tended to be replaced by other detection devices having a longer integration time as well as a longer memory that provides a record of past events. Yet the ear, as a detector and analyzer of tonal signals in noise backgrounds, is a remarkably compact and efficient device that still has a place in modern sonar. Surprisingly, its performance as a detector of sinusoidal signals approaches that of the optimum detector. In addition, it has other, and notably complex, characteristics that are still under intensive research.* Here

* In the Journal of the Acoustical Society of America for the year 1965, over 60 papers and Letters to the Editor were published on various aspects of physiological and psychological acoustics.

TABLE 12.1 Detection Threshold for Single and Multiple Inputs

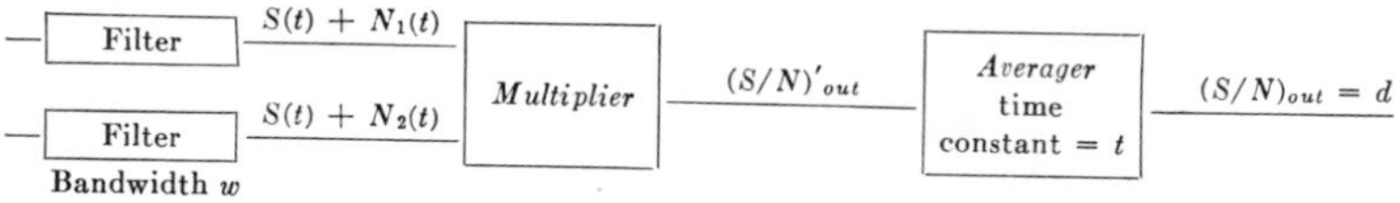

— Filter — $S(t) + N_1(t)$ → Multiplier → $(S/N)'_{out}$ → *Averager* time constant = t → $(S/N)_{out} = d$

— Filter — $S(t) + N_2(t)$ →

Bandwidth w

Case	Type of detector	$(S/N)'_{out}$	$(S/N)_{out} = d = (S/N)^1\, 2wt$	Detection threshold $= 10 \log [S/N/w]_{in}$
Single hydrophone:				
Signal known completely $N_2(t) = 0$	Coherent detector	$(S/N)_{in}$	$2wt(S/N)_{in}$	$10 \log \frac{d}{2t}$
Signal unknown $N_1(t) \equiv N_2(t)$	Autocorrelator (square-law detector)	$\frac{1}{2}\left(\frac{S}{N}\right)^2_{in}$	$wt\left(\frac{S}{N}\right)^2_{in}$	$5 \log \frac{dw}{t}$
Two hydrophones: $N_1(t) \neq N_2(t)$, but of same variance				
Outputs multiplied	Crosscorrelator	$\left(\frac{S}{N}\right)^2_{in}$	$2wt\left(\frac{S}{N}\right)^2_{in}$	$5 \log \frac{dw}{2t}$
Outputs added first and then multiplied (squared)	Adder + square-law detector	$2\left(\frac{S}{N}\right)^2_{in}$	$4wt\left(\frac{S}{N}\right)^2_{in}$	$5 \log \frac{dw}{4t}$

Limitations

1. Small S/N, i.e., $\overline{S(t)^2}/\overline{N(t)^2} \ll 1$.
2. Stationary white Gaussian noise, though not much change may be expected for other reasonable noise statistics.
3. $N_1(t)$, $N_2(t)$, and $S(t)$ are all uncorrelated, that is, $\overline{N_1(t)N_2(t)} = 0$.
4. $2wt \gg 1$. The number of sample points represented by the averager is large.
5. An alert observer, if one is used.

Continuation to m hydrophones (compiled from Ref. 20)

Case	$(S/N)_{out} = d$	Detection threshold $= 10 \log [S/N/w]_{in}$
m hydrophones (incoherent noise):		
Added and square-law detected	$m^2 wt\left(\frac{S}{N}\right)^{2*}_{in}$	$5 \log \frac{dw}{m^2 t}$
Correlated in pairs in $m(m-1)/2$ correlators, and outputs added	$m(m-1)\,wt\left(\frac{S}{N}\right)^2_{in}$	$5 \log \frac{dw}{m(m-1)t}$
Split into two halves and correlated	$\frac{m^2}{2}\,wt\left(\frac{S}{N}\right)^2_{in}$	$5 \log \frac{2dw}{m^2 t}$
Multiplied all together, m even	$\frac{m!m}{2}\,wt\left(\frac{S}{N}\right)^m_{in}$	$\frac{10}{m} \log \frac{2d}{m!m\,wt} + 10 \log w$

* Applies for a keyed signal. For an unkeyed signal, replace m^2 by $(m-1)^2$.

our concern will be restricted to but one aspect of the behavior of this wonderful sensory organ—its ability to detect sinusoidal signals in noise and reverberation backgrounds.

In the literature, the signal-to-noise ratio for auditory detection is frequently called *recognition differential.* This term is defined (25) as the "amount by which the signal level exceeds the noise level presented to the ear when there is a 50 percent probability of detection of the signal." Because there is no specification concerning false alarms, the term "recognition differential," as thus defined and as referred to in the older literature, is quantitatively almost meaningless.* In the current psychoacoustic literature, the use of recognition differential has all but disappeared, and has been replaced by a specification of the detection index d' $(= d^{1/2})$, equivalent to a pair of probabilities, needed for detection under stated conditions of signal and noise.

Physical Model of Audition Of the many component parts of the human hearing mechanism, one particular structure, called the *basilar membrane,* is of particular importance. This is a thin membrane lying coiled up inside a bony tube termed the *cochlea*—so called from its resemblance to a small snail shell. Embedded in the basilar membrane throughout its length are a great number of fine fibers, called the *auditory strings,* that initiate the complex sensory and nervous process resulting in the tonal sensation. According to the resonance theory of hearing, the auditory strings are analogous to the strings of a piano or harp; different portions of the basilar membrane and the corresponding embedded auditory strings resonate at different frequencies of the impressed sound. The ear is thus imagined to be like a comb filter consisting of a large number of narrow adjacent filter bands, called the *critical bands* of hearing. The bandwidth of the critical bands is the *critical bandwidth.*

The simple model of the ear for detection accordingly consists of a series of adjacent input filter bands corresponding to the various critical bands. These are followed, in the model, by a square-law detector and an integrator, or low-pass filter, to model the integration, or energy summation, time of the hearing process. Figure 12.9 shows the energy-detection model of the ear for the critical band centered at frequency f_n; the entire model would consist of a large number of such adjacent bandpass filters, each corresponding in the ear itself to a particular portion of the basilar membrane and its associated neural structure.

This simple model of the auditory detection process by no means accounts for all of the phenomena of audition; moreover, its parameters depend on many factors, including the input level and the immediate

* It appears likely that a false-alarm rate somewhere between 1 and 10 percent prevailed during many early laboratory measurements of recognition differential.

past stimulus of the ear, in a way unfamiliar to electronic circuitry. Nevertheless the model appears to account (26, 27) for the principal phenomena of auditory detection and, as will be seen, yields reasonable agreement with observed detection thresholds.

The critical-band concept was first proposed by Fletcher (28) to explain certain observations of the masking of tones by broadband noise. A variety of methods have been devised to measure the critical bandwidth. One method, first used by Fletcher and later by Schafer et al. (29), employs a tone in a narrow, adjustable band of noise. Observations are made of the bandwidth of the noise above which the detectability of the tone is constant; this bandwidth equals the critical band of the ear. Figure 12.10 is a compilation of measurements by this and other methods of the critical bandwidth at various frequencies. In the vicinity of 500 Hz, the critical bandwidth lies in the vicinity of 50 Hz. Measurements by some methods, such as by loudness summation, the detection of multiple tones, and the masking of noise by tones (the reverse of the method just mentioned), yield bandwidths appreciably larger than this, although the variation with frequency is the same. The contradictory findings as to the critical bandwidth of the ear have been reviewed by Zwicker, Flottorp, and Stevens (30), DeBoer (31), and Swets, Green, and Tanner (32). The dashed curve of Fig. 12.10 is an estimated eye average of the smaller bandwidth values.

The other parameter of the model—the integration time for audition—has received much less attention. Some measurements (33) indicate that the detectability of tone pulses is constant for durations greater than 1 sec, indicating that the ear integrates energy up to, but not beyond, an interval of 1 sec. More recent data (34) show an integration time of the order of 0.2 sec. Apparently this parameter lies somewhere in the

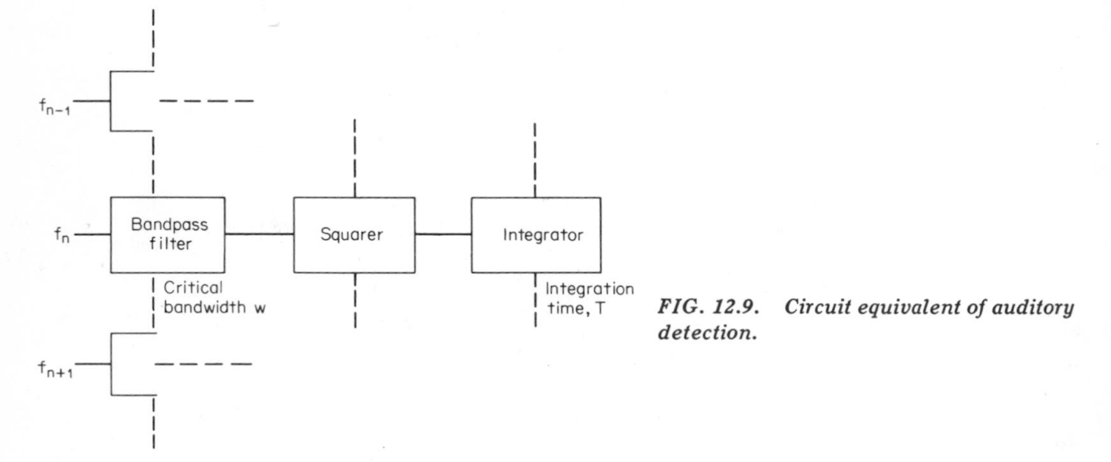

FIG. 12.9. Circuit equivalent of auditory detection.

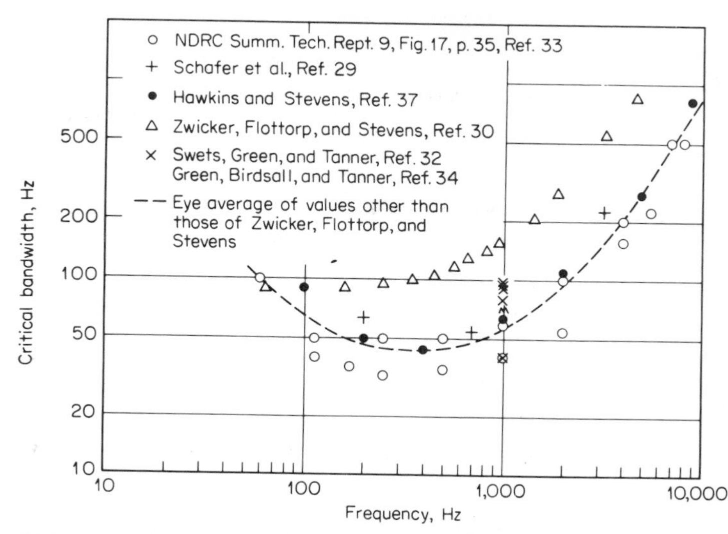

FIG. 12.10. Measured values of the critical bandwidth of the ear.

decade 0.1 to 1.0 sec and is the time interval over which the auditory process, according to the physical model, effectively integrates acoustic energy.

Detection Threshold for Tones in Broadband Noise The detectability of tone pulses masked by broadband Gaussian noise varies with the duration of the tone. Figure 12.11 shows the detection threshold of tonal pulses as compiled from measured values in the literature. The values plotted refer to a detection probability of 0.5 with a false-alarm rate of the order of 0.05; in all but the most recent experiments, the false-alarm rate was not stated. In the work of Green, Birdsall, and Tanner (34), the plotted thresholds correspond to a value of $d' = 1.8$ or a detection index $d = (d')^2 = 3.3$; which, according to the ROC curves, implies a false-alarm probability of 0.05 at the 50 percent level of detection probability.

Detectability of Echoes in Reverberation Against a background of reverberation, the detectability of tones (echoes) is remarkably different. In echo ranging with sinusoidal pulses, the governing parameter is the frequency difference between the echo and the reverberation background. This difference in frequency is the Doppler shift caused by the motion of the sonar ship and target relative to the reverberation-producing scatterers. It amounts to 0.69 hertz per knot of relative velocity per kilohertz (Sec. 9.12).

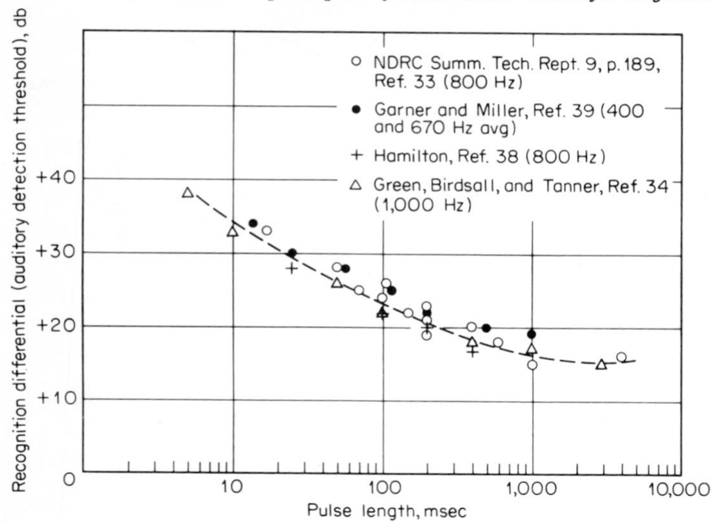

FIG. 12.11. Recognition differential (auditory DT) for sinusoidal pulses in broadband noise reduced to 1-Hz bands.

Much attention was given in the World War II years to the detectability of recorded echoes of adjustable level in recorded reverberation backgrounds. Figure 12.12 shows smoothed curves of the aural echo-to-reverberation ratio required for the detection of echoes of different durations, as a function of the frequency difference between echo and reverberation. In the experiments summarized by these curves, the echoes and reverberation obtained with sonar operating near 25 kHz were heterodyned down to a frequency of about 800 Hz. Of particular note is the rapid improvement of detectability with frequency difference. Also noteworthy is the asymmetry of these curves; "down Doppler" echoes having a frequency lower than the reverberation are more readily detected than are "up Doppler" echoes of the same frequency difference. This suggests an asymmetry in the response characteristic of the critical bands of the ear—an effect that has been

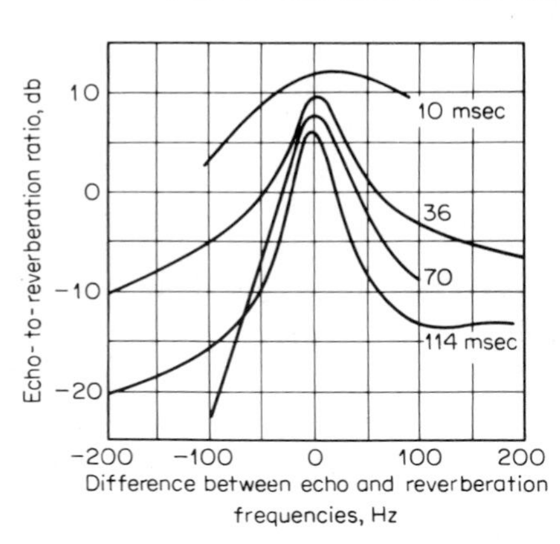

FIG. 12.12. Echo-to-reverberation ratio needed to detect echoes of various lengths in milliseconds against a reverberation background. (Ref. 9, Fig. 14, p. 254.)

observed in later experiments (35) on the masking of one tone by another tone of slightly different frequency.

Comparison with the Model Using the model of Fig. 12.9 and with some knowledge of its parameters, it is of interest to compare the performance of the ear as a detector of tones in broadband noise backgrounds with the performance of the theoretical model. We select a detection index d equal to 4, corresponding to $p(D) = 0.5$ and $p(\text{FA}) = 0.03$ and assume a critical bandwidth w of 50 Hz (Fig. 12.10) and an integration time T of 0.5 sec. Then using the expression

$$\text{DT} = 5 \log \frac{dw}{t} + \left| 5 \log \frac{T}{t} \right|$$

we may compute DT as a function of the pulse duration t. The result is shown in Fig. 12.13, together with the average observed curve taken from Fig. 12.11. Although similar in shape, the computed curve lies lower than the observed curve, indicating that the ear falls somewhat short, by about 4 db, of what would be expected from the model; the difference can be lessened by a more favorable choice of parameters (such as $w = 100$ Hz, and $T = 1.0$ sec). In any case, it appears that the simple physical model provides a reasonable replica of the detection performance of the ear. In terms of the model, the ear is a remarkably efficient detector of tonal signals in Gaussian noise backgrounds. As has been shown by laboratory comparison of a human and an electronic detector (36), the ear approaches the performance of the theoretical energy detector having the same bandwidth and integration time.

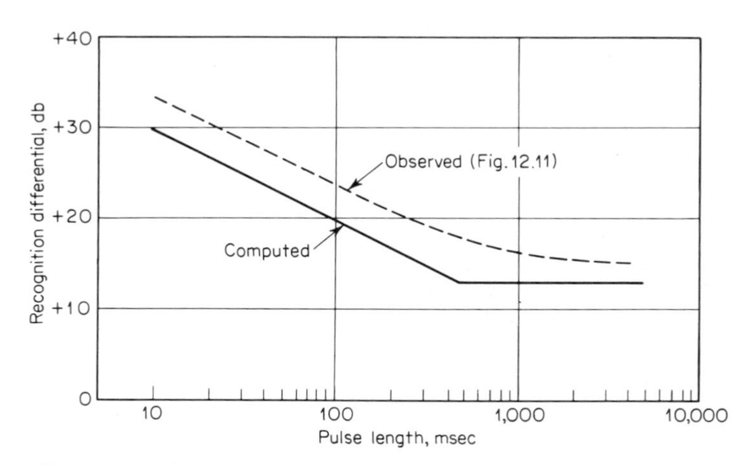

FIG. 12.13. Comparison of observed and computed recognition differentials.

REFERENCES

1. Urick, R. J., and P. L. Stocklin: A Simple Prediction Method for the Signal Detectability of Acoustic Systems, *Naval Ordnance Lab. Tech. Rept.* 61-164, 1961.
2. Helstrom, C. W.: "Statistical Theory of Signal Detection," Pergamon Press, New York, 1960.
3. Bennett, W. R.: Methods of Solving Noise Problems, *Proc. IRE,* **44**:609 (1956).
4. Woodward, P. M.: "Probability and Information Theory," McGraw-Hill Book Company, New York, 1953.
5. Laning, J. H., and R. H. Battin: "Random Processes in Automatic Control," McGraw-Hill Book Company, New York, 1956.
6. Davenport, W. B., and W. L. Root: "Random Signals and Noise," McGraw-Hill Book Company, New York, 1958.
7. Freeman, J. J.: "Principles of Noise," John Wiley & Sons, Inc., New York, 1958.
8. Bendat, J. S.: "Principles and Applications of Random Noise Theory," John Wiley & Sons, Inc., New York, 1958.
9. Schwartz, M.: "Information Transmission Modulation and Noise," McGraw-Hill Book Company, New York, 1959.
10. Lee, Y. W.: "Statistical Theory of Communication," John Wiley & Sons, Inc., New York, 1963.
11. Harman, W. W.: "Principles of the Statistical Theory of Communication," McGraw-Hill Book Company, New York, 1963.
12. Stewart, J. L., and W. B. Allen: Pseudorandom Signal Correlation Methods of Underwater Acoustic Research, *J. Acoust. Soc. Am.,* **35**:810(A) (1963).
13. Allen, W. B., and E. C. Westerfield: Digital Compressed-Time Correlators and Matched Filters for Active Sonar, *J. Acoust. Soc. Am.,* **36**:121 (1964).
14. Persons, C. E., M. K. Brandon, and R. M. Zarnowitz: Linear and Nonlinear Correlators for Pseudo Random Signal Detection, *U.S. Navy Electron. Lab. Rept.* 1219, 1964.
15. Allen, W. B.: Pseudorandom Signal Correlation in Sonar Systems, *U.S. Navy Electron. Lab. Rept.* 1299, 1965.
16. Becken, B. A.: Sonar, "Advances in Hydroscience," vol. 1, Academic Press Inc., New York, 1964.
17. Peterson, W. W., and T. G. Birdsall: The Theory of Signal Detectability, *Univ. Mich. Eng. Res. Inst. Rept.* 13, 1953.
18. Lindgren, B. W.: "Statistical Theory," pp. 145–150, The Macmillan Company, New York, 1962.
19. Peterson, W. W., T. G. Birdsall, and W. C. Fox: The Theory of Signal Detectability, *Trans. IRE,* **PGIT-4**:171 (1954).
20. Faran, J. J., and R. Hills: Application of Correlation Techniques to Acoustic Receiving Systems, *Harvard Univ. Acoust. Res. Lab. Tech. Memo* 28, 1952.
21. Lawson, J. L., and G. E. Uhlenbeck: "Threshold Signals," M.I.T. Radiation Laboratory Series, vol. 24, sec. 8.10, McGraw-Hill Book Company, New York, 1950.
22. Middleton, D.: "Introduction to Statistical Communication Theory," chaps. 18–23, McGraw-Hill Book Company, New York, 1960.
23. Faran, J. J., and R. Hills: Correlators for Signal Reception, *Harvard Univ. Acoust. Res. Lab. Tech. Memo* 27, 1952.
24. Weston, D. C.: Correlation Loss in Echo Ranging, *J. Acoust. Soc. Am.,* **37**:119 (1965).
25. Acoustical Terminology, sec. 12.27, USA Standards, New York, 1960.
26. Sheeley, E. C., and R. C. Bilger: Temporal Integration as a Function of Frequency, *J. Acoust. Soc. Am.,* **36**:1850 (1964).

27. Jeffress, L. A., and A. D. Gaston: ROC Curves from an Ear Model, *J. Acoust. Soc. Am.*, **38**:928(A) (1965).
28. Fletcher, H.: Auditory Patterns, *Rev. Mod. Phys.*, **12**:47 (1940).
29. Schafer, T. H., R. S. Gales, C. A. Shewmaker, P. O. Thompson: Frequency Selectivity of the Ear as Determined by Masking Experiments, *J. Acoust. Soc. Am.*, **22**:490 and (1950).
30. Zwicker, E., G. Flottorp, and S. S. Stevens: Critical Bandwidth in Loudness Summation, *J. Acoust. Soc. Am.*, **29**:548 (1957).
31. DeBoer, E.: Note on the Critical Bandwidth, *J. Acoust. Soc. Am.*, **34**:985 (1962).
32. Swets, J. A., D. M. Green, and W. P. Tanner: On the Width of the Critical Bands, *J. Acoust. Soc. Am.*, **34**:108 (1962). Also, Swets, J. A.: "Signal Detection and Recognition by Human Observers," chap. 23, John Wiley & Sons, Inc., New York, 1964.
33. Recognition of Underwater Sounds, Office of Scientific Research and Development, *Natl. Defense Res. Comm. Div. 6 Sum. Tech. Rept.* 9, 1946.
34. Green, D. M., T. G. Birdsall, and W. P. Tanner: Signal Detection as a Function of Signal Intensity and Duration, *J. Acoust. Soc. Am.*, **29**:523 (1957). Also, Swets, J. A.: "Signal Detection and Recognition by Human Observers," chap. 11, John Wiley & Sons, Inc., New York, 1964.
35. Greenwood, D. G.: Auditory Masking and the Critical Band, *J. Acoust. Soc. Am.*, **33**: 484 (1961).
36. Sherwin, C. W., et al.: Detection of Signals in Noise: A Comparison between the Human Detector and an Electronic Detector, *J. Acoust. Soc. Am.*, **28**:617 (1956).
37. Hawkins, J. E., and S. S. Stevens: Masking of Pure Tones and of Speech by White Noise, *J. Acoust. Soc. Am.*, **22**:6 (1950).
38. Hamilton, P. M.: Noise Masked Thresholds as a Function of Tonal Duration and Masking Bandwidth, *J. Acoust. Soc. Am.*, **29**:506 (1957).
39. Garner, W. A., and G. A. Miller: Masked Threshold of Pure Tones as a Function of Duration, *J. Exptl. Psychol.*, **37**:293 (1947).

THIRTEEN

Design and Prediction in Sonar Systems

13.1 Sonar Design

The various applications of the sonar equations fall into two general classes. One class of applications involves *sonar design*, where a sonar system is to be designed to accomplish a particular purpose. In a sonar design problem, a set of sonar parameters that will provide the desired performance must be found. This performance can usually be expressed in terms of *range*, through its counterpart, by some assumed propagation condition, the parameter *transmission loss*.

This selection of parameters in sonar design is beset with difficulties arising from constraints that are of economic, mechanical, or electrical origin. Sonar systems must sometines be primarily inexpensive, as in expendable units such as sonobuoys. Sometimes they must fit in a confined space, as in a torpedo, where the maximum size of the transducer to be used is dictated by dimensions over which the design engineer has no control. Sonar systems may also have to be designed to consume only a limited amount of electric power, as in a battery-powered underwater acoustic beacon, where a limitation is placed on the available acoustic power output and the pulse duration. Generally speaking, one or more of the parameters related to the system itself, such as directivity index or source level, may be fixed or limited by practical considerations not under the designer's control. The final design is achieved by "trade offs" and compromises between performance and achievable values of the equipment parameters. It is reached by what amounts to repeated solutions of the sonar equations—by a trial-and-error process wherein successive adjustments of parameters and performance are made until a reasonably satisfactory compromise is reached. Complications arise when the desired performance involves two or more of the variables. For example, a certain search rate, or area searched for a target in a given time, may be desired; this is a function of both range and beam width. In such problems, a number of trial solutions of the sonar equations will be needed to give a "feel" for the best set of conditions.

Sometimes the fortunate design engineer has a free choice of the operating frequency, or the operating frequency band, of the sonar under design. In such cases, the choice will be influenced by the optimum frequency appropriate to the desired maximum range of the sonar. This choice will be considered in the section to follow.

In an active-sonar design problem, the design will depend in part on whether the echoes occur in a background of noise or reverberation. In active-sonar systems, the range increases with acoustic power output until the echoes begin to occur in a reverberation background. When this occurs, the range is said to be *reverberation-limited.* Beyond this value of output power, no increase of range is had, since both echo level and reverberation increase together with increasing power. It follows, as a precept in active-sonar design, that the acoustic output power should be increased until the *reverberation level is equal to the level of the noise background at the maximum useful range of the system.* Unfortunately, although this is a useful general rule, it cannot always be followed because of limitations imposed by the amount of available power or because of interaction effects and cavitation at the sonar projector.

13.2 Sonar Prediction

The other broad class of problems has to do with *performance prediction.* Here the sonar system is of fixed design—and, indeed, may already be in operational use—and it is desired to predict its performance under a variety of conditions. Alternatively, if field trials of a system have already been made, it may be necessary to account for the performance that has been achieved—a kind of "postdiction," wherein a numerical explanation is required for the results obtained. This class of problems normally requires solving the appropriate form of the sonar equation for the parameter containing the range. The passive-sonar equation may be written

$$\begin{aligned} \mathrm{TL} &= \mathrm{SL} - \mathrm{NL} + \mathrm{DI} - \mathrm{DT} \\ &= \mathrm{FM} \end{aligned}$$

where the sum of the parameters on the right is called (Table 2.2) the *figure of merit* FM for the particular target referred to in the parameter SL. Similarly, the active-sonar equation for a noise background may be written

$$\begin{aligned} \text{TL} &= \tfrac{1}{2}(\text{SL} - \text{TS} - \text{NL} + \text{DI} - \text{DT}) \\ &= \text{FM} \end{aligned}$$

where FM is the figure of merit for the target implied by the value used for the parameter TS. The prediction of range requires the conversion into range of the value of transmission loss that is equal to the figure of merit. The conversion demands a specification of the propagation conditions, such as layer depth and propagation path, under which the equipment will be, or has been, used. With reverberation backgrounds, the transmission loss is usually the same for both the target and reverberation, and the range occurs implicitly in the terms 10 log A or 10 log V, representing in decibel units the reverberating area or volume, respectively.

13.3 The Optimum Sonar Frequency

Existence of an Optimum Frequency When range calculations at different frequencies are made for a sonar set of a particular design and for some specified propagation and target conditions, it is often found that the range has a maximum at some particular frequency. This frequency is the *optimum frequency* for the particular equipment and target characteristics being considered. At the optimum frequency a minimum figure of merit is required to reach a given range. Hence, the optimum frequency is a function of the detection range as well as the specified set of medium, target, and equipment parameters. If the operating frequency is made much higher than optimum, the absorption of sound in the sea reduces the range; if the operating frequency is made much lower than optimum, a number of other parameters become unfavorable and act to reduce the range. Examples of such parameters are the directivity index, background

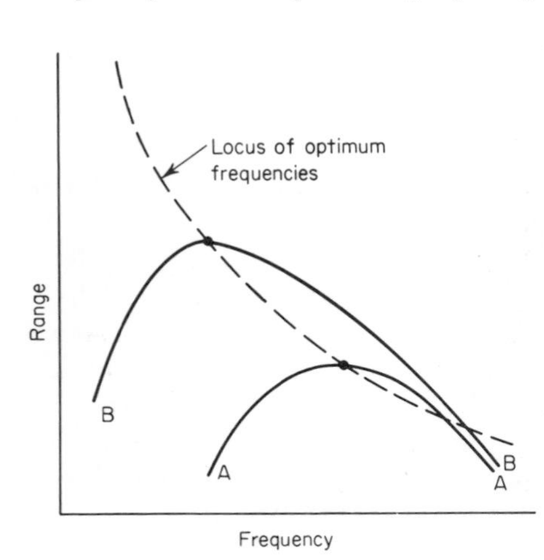

FIG. 13.1. Range as a function of frequency for two sonar systems of different figures of merit.

noise, and detection threshold (through a necessarily smaller bandwidth at the lower frequencies), all of which conspire to reduce the system figure of merit at low frequencies.

Curve *AA* of Fig. 13.1 shows a range-versus-frequency plot for a hypothetical sonar. If, by some means, the figure of merit of this sonar is raised by an amount that is the same for all frequencies, the range-frequency curve is shifted to *BB*. Although the range is increased at all frequencies, the optimum frequency, at which the maximum range occurs, has become lower. The locus of the peak values of a series of such curves gives the best frequency to use to obtain a desired range. Its shape and position depend on the system figure of merit and on the transmission loss and, more importantly, on how both vary with frequency.

Illustrative Example The determination of the optimum frequency can best be illustrated by an example. Consider a passive listening system that employs a line hydrophone 5 ft long. It is desired to find the optimum frequency for the detection of a freighter traveling at a speed of 10 knots. The noise background is taken to be the ambient noise of the sea in sea state 3 (wind speed 11 to 16 knots), and the detection threshold is zero db. Let the transmission loss be determined by spherical spreading plus absorption according to the relationship $\mathrm{TL} = 20 \log r + 0.01 f^2 r \times 10^{-3}$, where f is the frequency in kilohertz and r is the range in yards. Based on this expression, Fig. 13.2 shows curves of TL as a function of frequency for a number of different ranges. Superposed on the same plot is the line *AA*, equal to FM at different frequencies for the particular problem at hand, using appropriate values of the parameters.* At any frequency, the detection range is that for which TL = FM. This range has a maximum, for the curve *AA*, of 6,000 yd and occurs at 5 kHz. This is the optimum frequency for the assumed conditions. At this frequency, the slopes of the line *AA* and of an interpolated member of the family of TL curves are equal. At frequencies different from 5 kHz, the range is less, becoming reduced to 5,000 yd at both 2 and 10 kHz. If, by redesigning the system, the FM is increased by an amount that is constant with frequency, the line *AA* might be shifted to *BB*; the range will be increased to 19,500 yd and the optimum frequency lowered to 1.7 kHz. If the redesign is such as to change the slope of the FM curve, an altogether new optimum frequency will be obtained.

Analytic Method When, as in the example just given, the transmission loss can be expressed as a particular function of range for the conditions of interest, the optimum frequency can be found analytically. In the equality TL = FM, the maximum (or minimum) range is obtained by

* SL: Table 10.2; NL: Fig. 7.5; DI: Fig. 3.11; DT = 0.

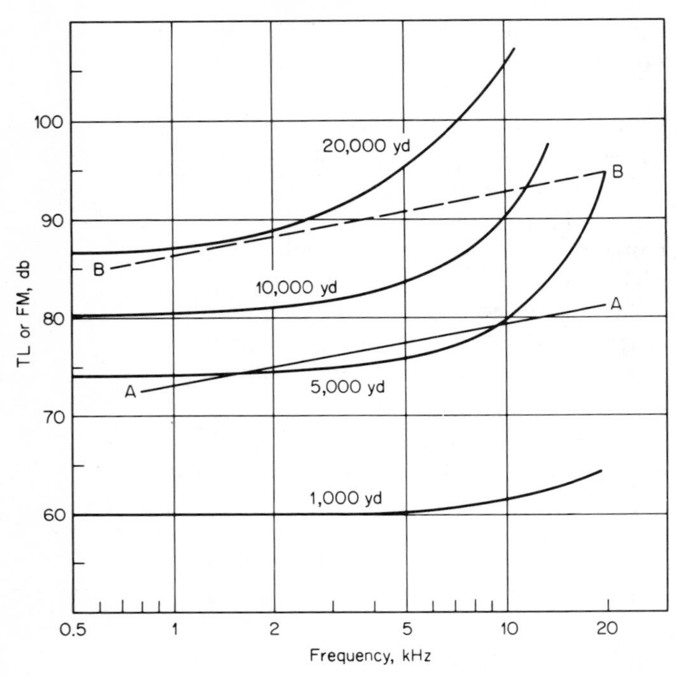

FIG. 13.2. Curves of transmission loss and figure of merit as a function of frequency.

differentiating both sides with respect to frequency and setting dr/df equal to zero. With the preceding expression for TL, we would have

$$\text{TL} = 20 \log r + 0.01 f^2 r \times 10^{-3} = \text{FM}$$

On differentiating and placing $dr/df = 0$, we obtain

$$0.02 f_0 r_0 \times 10^{-3} = \frac{d(\text{FM})}{df}$$

where f_0 = optimum frequency
r_0 = maximum range
$d(\text{FM})/df$ = rate of change of FM with frequency, db/kHz

In terms of the more conventional unit of decibels per octave of frequency, we can write

$$\left.\frac{d(\text{FM})}{df}\right|_{\text{db/octave}} = \frac{f_0}{\sqrt{2}} \left.\frac{d(\text{FM})}{df}\right|_{\text{db/kHz}}$$

since the octave whose geometric mean frequency is f_0 is $f_0/\sqrt{2}$ Hz wide.

It therefore follows that

$$\frac{0.02}{\sqrt{2}} f_0{}^2 r_0 = \frac{d(\text{FM})}{df}$$

and the optimum frequency becomes

$$f_0 = \left[\frac{70.7}{r_0} \frac{d(\text{FM})}{df} \right]^{1/2}$$

where $d(\text{FM})/df$ is the rate of change of FM with frequency in units of decibels per octave, and f_0 and r_0 are in units of kilohertz and kiloyards, respectively. Since for a passive system

$$\text{FM} = \text{SL} - \text{NL} + \text{DI} - \text{DT}$$

it follows that

$$\frac{d(\text{FM})}{df} = \frac{d(\text{SL})}{df} - \frac{d(\text{NL})}{df} + \frac{d(\text{DI})}{df} - \frac{d(\text{dT})}{df}$$

so that the quantity $d(\text{FM})/df$ is the sum, with due regard for sign, of the rates of change with frequency of the sonar parameters of which it is composed. Considering the example given above (Fig. 13.2), $d(\text{FM})/df$ would be found to be approximately equal to $-6 + 5 + 3 + 0 = +2$ db/octave. On substituting in the above expression for f_0 and taking $r_0 = 6{,}000$ yd, f_0 becomes equal to 5.3 kHz. In echo ranging, where the two-way transmission loss is involved, the expression for f_0 becomes

$$f_0 = \left[\frac{35.4}{r_0} \frac{d(\text{FM})}{df} \right]^{1/2}$$

The optimum frequency accordingly depends upon the frequency variation of all the sonar parameters and is especially sensitive to the frequency variation of the absorption coefficient. It is not sharply defined, but is the peak of a broad maximum extending over a frequency range of several octaves. The optimum frequency may be defined in terms other than range, as, for example, search rate or processing time, as discussed by Stewart, Westerfield, and Brandon (1). For reverberation backgrounds, the figure of merit is itself a function of range, and the optimum frequency is not as easily determined. Normally, an optimum frequency does not exist in reverberation-limited systems since the frequency-dependent absorption coefficient is ordinarily the same for the echo and for the reverberation background.

An extended discussion of the subject is given by Horton (2), who can be credited with having first recognized the existence of optimum frequencies in sonar applications.

13.4 Applications of the Sonar Equations

Sonar problem solving

The following are some examples of how the sonar equations may be used to solve problems in a number of different applications of sonar. The examples given and the conditions assumed do not necessarily have any practical significance, but are selected more or less at random to illustrate how the equations are used in some specific problems concerning the many modern uses of sonar.

The approach to problem solving by means of the sonar equations is to select the equation appropriate to a particular problem and then to solve it for the unknown parameter in terms of the other parameters which are either specifiable or can be selected, with more or less uncertainty, from specified conditions on which they depend. Typical values for nearly all conditions of interest can be found in curves or tables given in earlier chapters.

In an actual design problem the usually straightforward computation should be accompanied by a plot of echo or signal level, together with the reverberation and noise masking levels, as a function of range. Such a plot will indicate most strikingly how the range, determined by intersection of the curves of signal and background, will vary with changes in level. This plot will lend confidence to the numerical computations. Once the range is determined, other quantities of perhaps greater significance, such as area searched per unit time, can be readily computed.

Active Submarine Detection

PROBLEM: An echo-ranging sonar mounted on a destroyer has a power output of 1,000 watts at a frequency of 8 kHz. Its DI is 20 db and it uses a pulse duration of 0.1 sec, with a receiving bandwidth of 500 Hz. Find the range at which it can detect a beam-aspect submarine at a depth of 250 ft in a mixed layer 100 ft thick when the ship is traveling at a speed of 15 knots. Detection is required 50 percent of the time, using incoherent processing, with a probability of 0.01 percent of occurrence of a false alarm during the echo duration.

SOLUTION: The active-sonar equation, solved for TL, is

$$\mathrm{TL} = \tfrac{1}{2}(\mathrm{SL} + \mathrm{TS} - \mathrm{NL} + \mathrm{DI} - \mathrm{DT})$$

SL is given by Fig. 4.3, using $\mathrm{DI}_T = 20$ db, as 121; by Table 9.2, TS = 25; by Fig. 11.10 and reducing from 25 kHz by assuming −6 db/octave spectral slope, $\mathrm{NL} = -47 + 20 \log (25/8) = -37$; DI = +20; by Fig. 12.6, $d = 15$ and $\mathrm{DT} = 5 \log (15 \times 500/0.1) = +24$. Therefore $\mathrm{TL} = \tfrac{1}{2}(179) = 90$. Referring to Fig. 6.7*b*, for a layer depth of 100 ft, and assuming that the transmission is the same as that for a source depth of 50 ft, the range corresponding to this value of TL is 5,500 yd.

Passive Submarine Detection

PROBLEM: A submarine radiating a 500-Hz line component at a source level 60 db crosses a convergence zone. Another submarine, located 30 miles away, listens with a nondirectional hydrophone. Assuming a noise background equivalent to that of the deep sea in sea state 3, how long an observation time will the second submarine need to detect the first if it uses incoherent (energy) processing in a receiver band 100 Hz wide and if a detection probability of 50 percent, with a 1 percent false-alarm probability, is satisfactory?

SOLUTION: When solved for the parameter of interest, the passive-sonar equation is

$$\mathrm{DT} = \mathrm{SL} - \mathrm{TL} - \mathrm{NL} + \mathrm{DI}$$

SL is given as +60; TL is taken as being equal to spherical spreading to 30 miles plus a convergence gain of 10 db, or TL = 20 log (30 × 2,000) − 10 = +86; by Fig. 7.5, NL = −34; DI = 0 for a nondirectional hydrophone. Therefore DT = +8. By the formula, DT = 5 log (dw/t), with $w = 100$ (given) and $d = 6$ (Fig. 12.6), we find an observation time of $t = 15$ sec. The signal energy must therefore be integrated for this length of time in order for detection to occur at the required probability levels.

Minesweeping

PROBLEM: A minesweeper tows behind it, for the purpose of sweeping acoustic mines, a broadband sound source having a source spectrum level of 50 db in a 1-Hz band. The mines to be swept are sensitive to noise in the band 100 to 300 Hz, and are suspected to be set to be actuated when the level of noise in this frequency band is 40 db above the spectrum level of the ambient-noise background in coastal waters at a wind speed of 30 to 40 knots. If spherical spreading describes the transmission loss, at what range will the minesweeper sweep (actuate) these mines?

SOLUTION: Solving the passive equation for TL, we have

$$\mathrm{TL} = \mathrm{SL} - \mathrm{NL} + \mathrm{DI} - \mathrm{DT}$$

Since the spectrum level of the broadband source is 50 db, the level in the sensitive frequency band of the mines is SL = 50 + 10 log 200 = 73; by Fig. 7.7, NL = −16; DI = 0 is implied by the nature of the problem; DT = 40 is given. Therefore, TL = 49. For spherical spreading, this corresponds to a swept range of 280 yd.

Depth Sounding

PROBLEM: A fathometer transducer is mounted on the keel of a destroyer and is pointed vertically downward. It has a DI of 15 db with a source level of 100 db at a frequency of 12 kHz. Assuming that reflection takes place at the sea bottom with a reflection loss of 20 db, at what speed of the destroyer will the echo from the bottom in 15,000 ft of water be equal in the self-noise level of the ship in the 500-Hz receiving bandwidth of the fathometer receiver?

SOLUTION: Because reflection at the sea bottom has been postulated, the actual source can be replaced by an image source in the bottom at a range equal to twice the water depth. The transmission loss then will be

$$\text{TL} = 20 \log 2d + \alpha 2d \times 10^{-3} + 20$$

where d is the water depth in yards. With $d = 5{,}000$ yd and α taken at 1 db/kyd, TL = 110 db. The appropriate form of the sonar equation is

$$\text{SL} - \text{TL} = \text{NL} + 10 \log w - \text{DI}$$

where NL + log w is the noise level in the bandwidth w of the receiver. Solving for the unknown parameter,

$$\text{NL} = \text{SL} - \text{TL} - 10 \log w + \text{DI}$$

With SL = 100 db (given), $10 \log w = 10 \log 500 = 27$ db, DI = 15 db, we find NL = -22 db at 12 kHz. This would correspond at 25 kHz to a value of NL = $-22 - 20 \log (25/12) = -28$ db. Referring to Fig. 11.10, the ship speed at which the 25-kHz isotropic self-noise level is -28 db is 25 knots.

Minehunting

PROBLEM: A mine of average aspect lies on a sand bottom. It is desired to detect the mine at a slant distance of 100 yd by means of an active sonar located 20 yd from the bottom. If a pulse duration of 10 msec is used, what horizontal beam width will be required if detection can be achieved at a detection threshold of zero db?

SOLUTION: The sonar equations for a reverberation background are

$$\text{SL} - 2\text{TL} + \text{TS} = \text{RL} + \text{DT}$$
$$\text{RL} = \text{SL} - 2\text{TL} + S_s + 10 \log A$$
$$A = \Phi r \frac{ct}{2}$$

Solving for A and eliminating common terms from the first two expressions, we obtain

$$10 \log A = \text{TS} - S_s - \text{DT} \log A$$

By Table 9.2, we estimate TS = -17 db; by Fig. 8.21 and estimating for a grazing angle equal to $\sin^{-1} (20/100) = 12°$, $S_s = -37$ db; DT is given as zero. Therefore $10 \log A = 20$ db and $A = 100$ yd^2. Solving the third equation for Φ, with $A = 100$, $r = 100$, and $ct/2 = 1{,}600 \times 0.01/2 = 8$ yd, we find $\Phi = 1/8$ rad $= 7.2°$. By Table 8.1, this would require a horizontal line transducer 11 wavelengths long.

Explosive Echo Ranging

PROBLEM: A 1-lb charge is used as a sound source for echo ranging on a submarine. Find the detection range of a bow-stern-aspect submarine target in a background of deep-sea ambient noise in sea state 6. Detection is required 90 percent of the time with a 0.01 percent chance of a false alarm in the echo duration of 0.1 sec. A nondirectional hydrophone with a 1-kHz bandwidth centered at 5 kHz is used for reception. Let the source and receiver depths be 50 ft in a mixed layer 100 ft thick, and let the target depth be 500 ft.

SOLUTION: For short transient sources, the source level is

$$\text{SL} = 10 \log E - 10 \log t_e$$

where E = source level in terms of energy density
t_e = echo duration

Solving the active sonar equation for TL, we obtain

$$\text{TL} = \tfrac{1}{2}(10 \log E - 10 \log t_e + \text{TS} - \text{NL} + \text{DI} - \text{DT})$$

The quantities t_e and DI are given in the problem statement. Since the source is broadband, and using Fig. 4.15 at 5 kHz, $E = +80 + 10 \log 1{,}000 = 110$ db in the 1-kHz receiver bandwidth; by Table 9.2, TS = 10 db; by Fig. 7.5, NL = -43 db; by formula, DT = $5 \log (dw/t)$, using $d = 25$ (Fig. 12.6), $w = 1{,}000$ and $t = 0.1$ DT = 27. DI = 0 (given). With these values TL is found to be 73 db. By Fig. 6.6*c*, the range is 2,600 yd at 2 kHz; by Fig. 6.7*c*, the range is 2,200 yd at 8 kHz; on interpolating for 5 kHz, the estimated range becomes 2,400 yd. However, it should be remarked that in this problem the range is apt to be reverberation-limited rather than noise-limited.

Torpedo Homing

PROBLEM: In an active homing torpedo, a detection range of 3,000 yd is required on an average-aspect submarine. A detection threshold of 30 db is needed to cause the torpedo to "home" on its target. If the torpedo transducer is a plane-piston array restricted to a diameter of 15 in., how much acoustic power output is needed at an operating frequency of 40 kHz? The transmission loss is assumed to be adequately described by spherical spreading and absorption at a temperature of 60°F, and the self-noise is to be taken equal to the ambient noise of the deep sea in sea state 6.

SOLUTION: Solving the active-sonar equation for SL, we obtain

$$\text{SL} = 2\text{TL} - \text{TS} + \text{NL} - \text{DI} + \text{DT}$$

From Fig. 5.5, TL = 85 db; by Table 9.2, TS = 15 db; by Fig. 7.5, NL = -59 db; by Fig. 3.11, DI = 30 db; DT = 30, given. We therefore find SL = 116 db, and by Fig. 4.3, with $\text{DI}_T = 20$, the required power output is found to be 30 watts.

Fish Finding

PROBLEM: A compact school of fish containing 1,000 members, each averaging 1 ft in length, lies 100 yd from a fishing boat equipped with a fish-finding sonar. What will be the level of the echo from this school of fish at a frequency of 60 kHz, assuming that the transducer has a beam pattern broad enough to contain the entire school? The sonar projector radiates 100 acoustic watts of power and is a circular plane array 10 in. in diameter.

SOLUTION: The echo level is the left-hand side of the active-sonar equation and is equal to SL − 2TL + TS. By Fig. 3.11, DI = 30 db; by Fig. 4.3, SL = 121 db; with spherical spreading and absorption, using $\alpha = 19$ db/kyd (Fig. 5.3), TL = 42 db; by Table 9.2, TS = −31 for a single fish 1 ft long, and for 1,000 fish, TS = −31 + 10 log 1,000 = −1 db. The echo level becomes +36 db re 1 dyne/cm^2. If the transducer has a receiving sensitivity of −70 db, the echo would appear as a voltage equal to −34 db re 1 volt across the transducer terminals.

Communication

PROBLEM: In the sofar method of aviation rescue, a downed aviator drops a 4-lb explosive charge set to detonate on the axis of the deep sound channel. How far away can the detonation be heard by a nondirectional hydrophone, also located on the axis of the deep sound channel, at a location of moderate shipping in sea state 3? The receiving system uses a frequency band centered at 150 Hz and squares and integrates the received signals for an interval of 2 sec—an interval estimated to be sufficiently long to accommodate all the energy of the signal. A signal-to-noise ratio of 10 db is required for detection.

SOLUTION: Solving the passive equation for TL, we obtain TL = SL − NL + DI − DT. Recognizing the existence of severe signal distortion, we convert to energy-density and obtain TL = 10 log E_0 − (NL + 10 log t) + DI − DT, where E_0 is the source energy-density and t is the integration time. From Fig. 4.15, 10 log E_0 for a 4-lb charge at 150 Hz = 107 db; by Fig. 7.5, NL = −32; DI = 0 db, DT = 10 db, and 10 log t = 3 db are given in the problem statement. Therefore, TL = 126 db. To convert to range, we write (Sec. 6.2) TL = 10 log r + 10 log r_0 + $\alpha r \times 10^{-3}$. Assume $r_0 = 10{,}000$ yd. Using the formula (Sec. 5.3) $\alpha = 0.1f^2/(1 + f^2)$, where f is in kilohertz, we find $\alpha = 0.00225$ db/kyd. Drawing a curve of TL against r, we read off, for TL = 126, the value of r = 8,000 kyd, or 4,000 miles.

Cautionary Remarks A few words of caution must be said regarding the "pat" solutions of the problems just given. Everything depends upon the values of parameters assumed in their solution. These values are always accompanied by uncertainties arising from two sources: first, uncertainty that the conditions assumed are really those of actual interest and importance; and second, uncertainty that, under these conditions, the chosen values of the parameters are valid.

The first of these two sources of uncertainty involves the specification of the conditions, some natural, some man-made, that the engineer feels will be representative of the environment and the target in and against which the system must operate. Here extreme cases will often have to be worked out in the hope that conditions beyond the selected limits will not be of practical significance. The second uncertainty arises from the presently crude state of underwater sound as a body of quantitative knowledge. Even when all the necessary nature and target conditions are specified, the associated acoustic parameter is apt to be uncertain by several decibels or more, simply because of insufficient quantitative information. As a result of the uncertainties in the parameters, the solution to a particular problem is often valid only as to order of magnitude and is useful only as a rough guide to the likely performance of the system under consideration. Increasingly dependable predictions will become possible in the future as more reliable values of the sonar parameters are obtained by research and measurement.

REFERENCES

1. Stewart, J. L., E. C. Westerfield, and M. K. Brandon: Optimum Frequencies for Sonar Detection, *J. Acoust. Soc. Am.*, **33**:1216 (1961).
2. Horton, J. W.: Fundamentals of Sonar, *U.S. Naval Inst., art.* 7C-3, 1957.

Index

Absorption of sound in the sea, 86–93
causes, 87
depth variation, 90
frequency variation, 88
measurement methods, 86
Absorption cross section, 202
Acoustic axis, 39
Active sonar, 1, 332
Afternoon effect, 102
Alberich (nonreflecting coating for submarines), 6
Ambient noise, 160–184
amplitude distribution, 177
depth effect, 175–177
directionality, 180
sources, 161–166, 171
spectra, 166
variability, 170
Ambient noise levels, in coastal waters, 168
in deep water, 166
defined, 160
in ice-covered waters, 178
ship traffic in, 164
Anomalous depth effect, 272
Arctic, ambient noise, 178–180
propagation, 139–141
reverberation, 225
Array gain, 49–54
defined, 51
dependence on coherence, 53
against flow noise, 296
Arrays, 39–58
advantages of, 30
examples of, 30
"Asdic," explanation of term, 4
Attenuation coefficient, 91
Attenuation loss, 83
Audition, 317–325

Backscattering cross section, 18, 189
Band level, 14, 265
Bathythermograph, description, 91
development of, 6
Beam patterns, 39–44
of ambient noise, 180
Beam patterns, of line and plane arrays, 42
nomogram for, 43
of shaded arrays, 46
Beamwidth of arrays, 44
for reverberation, 193
Binomial shading, 46
Biological sources, of ambient noise, 171–173
of reverberation, 204
Bistatic sonar, 20, 189, 234
Bottom-bounce paths, 154
Bubble pulses, 71, 74
Bubbles in the sea, 199–204
attenuation by, 203
cross sections of, 202
damping of, 201
resonance, 201
sound velocity, 200
Butterfly pattern, 251

Calibration of transducers, 31–39
comparison method, 31
other methods, 32
reciprocity method, 31–39
Caustics in the sea, 107, 137
Cavitation, on domes, 298
index, 271
noise generation by, 269–272
of projectors, 64–69
suppression by depth, 272
threshold, 64
Channeling of sound, in Arctic, 139
in deep sea, 132–141
by mixed layer, 122–132
in shallow water, 141–149
Chebyshev polynomials, 56
Coherence, of ambient noise, 182
in array design, 51–54
of isotropic noise, 52
of reverberation, 227
of turbulent pressures, 295
Continuous spectrum, 14, 264, 267
Convergence gain, 83, 137
Convergence zones, 137, 154

Correlation, of ambient noise, 182
in flow noise, 295
of reverberation, 288
of signals and noise, 50–54
of temperature in the sea, 150
of transmitted sound, 153
Correlators, in arrays, 55
for detection, 310
Critical angle, 115
Critical bands, 319
Critical speed, 271
Cross section, for reverberation, 189
for target strength, 236
Cutoff frequency, in the mixed layer, 125
in shallow water, 144
Cylindrical spreading, 84

Decibel, 14
Deep scattering layer (DSL), 204–208
characteristics, 207
effect on shadow zones, 114
migration, 205
organisms in, 204
resonance effects, 206
Deep sound channel, 132–141
convergence zones in, 137
ray paths in, 136
signal characteristics, 132
transmission model, 135
Deflagration, 70
Depth sounding, 11, 333
Design of sonars, 17, 326–327
Detection index, 309
Detection probability, 306
Detection threshold, 304–325
defined, 305
effect of, averaging, 312
duration and bandwidth, 314
repeated signals, 313
formulas for, 310–311
nomogram for, 312
Directionality, of ambient noise, 180
of arrays, 39–45
of radiated noise, 273
of reverberation, 194
Directivity index, defined, 47, 60
nomogram for, 49
receiving, 47
for reverberation, 194
of simple transducers, 48
transmitting, 60
Dolph-Chebyshev shading, 46
Domes, 298–300
Doppler shift, of echoes, 259
of reverberation, 228
DSL (*see* Deep scattering layer)

Echo duration, 25
Echo excess, 22
Echo level, 22, 196
Echoes, characteristics of, 259–261
detection in reverberation, 321–323
Echoes, formation processes of, 256–259
ECR (Eyring-Christensen-Raitt) layer (*see* Deep scattering layer)
"Eel" (listening device), 4
Efficiency of projectors, 64
Energy flux density, defined, 13
of explosives, 76–78
in reverberation, 198
in sonar equations, 23–25
Explosive echo ranging, 11, 335
Explosives as sound sources, 69–80
advantages of, 78
energy spectra, 76
pressure signature, 70–71
Extinction cross section, 199, 202

False alarm probability, 306
Fathometer, 4
Figure of merit, 22, 328
Fish, target strength of, 255
Fish finding, 11, 336
Flow noise, discrimination against, 296
as radiated noise, 274
as self noise, 293–298
Fluctuation of transmitted sound, causes of, 150
magnitude of, 151
in shallow water, 153
surface reflection as a source of, 110, 153

GHG (array listening equipment) sonar, 6

Hull drone, 268
Hydrodynamic noise, as flow noise, 291–293
as radiated noise, 266, 274
Hydrophone, 1, 28

Image interference, 110
Impedance, specific acoustic, 12
Intensity, defined, 13
in transmission loss, 82
Interaction effects, 68
Internal waves, 128
Intromission, angle of, 115
Ionic relaxation, 87

JP sonar, 5

Knudsen spectra, 167

Lambert's law, 223
Layered reverberation, 197
Leakage coefficient, 123
Leonardo's air tube, 4
Level of a sound wave, 14
Limiting ray, 127

Line arrays, 41–49
  continuous, 42–43
  of equally spaced elements, 41–42
  nomogram for, 44
Line components in noise, 264
Lloyd mirror effect, 110, 114
Log, acoustic, 12

Machinery noise, as radiated noise, 267–268
  as self noise, 291
Magnetostriction, 28, 61
Matched filter, 311
Microstructure of the sea, 150
Minehunting, 9, 334
Mines, acoustic, 7
  pressure, 7
  target strength of, 254
Minesweeping, 7, 333
Mixed layer channeling, 122–132
  depth variation, 123
  leakage out of, 127
  model for, 126
  occurrence of, 100, 128
  transmission loss in, 128–131
Monostatic sonar, 20, 234
Multiplicative arrays, 54–58
  advantages of, 56
  types of, 55
MV device, 4

Noise (*see* Ambient noise; Flow noise; Radiated noise; Rain noise; Self noise)
Noise level, of ambient noise, 160
  of self-noise, 286
  in sonar equations, 19
Noise-limited range, 23
Normal-mode theory, 103, 143

Optimum frequency, 328–331

Passive sonar, 1, 333
Performance figure, 22
Period meter, 229
Piezoelectricity, 28, 61
Prediction of performance, 16, 327
Processing gain, 306
Product theorem, 44
Projectors, beam patterns of, 61
  cavitation limitations, 64–68
  efficiency of, 64
  interaction effects, 68
  source level of, 60
Propeller, singing, 273
Propeller beats, 273
Propeller noise, 266, 268–274

QC sonar, 5

Radiated noise, 263–285
Rain noise, 173
Ray diagrams, examples of, 113, 114, 138
  ray theory in, 104
  transmission loss from, 107
Ray theory, compared with normal mode theory, 145
  for shallow water, 142
Ray tracing, 107
Rayleigh distribution, 226
RBR (refracted bottom-reflected) rays, 136
Reciprocity, calibration, 31–39
  parameter, 35–37
  principle, 34
Recognition differential, 319
Reflection, from sea bottom, 114
  from sea surface, 109, 128
  from targets, 237–240
Reflection loss, defined, 109
  of sea bottom, 115–118
  of sea surface, 109
Reliable acoustic path, 155
Resonance, in the deep scattering layer, 206
  in noise generation, 268, 298
  of targets, 258
Response, open circuit, 31
  receiving, 30
  transmitting-current, 31
Reverberating area, 195
Reverberating volume, 193
Reverberation, 187–234
  bottom, 217–225
  characteristics, 226–228
  detection of echoes in, 321
  frequency spread of, 228
  under ice-cover, 225
  layered, 197
  prediction, 229
  surface, 194–196, 209–217
  volume, 190–194, 204–208
Reverberation level, equivalent plane wave, defined, 190, 191
  for short transients, 198
  as sonar parameter, 21
Reverberation-limited range, 23
ROC (receiver-operating-characteristic) curves, 307–309
RSR (refracted surface-reflected) rays, 136

Scanning sonar, 8
Scattering cross section, 202
Scattering strength, compared with target strength, 196
  defined, 188
  for surface reverberation, 194
  for volume reverberation, 191
Sea bottom, backscattering from, 217–225
  reflection by, 115, 117, 217
  shadow cast by, 114
Sea surface, ambient noise from, 165
  fluctuation caused by, 110, 153
  reflection by, 110–113, 128